Mobile Networks: Enabling Technologies and Services

Mobile Networks: Enabling Technologies and Services

Edited by **Benito Olin**

WILLFORD PRESS

New York

Published by Willford Press,
118-35 Queens Blvd., Suite 400,
Forest Hills, NY 11375, USA
www.willfordpress.com

Mobile Networks: Enabling Technologies and Services
Edited by Benito Olin

International Standard Book Number: 978-1-68285-145-6 (Hardback)

Printed in the United States of America.

Contents

Preface

This book has been an outcome of determined endeavour from a group of educationists in the field. The primary objective was to involve a broad spectrum of professionals from diverse cultural background involved in the field for developing new researches. The book not only targets students but also scholars pursuing higher research for further enhancement of the theoretical and practical applications of the subject.

Mobile networks are very crucial to the overall communication systems in the present scenario. There is an increasing demand for improved mobile networks both by business enterprises and individual consumers. This book attempts to provide a detailed explanation of the vital concepts, applications and analysis of mobile technologies and services for the present telecommunication industry. It compiles contributions from internationally acclaimed panel of experts and scientists from across the globe that discusses some of the key advancements and innovations made in the field of mobile computing and networks. The book examines some of the most significant topics in mobile network technology like network transmission, signal processing, area coverage, etc. This book will aid all the students, researchers, scientists and industry experts in gaining a thorough understanding of latest developments and applications of mobile networks.

It was an honour to edit such a profound book and also a challenging task to compile and examine all the relevant data for accuracy and originality. I wish to acknowledge the efforts of the contributors for submitting such brilliant and diverse chapters in the field and for endlessly working for the completion of the book. Last, but not the least; I thank my family for being a constant source of support in all my research endeavours.

Editor

Condensation-based routing in mobile ad-hoc networks

Francesco Palmieri[a,*] and Aniello Castiglione[b]

[a]Dipartimento di Ingegneria dell'Informazione, Seconda Università degli Studi di Napoli, Aversa (CE), Italy

[b]Dipartimento di Informatica, Università degli Studi di Salerno, Fisciano (SA), Italy

Abstract. The provision of efficient broadcast containment schemes that can dynamically cope with frequent topology changes and limited shared channel bandwidth, is one of the most challenging research topics in MANETs, and is crucial to the basic operations of networks serving fully mobile devices within areas having no fixed communication infrastructure. This problem particularly impacts the design of dynamic routing protocol that can efficiently establish routes to deliver data packets among mobile nodes with minimum communication overhead, and at the same time, ensure high throughput and low end-to-end delay. Accordingly, this work exploits and analyzes an adaptive probabilistic broadcast containment technique based on a particular condensation phenomenon borrowed from Quantum Mechanics and transposed in self-organizing random networks, that has the potential to effectively drive the on-demand route discovery process. Simulation-based performance analysis has shown that the proposed technique can introduce significant benefits on the general performance of broadcast-based reactive routing protocols in MANETs.

Keywords: Mobile ad-hoc networks, probabilistic broadcast, Bose-Einstein condensation, reactive routing protocols

1. Introduction

The widespread deployment of wireless technologies and mobile computing devices has stimulated a considerable interest in mobile ad-hoc networks (MANETs). These are self-organizing and adaptive wireless communication infrastructures consisting of autonomous mobile nodes, operating in a distributed and asynchronous peer-to-peer fashion, that play both the role of connected hosts and routers by simultaneously running user applications and enabling communications between nodes within their wireless coverage areas through the relay of packet traffic on behalf of them. Such nodes can dynamically move around arbitrarily at specific speeds in any direction, so that the resulting network lacks of any stable communication architecture and management structure and is characterized by a randomly evolving topology in both space and time.

Real-time distributed applications, such as "command and control" in mobile sensor environments [24], communications for battlefield operations and life-critical vehicular interactions or road condition related systems [22], multimedia entertainment, group-working and cooperative strategy playing, multiplayer games, etc., are the most attractive candidates to be used over such network infrastructures due to the

*Corresponding author: Francesco Palmieri, Dipartimento di Ingegneria dell'Informazione – Seconda Università degli Studi di Napoli, Via Roma 29, I-81031 Aversa (CE), Italy. E-mail: mailto: francesco. palmieri@unina.it.

enormous potentials offered by this technology in enabling ad-hoc opportunistic real-time interactions among mobile networked entities.

Unfortunately, providing effective real-time services in such a complex and unstable environment may be a really challenging task because of the high degree of dynamism characterizing the involved mobile nodes, together with their energy, connection bandwidth, processing power and memory limitations as long as possible impairments and oddities (interference phenomena, etc.) affecting the wireless channels used to ensure communications and the lack of any kind of central control or hierarchical organization.

Consequently, the routing mechanisms and algorithms used to forward packets from their source to the final destination, via multiple continuously changing node paths, must cope with all the above problems and be extremely lightweight in terms of computational and storage resources requirements. In particular, the lack of a fixed topology makes *flooding* the fundamental dissemination technique for network-wide broadcast propagation, serving as an unavoidable building block for implementing routing protocols in MANETs. Unfortunately, flooding generates a large amount of redundant packets that may rapidly drive to exhaustion some critical resources such as bandwidth and energy as well as cause contention, collisions, and hence, additional packet loss. On the other hand, reducing the number of redundant broadcasts may reduce the degree of reliability. Thus, we have to cope with the challenge of striking a balance between the introduced level of redundancy and reliability. Probabilistic broadcast techniques, especially adaptive ones, may be effective for selective broadcast containment in all the flooding-based protocols.

Accordingly, some fundamental properties of random peer-to-peer organizations, characterized by an almost totally "flat" scheme, *Power Laws* and large-scale scalability, have been exploited to model a broadcast-storm resistant flooding scheme that revealed to be particularly effective in evolving ad-hoc networks characterized by competition for links. Here, when new nodes are continuously added to the network, a single node (typically the best connected one) or a few nodes (one for each specific area of coverage) acquire a macroscopic fraction of all the links (i.e. they condense on it). In fact, despite the non-equilibrium nature of these ad-hoc self-organizing networks, whose unstructured topology reflects the competition for links, their dynamic evolution follows the Bose statistics [10] and can experience *Bose-Einstein condensation* [4] phenomena. Starting from these considerations, this work investigates the potential performance improvements that can be achieved in the on-demand route discovery context for reactive ad-hoc routing protocols, by applying adaptive probabilistic broadcast containment techniques based on the exploitation of the aforementioned condensation phenomena "transposed" in the field of dynamic random networks.

The *Ad-hoc On-Demand Distance Vector* (AODV) routing protocol [18] has been selected in order to assess the effectiveness of the proposed technique, because it is one of the most used and studied routing protocols in the mobile ad-hoc environment and its route discovery facility is widely known to lead to significant delays in networks with a large diameter. Performance evaluation results indicate that the proposed technique enables AODV to achieve an higher data packets delivery ratio and a better reachability, while reducing the routing overhead due to the saved re-broadcasts.

2. Related work

In the recent mobile networking literature there is a considerable attention to performance issues in MANET routing [2,12,13,21] and in particular to the problem of limiting the communication over-head associated with broadcast propagation in on-demand route discovery and maintenance [3]. A lot of approaches have been proposed to face with this challenge, often based on probabilistic routing mechanisms [20] aiming at properly adjusting the forwarding probability according to local topological considerations in such a way to avoid, if possible, the unnecessary and redundant propagation of broadcast messages. Several broadcasting scenarios have been studied in [17] with the sake of mitigating propagation redundancy by using a fixed probabilistic scheme. In addition, a more dynamic probabilistic on-demand route discovery method has been proposed in [23], where the forwarding probability for route request message is estimated by using the number of duplicate requests received on each node.

In [7], another adaptive probabilistic scheme is proposed, where the probability for a node to retransmit a broadcast packet is directly associated to the local node degree combined with a fixed efficiency parameter controlling the reachability of the broadcast. Similarly, an on-demand route discovery scheme combining traditional probabilistic broadcasting with more specific considerations about the coverage area interested in the broadcast propagation, has been presented in [11], whereas an hybrid approach based on both probabilistic and counter-based propagation has been presented in [16]. Different solutions, alternative to probabilistic broadcasting, have also been exploited, such as the one based on Multi-Point Relays (MPR) selection proposed in [14], or the one based on gossiping and presented in [9].

At the best of our knowledge, the adaptive probabilistic scheme reported in this article is the first experience using concepts derived from Quantum Mechanics, such as the Bose-Einstein condensation, to drive and control the propagation of route discovery messages across large scale MANET infrastructures.

3. Effective routing in MANETs

In traditional fixed-topology networks, taking routing decisions is a quite straightforward task: packets to be forwarded are usually sent to the neighbor located on the best path, that is the one with the lowest cost to the final destination. Due to the relatively high reliability of paths/routes, no duplication is needed, and hence each packet on every hop is only sent to a single node. On the contrary, in MANETs' things are completely different and the dynamic nature of nodes, that continuously join and leave the network or change their attachment connections by moving between different coverage areas, may introduce frequent path breaks/failures and recalculations. In this environment, a route consists of an ordered set of intermediate nodes, determined on-the-fly and subject to rapid changes, that transport a packet from source to destination by forwarding it from one node to the next one (see Fig. 1).

Each route is a short-lived object resulting from a continuously evolving topology composition process driven by point-to-point or multi-point ad-hoc *peer-to-peer* associations dynamically established by the moving nodes. That is, when a new node enters into the transmission range of another one, this creates the possibility of having new routes for a destination being available. The detection of a newly moved or powered on node is implemented through a traditional beaconing mechanism. The neighbor discovery single-hop *HELLO* messages, propagated in broadcast by each node on all its interfaces and retransmitted at regular intervals, give information regarding the availability of a node as a next hop, allowing neighbor adjacency lists to be dynamically built on all the network nodes. Suppose that a node X moves into the neighborhood (wireless coverage range) of a node Y. When the nodes start exchanging HELLO packets, the IP address of X is immediately added to the list of neighbors of Y and vice versa (see Fig. 2). Analogously, when Y detects that periodic HELLO packets are no more received from an adjacent node X, it is removed from the neighbor list, assuming that it has moved away or has been powered off.

Thus, establishing and maintaining network connectivity will require greater information exchange efforts and some duplication in traffic relay, leading to a noticeable increase in communication overhead. More precisely, when a packet to be forwarded arrives to a node, there might not be an available path to its destination so that the node may have the necessity of temporarily buffering it while searching

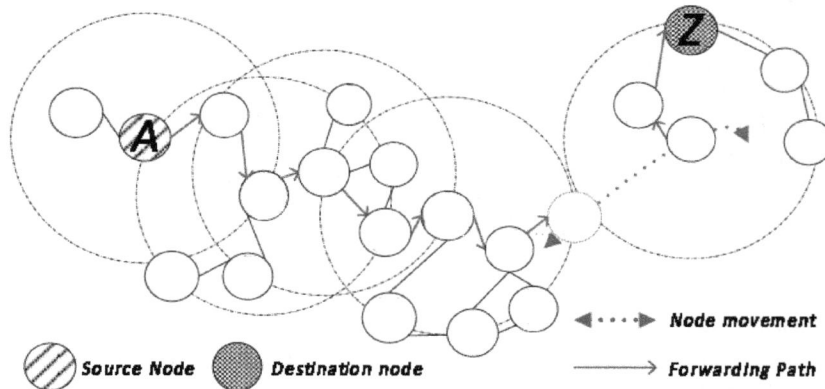

Fig. 1. Hop-by-hop routing in MANETs.

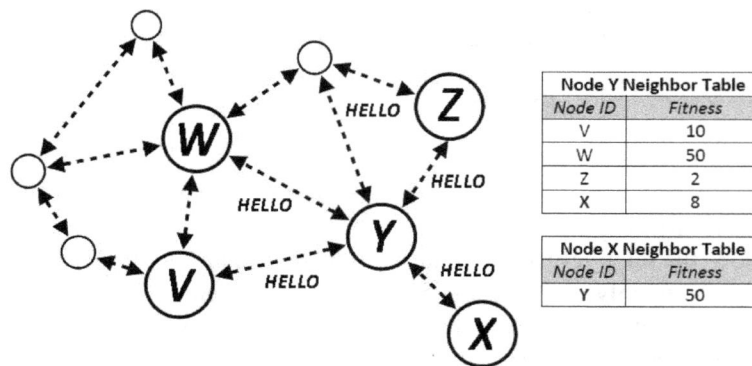

Node Y Neighbor Table	
Node ID	Fitness
V	10
W	50
Z	2
X	8

Node X Neighbor Table	
Node ID	Fitness
Y	50

Fig. 2. Neighbor discovery in MANETs.

for a feasible route to the involved destination. The decision on whether forwarding a packet to more than a node is also a concern, since multiple forwarding actions drastically increase the probability of a successful delivery, also in an inherently unreliable environment such as the MANET one.

Such decisions are, unluckily, somewhat difficult to take because only in some situations it could be reasonable to establish an hard threshold and only forward a message to a neighbor when its delivery probability exceeds such threshold. In fact, when excluding a node from the propagation tree, no one can be sure that another neighbor having higher delivery chances will be selected in a realistic time frame. Consequently, in several situations, a greater flexibility is needed during the decision on which neighbor node will be the destination of a broadcast packet. In addition, also the determination of the number of nodes to which a certain packet should be propagated can be an hard choice. In fact, the distribution of packets to the largest possible number of neighbors will extend the probability of a positive delivery, while, simultaneously, it will consume more resources in the entire system. On the contrary, the ability to forward packets only to a limited subset of nodes, and asymptotically to a single one, makes certainly use of few system resources, but decreases the probability of positive delivery and increases the overall end-to-end delay. Furthermore, it should be considered that wireless connections, with respect to their wired counterparts, supply to the involved communicating nodes a lower and less stable bandwidth.

Consequently, proper extensions are needed at the routing protocols level to support adaptiveness in following the evolving network status as well as to improve the overall efficiency by containing the protocol overhead for maximizing the real bandwidth available for communication.

At the state-of-the-art, MANET routing protocols can be classified into two main classes, depending on their basic operating features:

- *Proactive routing protocols* strive to keep mobile routes constantly up-to-date as well as to ensure their consistency among the different nodes in a network. In order to have a comprehensive control/view of the entire network, changes in the topology need to be spread over the whole communication infrastructure. Keeping "fresh" routing information for every destination in the network ensures a very limited delay in end-to-end communication between each pair of nodes. Unfortunately, this family of protocols exhibits the drawback of introducing further control traffic to dynamically update route entries that are no more valid (i.e., stale ones). In particular, in scenarios characterized by an high degree of mobility, the implementation of a proactive routing protocol introduces a communication overhead that makes its deployment cost excessively high. For that reason, in presence of ad-hoc communication networks with an high quantity of mobile nodes, reactive routing protocols can be considered a better choice than the proactive one.
- *Reactive routing protocols* create a route only when it is required, according to an on-demand paradigm. When a new route is needed towards a specific destination, the originating node starts a route discovery process by starting the propagation of a RREQ (Route Request) broadcast packet. The target nodes, or just the intermediate ones (if they own the specific information needed), when replying to the RREQ messages, deliver a backward RREP message (Route Reply) reporting how many intermediate hops are needed to reach the destination. Every node that receives a message of type RREP stores a forward route to the destination node in its routing table. After the complete path has been successfully set up after the source node has received a RREP reply packet, a route management procedure is started to preserve the obtained routing information until the target is no more accessible or the path to it is no longer needed. Reactive protocols require less bandwidth than the proactive ones because they avoid, as possible, the delivery of unneeded recurrent routing information and, at the same time, reduce the network load when just a limited subset of all the available routes is used in a given circumstance. On the other hand, the latency of the on-demand route discovery process can introduce a significant packet delay prior to any communication can take place.

In both cases, broadcasting is used to propagate information that need to be unconditionally disseminated throughout the network (e.g. topology updates or RREQ packets for route discovery). However, RREQ packet does not need to reach all the nodes in the network once a particular path has been discovered that leads to the desired destination. Hence, significant optimization of the routing protocol performance can be achieved by reducing the unnecessary propagation of the above packets with some kind of broadcast control/containment technique. From now on, all the reported considerations will refer to reactive protocols supporting route discovery mechanisms that is the routing technique of choice for this proposal.

Simple flooding, also called *pure* or *blind flooding*, is the basic way for implementing the broadcast mechanism, where each node replicates a packet by forwarding it to all its neighboring nodes. Each node receiving a broadcast packet should re-propagate outwards it throughout its neighborhood, eventually stopping the process when every node in the network has already received the broadcast packet. Such propagation strategy, despite its inherent simplicity and reliability, presents a significant overhead due to the incontrollable proliferation of delivered messages, with the obvious consequences in terms of redundancy and collisions. Such phenomenon is commonly known as the *Broadcast Storm Problem* (BSP) [17].

Starting from these considerations, it becomes clear that the propagation mechanisms used in the on-demand route discovery process is fundamental to ensure an acceptable scalability and performance to the involved communication infrastructure. More precisely, the main issue in propagating on-demand routing requests in reactive routing scenarios is directly associated to the minimization of the number of nodes propagating the RREQ broadcast packets while keeping a satisfactory degree of reachability. Clearly, decreasing the number of broadcast retransmissions, immediately saves more bandwidth and reduces the overall node contention and power consumption, with the obvious consequences on the network performance. Thus, a discovery method that provides a good compromise between the minimization of critical resource usage and the maximization of network performance (e.g. throughput and end-to-end delay), is highly desirable.

3.1. Probabilistic routing

Several schemes based on probabilistic approaches have been proposed in [7,11,20,23], to overcome the drawbacks associated with simple flooding-based broadcast techniques. Such schemes are really simple to implement, and hence fast and computationally cheap. In fact, they are not adversely affected by node mobility because they require little or no topological information in order to make forwarding/replication decisions and do not need the support of any additional special-purpose hardware.

In traditional approaches based on probabilistic decision techniques, every node involved in broadcast propagation floods a packet throughout its neighborhood, according to a fixed forwarding probability p. This can imply that when the value of p is too low, most of the network nodes do not receive the broadcasts or, alternatively, in presence on a excessively high value of p, a proliferation of unnecessary redundant transmissions can be experienced. In detail, in presence of n mobile nodes, the maximum (worst case) number of messages needed to ensure the propagation of an RREQ message during a route discovery process will be $p \cdot (n - 2)$, when no intermediate node already knows a valid route to the target.

In presence of randomly distributed mobile nodes dynamically establishing ad-hoc connections between each other, several topological regions, characterized by different node density (e.g. sparse and dense regions) and degree of connectivity, can continuously emerge and reorganize themselves within the network. In a highly dense region, multiple nodes may share the same transmission channel coverage being organized in some clustered fashion around some hub/aggregation nodes. In these regions, significant resources can be saved with limited impacts on the broadcast propagation effectiveness, if several nodes do not forward the packet to some randomly chosen neighbors. Conversely, in sparse regions, characterized by a limited degree of coverage, some areas may be completely cut off from broadcast propagation unless the forwarding probability threshold is set high enough. Thus, to cope with these much different situations, such probability need to be adaptively set on all the nodes depending on their local topological and coverage characteristics. Analogously, the distribution of ad-hoc connections within a MANET, and hence the number of connections of each node (its connectivity degree), can significantly influence the broadcast probability, since forwarding the packets towards highest-degree nodes, connected to a large number of peers, greatly enhances the successful delivery chances and reduces the length of the chosen routes/paths throughout the network.

All the approaches based on a static forwarding probability p are adversely affected by unfair broadcast distribution phenomena due to the fixed value of p that is used to take forwarding decisions on each node, independently from its specific topological features. Conversely, mobile nodes should be properly categorized according to their communication capabilities (bandwidth, coverage etc.), position and role within the network so that the forwarding probability can be dynamically adjusted to drive the broadcast propagation towards the best connected areas (the denser regions). Furthermore, the limitation of redundant broadcasts should ideally not affect the overall network reachability, and introduce only a minimum dynamic topology management overhead also in presence of a significant node mobility degree, ensuring the needed scalability to the MANET infrastructure.

3.2. Adaptive probabilistic schemes

To cope with the above problems, a more adaptive and flexible probabilistic route discovery scheme is needed. It should work by dynamically adjusting the forwarding probability p on each node according to its topological features, mainly characterized by local density considerations. Such density can be easily estimated from the number of adjacent neighbors, that is the node degree.

In typical MANET organizations we can observe the presence of a significant number of nodes connected with a few links (leaf, or lower degree nodes), whereas only a small number of nodes (known as *connection hubs*) with a great number of links ensure the overall connectivity to all their neighbors. This is due to an implicit preferential attachment mechanism, according to which each node joining the network, in presence of multiple choices and when constrained by the involved transmission technology to a limited number of possible associations, aims at connecting to the neighbors characterized by the best communication capabilities (physical link, aggregated bandwidth, signal health/power, reliability, etc.). Thus, depending on their location, resource availability, link coverage and quality, some nodes tend to physiologically assume the role of hubs or super-nodes within the MANET dynamic peer-to-peer topology, where all the other ones, with lower communication capabilities, will limit their participation to the role of clients or peripheral routers serving a very limited network region. This implies some extent of compliance to the Power Laws, so that the number of hops between nodes is significantly reduced (*small-world* property) [1,8].

The resulting hierarchy is an extremely important factor to cope with the node heterogeneity, to ensure the scalability of the routing algorithms and to limit the impact of node mobility. In fact, the presence of a few highly connected "focal points" in the network, allows faster propagation of information and hence speeding up both the RREQ and RREP traffic, with significant performance gains for the overall route discovery process. However, due to the former characteristic, if several spoke nodes propagate some information at almost the same time, the involved messages tend to concentrate at the hub node. That is, the nodes with the highest connectivity are subject to most of the search/query traffic. This tendency is likely to increase with the number of connected nodes in the network since also the number of links connected to the hub node increases consequentially.

3.3. Exploiting condensation phenomena

When, as the result of moving from a previous location and coverage area to another one, mobile nodes are continuously disconnected and re-added to the network, joining, according to a *preferential attachment* principle, to m already existing nodes (with $m \geqslant 1$) characterized by different connectivity capabilities and hence competing for new communication links/channels, new interesting phenomena can be observed on the evolving network, from which we can take further advantage in the proposed adaptive probabilistic routing scheme. More precisely, we exploit the existence of a close link, presented in [4], between the theory of evolving ad-hoc networks and the equilibrium of a Bose gas (governed by the Bose-Einstein statistics [6]) in Quantum Mechanics according to which a dynamically growing network can experience specific clustering/condensation phenomena, due to the competition for links, that resemble the interactions between particles in the above gas. We are interested in modeling the transitions from equilibrium phases, in which all the nodes have the same potential of accepting new connections but some of them (the hubs) gradually tend to form clusters around them by acquiring edges at a higher rate than the others, and a condensation one, in which the best potentially connected nodes condensate most of the new links around them by acquiring a significant portion of all the ad-hoc connections within the network and maintains them over time. This perfectly resembles the analogous phenomenon observable in gases, known as Bose-Einstein condensation [5], in which a large fraction of the bosons occupy the lowest external potential quantum state and the resulting quantum effects become apparent on a macroscopic scale. These considerations and phenomena occurring in complex networks have been formally represented in [4,19] by using a precise model that will constitute the basis for driving the proposed probabilistic routing scheme.

To model the different ability of the network nodes to compete for new ad-hoc connections within their neighborhood, a "fitness" score η_i (chosen according to a specific distribution $\rho(\eta)$) is associated to each node $i \in N$, representing the differences in communication technology, aggregated bandwidth or available channels. In [4], a close mapping from a Bose gas to an evolving network has been defined by assigning an "energy level" ε_i to each node i, depending on its fitness η_i through the relation:

$$\varepsilon_i = -\frac{1}{\beta} \ln \eta_i \tag{1}$$

where β is a constant value conditioning its dependency from the fitness η. In particular, when β approaches to zero, all the nodes tend to acquire equal fitness, when instead $\beta \gg 1$ nodes with different "energy" assume very different fitness values. A connection between two nodes i and j characterized by the energy levels ε_i and ε_j and fitness scores η_i and η_j corresponds in the mapping to two non-interacting particles placed respectively on their own energy levels. Each new node i joining the network implies the introduction of a new energy level ε_i as long as $2 \cdot m$ new particles in the mapping abstraction. Only m of them remain associated to the energy level ε_i, corresponding to the new m outgoing connections established from the node i, whereas the other m particles are distributed among the other energy levels (associate the incoming connections to m already existing nodes).

The probability Π_i that any new node joining the network will connect one of its m links to an already existing node i, directly depends on the degree k_i and on the "energy" on its level ε_i, and is adversely conditioned from the same properties associated to the other nodes j within its neighborhood N:

$$\Pi_i = \frac{e^{-\beta \varepsilon_i} k_i}{\sum\limits_{j \in N} e^{-\beta \varepsilon_j} k_j} \tag{2}$$

In the aforementioned gas/growing network analogy, Π_i can also be viewed as the probability that a particle lands on level i, characterized by an energy ε_i and the number or particles on this level $n(\varepsilon_i)$, closely corresponding to the number of nodes actually connected to the node i, is known [4] to follow the Bose statistics:

$$n(\varepsilon_i) = \frac{1}{e^{\beta(\varepsilon_i - \mu)} - 1} \tag{3}$$

where the aforementioned constant β plays, in the Bose gas analogy, the role of an inverse temperature whereas μ corresponds to the chemical potential.

Usually, nodes that have been newly created, link to firmly connected nodes acting as "hub". Those nodes are distinguished by the fact of having an high level of fitness. In presence of nodes that are characterized by the same potential in concentrating connections, and hence have the same fitness, the above model coincides with the classic scale free one, described by the Power Laws, leading the network into an almost stable equilibrium.

On the other hand, if the nodes that are already present in the network begin to have a considerable different connection potential, nodes characterized by an higher fitness begin to create links on them in a way which creates clusters and letting to predominate only to those nodes having higher fitness ("fit-get-rich"). In the last phase of the evolution process, it can be observed the condensation of most of the links around the fittest and best connected nodes. In detail, such condensation phenomenon is driven, according to [4], by a fitness distribution $\rho(\eta)$ modeled as:

$$\rho(\eta) = (1 - \eta)^\lambda \tag{4}$$

for each $\lambda > 1$.

3.4. The proposed adaptive algorithm

We would like to make use of the aforementioned observations and their underlying properties to improve the performance of reactive routing protocols in MANETs by introducing into the AODV protocol a novel highly adaptive probabilistic routing strategy "conditioned" by the condensation phenomenon, assuming that the quantity of "energy" accumulated on a specific node (or its capacity to acquire connections) will indicate how likely this node will be able to deliver a packet to a destination through a reliable and sufficiently short route/path.

To accomplish this, we establish a probabilistic broadcast forwarding strategy based on Eq. (2) simplified as:

$$p_i = \frac{\eta_i k_i}{\sum\limits_{j \in N} \eta_j k_j}. \tag{5}$$

where p_i is the probability of forwarding a packet to a specific node i, with degree k_i and fitness η_i and N is the complete set of available neighbor nodes with their associated degree k_j and fitness η_j with $j \in N$.

The fitness η_i assigned to each node can depend on its maximum b_i and residual bandwidth r_i as long as its signal power s_i, noise level n_i and total/residual number of channels (respectively c_i and a_i) as defined in the following equation:

$$\eta_i = \frac{r_i \log(b_i) \cdot a_i \log(c_i) \cdot s_i}{n_i}. \tag{6}$$

The above criteria also derive from the consideration that, in the typical preferential attachment model, new nodes strive to establish new connections with nodes with higher degree and fitness, and hence these properties jointly determine the attractiveness of a node. Consequently, as the node fitness changes, so does its role in MANET routing, so that the fittest nodes are those that have greater chances to successfully respond (with RREP messages) to route requests queries (RREQ messages) in the route discovery process.

The whole probabilistic forwarding strategy, to be adopted on each mobile node for RREQ message delivery in the AODV routing protocol, with the sake of containing the adverse effects of broadcasts during the route discovery process, and driving the RREQ forwarding towards the fittest and best connected nodes, is described in detail by using pseudo-code in Algorithm 1.

Algorithm 1 Adaptive condensation-based probabilistic broadcast forwarding

---

Input:
m: *message/packet to be forwarded*
$N = \{n_1, \ldots, n_N\}$: *available neighbor nodes set*
    for all nodes $n_i \in N$ **do**
        $p_i = forwarding\_probability(n_i, N)$ $\{p_i$ is calculated as in Equation 5$\}$
        $x \leftarrow random \in [0, 1]$
        if $x \leqslant p_i$ **then**
          send m to n_i
        end if
    end for

Clearly, as the number of retransmissions required for broadcasting RREQ messages is reduced, more bandwidth is saved and also contention and node power consumption are reduced, by improving overall MANET performance.

4. Performance evaluation

The performance and effectiveness of the proposed scheme has been analyzed by using discrete event simulation. The NS-2 simulator [15] has been used in all the evaluation experiments, where the mobile nodes have been modeled as homogeneous (i.e. with the same transmission range and interfaces, simulating the default Lucent's WaveLAN 802.11b card with a 2Mb/sec bit rate), and the wireless channel is fully shared and can be accessed by any node at random times. Since the 802.11b MAC specification is based on CSMA/CA, broadcast collision, control is limited to basic collision avoidance at the carrier sensing level and thus some collisions in broadcast packet transmission may take place.

Several variable CBR traffic loads, relying on UDP transport and with 512 byte packet size, have been used in the analysis, within the context of a *random waypoint* mobility model in a square field of $1000 \times 1000\ m^2$ and a pause time of 0 seconds corresponding to continuous motion of the mobile nodes (with $20\ m/s$ maximum nodal speeds). Each simulation was run for 100 seconds. Unicast AODV routing with link layer support and HELLO messages enabled (needed to build the neighbor lists) has been used to discover all the destination nodes solicited by the test traffic. A modified protocol version (referred to as C-AODV in the figures), optimized to reduce the number of retransmission for the route request (RREQ) messages by using the previously presented condensation based broadcast containment techniques, has been compared against the traditional AODV implementation based on blind flooding.

In detail, the above performance comparisons have been performed in two different scenarios, whose goal was to study how the effectiveness of the proposed scheme is affected by the prevailing network conditions such as traffic load and node density:

- In the first one, a single random mobile ad-hoc network consisting of 100 nodes has been solicited with several variable traffic loads ranging from 10 to 100 32 Kbps CBR connections.
- In the second one, the network density has been varied from 50 to 150 nodes placed randomly on the aforementioned area, solicited by using a fixed 32 Kbps CBR load of 50 connections.

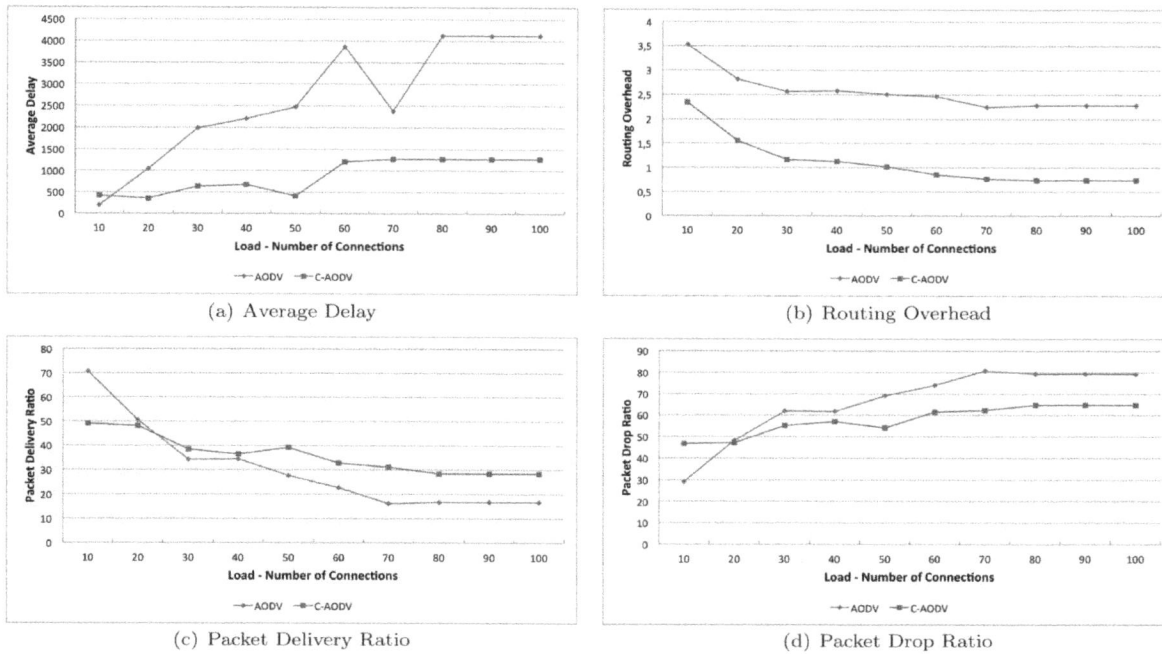

(a) Average Delay   (b) Routing Overhead

(c) Packet Delivery Ratio   (d) Packet Drop Ratio

Fig. 3. Performance results (a), (b), (c), (d) for a 100-nodes network and load varying from 10 to 100 connections.

In both the scenarios, the metrics used to evaluate the performance of the protocol included:

– The average *End-to-end Delay*, that is the interval between the instant a source generates a packet and the time at which the destination receives the packet;
– The *Routing Overhead* calculated as the sum of all routing packets transmitted during the simulation, or normalized routing load. For packets sent over multiple hops, each transmission over one hop is counted as one transmission;
– The *Packet Delivery Ratio*, representing the percentage of sent packets that have been correctly received;
– The *Packet Drop Ratio*, representing the percentage of sent packets that have been dropped.

The main result of all the simulation experiments is that the condensation-based protocol version (C-AODV) almost always outperform the conventional one. The improvements in end-to-end delay become progressively more appreciable when the connection load increases (compare Fig. 3(a) with Fig. 4(a)) whereas the gain observable in the reduction of *Routing Overhead* varies almost linearly with both the load and network density but its growth rate is significantly affected by the network dimensions and adversely conditioned by the increasing load (compare Fig. 3(b) with Fig. 4(b)). The benefits in packet delivery performance and packets dropped become evident only when the network load grows over 20 connections or more (see Figs 3(c) and 3(d)), whereas it generally maintains a slightly better performance for low density as long as high density networks (see Figs 4(c) and 4(d)).

5. Conclusions

A new broadcast containment strategy, where the forwarding probability at a node is dynamically computed based on its local density criteria and driven by condensation phenomena similar to those observable

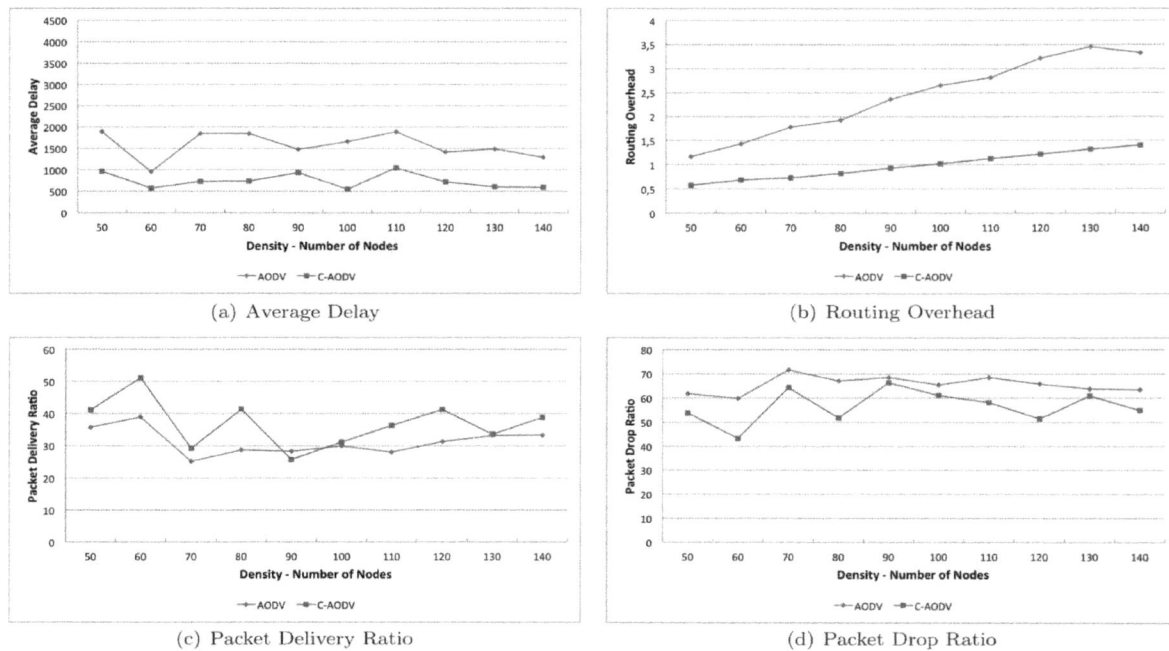

(a) Average Delay

(b) Routing Overhead

(c) Packet Delivery Ratio

(d) Packet Drop Ratio

Fig. 4. Performance results (a), (b), (c), (d) for a 50-connections load on networks varying from 50 to 140 nodes.

in a Bose gas in quantum mechanics, has been proposed and introduced within the route discovery framework of the AODV MANET routing protocol. The performance of the resulting condensation-based enhanced scheme (C-AODV) has been compared against the traditional one based on blind-flooding. The observed simulation results have shown that, for all the considered network densities and for almost all the offered loads, the proposed scheme outperforms the traditional blind-flooding based AODV routing protocol in terms of reduced routing overhead, packet delivery performance, end-to-end delay, and the achieved benefits become more appreciable when the number of connections or active nodes grows. The success of the presented adaptive probabilistic strategy is due to its ability of striking a good balance between exploration (expanding the RREQ propagation to specific areas of the search space that can bring to promising solutions) and exploitation (considering past experience factors and traffic properties to drive the search) within the network discovery/broadcast propagation process throughout the mobile ad-hoc network.

References

[1]   A.-L. Barabási and R. Albert, Emergence of scaling in random networks, *Science* **286**(5439) (15 October 1999), 509–512.
[2]   A. Barolli, E. Spaho, L. Barolli, F. Xhafa and M. Takizawa, QoS routing in ad-hoc networks using GA and multi-objective optimization, *Mobile Information Systems* **7**(3) (2011), 169–188.
[3]   S. Basagni, I. Chlamtac, V.R. Syrotiuk and B.A. Woodward, A Distance Routing Effect Algorithm for Mobility (DREAM), In *MOBICOM*, pages 76–84, 1998.
[4]   G. Bianconi and A.-L. Barabási, Bose-Einstein Condensation in Complex Networks, *Phys Rev Lett* **86** (Jun 2001), 5632–5635.
[5]   C.C. Bradley, C.A. Sackett, J.J. Tollett and R.G. Hulet, Evidence of Bose-Einstein Condensation in an Atomic Gas with Attractive Interactions, *Phys Rev Lett* **75** (Aug 1995), 1687–1690.
[6]   A.H. Carter, *Classical and Statistical Thermodynamics*, Prentice Hall, Upper Saddle River, NJ, 2001.

[7]   J. Cartigny and D. Simplot, Border Node Retransmission Based Probabilistic Broadcast Protocols in Ad-Hoc Networks, In *Proceedings of the 36th Annual Hawaii International Conference on System Sciences (HICSS'03) – Track 9 – Volume 9*, HICSS '03, Washington, DC, USA, 2003. IEEE Computer Society, pages 303–312.

[8]   M. Faloutsos, P. Faloutsos and C. Faloutsos, On power-law relationships of the Internet topology, *SIGCOMM Comput Commun Rev* **29** (August 1999), 251–262.

[9]   Z.J. Haas, J.Y. Halpern and L. Li, Gossip-based ad hoc routing, *IEEE/ACM Trans Netw* **14**(3) (2006), 479–491.

[10] K. Huang, *Statistical mechanics*, Wiley, New York, NY, USA, 1987.

[11] J.-S. Kim, Q. Zhang and D.P. Agrawal, Probabilistic broadcasting based on coverage area and neighbor confirmation in mobile ad-hoc networks, In *Global Telecommunications Conference Workshops, 2004. GlobeCom Workshops 2004. IEEE*, Nov. –3 Dec. 2004, pages 96–101.

[12] Y.-S. Kim, Y.-S. Shim and K.-H. Lee, A cluster-based web service discovery in MANET environments, *Mobile Information Systems* **7**(4) (2011), 299–315.

[13] E. Kulla, M. Hiyama, M. Ikeda, L. Barolli, V. Kolici and R. Miho, MANET performance for source and destination moving scenarios considering OLSR and AODV protocols, *Mobile Information Systems* **6**(4) (2010), 325–339.

[14] T.-H. Lin, H.-C. Chao and I. Woungang, An enhanced MPR-based solution for flooding of broadcast messages in OLSR wireless ad hoc networks, *Mobile Information Systems* **6**(3) (2010), 249–257.

[15] S. McCanne and S. Floyd, The network simulator – ns-2. http://www.isi.edu/nsnam/ns/.

[16] A. Mohammed, M. Ould-Khaoua and L.M. Mackenzie, An Efficient Counter-Based Broadcast Scheme for Mobile Ad Hoc Networks, In Katinka Wolter, editor, *EPEW*, volume 4748 of *Lecture Notes in Computer Science*, Springer, 2007, pages 275–283.

[17] S.-Y. Ni, Y.-C. Tseng, Y.-S. Chen and J.-P. Sheu, The Broadcast Storm Problem in a Mobile ad hoc Network, In *MOBICOM*, 1999, pages 151–162.

[18] C.E. Perkins, E. Belding-Royer and S. Das, Ad-hoc On-Demand Distance Vector (AODV) Routing, IETF RFC 3561, 2003.

[19] A. Réka and A.-L. Barabási, Statistical mechanics of complex networks, *Rev Mod Phys* **74** (Jan 2002), 47–97.

[20] Y. Sasson, D. Cavin and A. Schiper, Probabilistic broadcast for flooding in wireless mobile ad hoc networks, In *Wireless Communications and Networking, 2003. WCNC 2003. 2003 IEEE*, volume 2, March 2003, pages 1124–1130.

[21] E. Spaho, L. Barolli, G. Mino, F. Xhafa, V. Kolici and R. Miho, Implementation of CAVENET and its usage for performance evaluation of AODV, OLSR and DYMO protocols in vehicular networks, *Mobile Information Systems* **6**(3) (2010), 213–227.

[22] J. Zhang, C. Chen and R. Cohen, A Scalable and Effective Trust-Based Framework for Vehicular Ad-Hoc Networks, *J Wireless Mobile Networks, Ubiquitous Computing, and Dependable Applications* **1**(4) (2010), 3–15.

[23] Q. Zhang and D.P. Agrawal, Dynamic probabilistic broadcasting in MANETs, *J Parallel Distrib Comput* **65**(2) (2005), 220–233.

[24] T.A. Zia and A.Y. Zomaya, A Lightweight Security Framework for Wireless Sensor Networks, *J Wireless Mobile Networks, Ubiquitous Computing, and Dependable Applications* **2**(3) (2011), 53–73.

Francesco Palmieri is an assistant professor at the Engineering Faculty of the Second University of Napoli, Italy. His major research interests concern high performance and evolutionary networking protocols and architectures, routing algorithms and network security. Since 1989, he has worked for several international companies on nation-wide networking-related projects and, from 1997 to 2010 he has been the Director of the telecommunication and networking division of the Federico II University, in Napoli, Italy. He has been closely involved with the development of the Internet in Italy as a senior member of the Technical-Scientific Advisory Committee and of the CSIRT of the Italian NREN GARR. He has published more that 70 papers in leading technical journals/conferences and currently serves as Editor-in-Chief of an international journal and is part of the editorial board of several other ones.

Aniello Castiglione joined the Dipartimento di Informatica ed Applicazioni "R. M. Capocelli" of Università di Salerno in February 2006. He received a degree in Computer Science and his Ph.D. in Computer Science from the same university. He is a reviewer for several international journals (Elsevier, Hindawi, IEEE, Springer) and he has been a member of international conference committees. He is a Member of various associations, including: IEEE (Institute of Electrical and Electronics Engineers), of ACM (Association for Computing Machinery), of IEEE Computer Society, of IEEE Communications Society, of GRIN (Gruppo di Informatica) and of IISFA (International Information System Forensics Association, Italian Chapter). He is a Fellow of FSF (Free Software Foundation) as well as FSFE (Free Software Foundation Europe). For many years, he has been involved in forensic investigations, collaborating with several Law Enforcement agencies as a consultant. His research interests include Data Security, Communication Networks, Digital Forensics, Computer Forensics, Security and Privacy, Security Standards and Cryptography.

2

QoS routing in ad-hoc networks using GA and multi-objective optimization

Admir Barolli[a,*], Evjola Spaho[b], Leonard Barolli[c], Fatos Xhafa[d] and Makoto Takizawa[a]

[a]*Department of Computers and Information Science, Seikei University, Tokyo, Japan*
[b]*Graduate School of Engineering, Fukuoka Institute of Technology (FIT), Fukuoka, Japan*
[c]*Department of Information and Communication Engineering, Fukuoka Institute of Technology (FIT), Fukuoka, Japan*
[d]*Department of Languages and Informatics Systems, Technical University of Catalonia, Jordi Girona 1-3, Barcelona, Spain*

Abstract. Much work has been done on routing in Ad-hoc networks, but the proposed routing solutions only deal with the best effort data traffic. Connections with Quality of Service (QoS) requirements, such as voice channels with delay and bandwidth constraints, are not supported. The QoS routing has been receiving increasingly intensive attention, but searching for the shortest path with many metrics is an NP-complete problem. For this reason, approximated solutions and heuristic algorithms should be developed for multi-path constraints QoS routing. Also, the routing methods should be adaptive, flexible, and intelligent. In this paper, we use Genetic Algorithms (GAs) and multi-objective optimization for QoS routing in Ad-hoc Networks. In order to reduce the search space of GA, we implemented a search space reduction algorithm, which reduces the search space for GAMAN (GA-based routing algorithm for Mobile Ad-hoc Networks) to find a new route. We evaluate the performance of GAMAN by computer simulations and show that GAMAN has better behaviour than GLBR (Genetic Load Balancing Routing).

1. Introduction

The wireless mobile networks and devices are becoming increasingly popular and they provide users access to information and communication any-time and anywhere. The conventional wireless networks are often connected to a wired network (e.g. ATM or Internet). This kind of wireless network requires a fixed wire-line backbone infrastructure. All mobile hosts in a communication cell can reach a base station on the wire-line networks in one-hop radio transmission. In contrast, the class of Ad-hoc networks do not use any fixed infrastructure. The nodes of Ad-hoc networks intercommunicate through single-hop and multi-hop paths in a peer-to-peer fashion. Intermediate nodes between two pairs of communication nodes act as routers. Thus the nodes operate both as hosts and routers. The nodes are mobile, so the creation of routing paths is affected by the addition and deletion of nodes. The topology of the network may change rapidly and unexpectedly.

Much work has been done on routing in Ad-hoc networks [1,2]. Many protocols and algorithms such as Destination-Sequenced Distance-Vector (DSDV) protocol, cluster-based routing algorithms, Dynamic Source Routing (DSR) protocol, Ad-hoc On-demand Distance-Vector (AODV) protocol, Zone Routing Protocol (ZRP), Temporally Ordered Routing Algorithm (TORA), and Associative Bit Routing (ABR)

---

*Corresponding author. E-mail: admir.barolli@gmail.com.

have been proposed. The emphasis has been on providing the shortest path and achieving a high degree of availability in a dynamic environment where the network topology changes fast. However, all the above routing solutions only deal with the best effort data traffic.

The QoS routing has been receiving increasingly intensive attention in the wire-line network domain [3, 4]. The routing strategies can be classified into three classes: source, distributed and hierarchical routing. However, these QoS routing algorithms can not be applied directly to Ad-hoc networks, because of the bandwidth constraints and dynamic network topology of Ad-hoc networks.

Recently, because of the rising popularity of multimedia applications and potential commercial usages of Ad-hoc networks, QoS support in Ad-hoc networks has become an unavoidable task. To support QoS, the link state information such as delay, bandwidth, cost, loss rate, and error rate in the network should be available and manageable. But, getting and managing the link state information in Ad-hoc networks is very difficult because the quality of wireless link may change with the surrounding circumstances. Furthermore, the resource limitation and the mobility of hosts make things more complicated. The challenge we face is to implement complex QoS functionality with limited available resources in a dynamic environment.

In the literature, the research on QoS support in Ad-hoc networks includes QoS models [5,6], QoS resource reservation signalling [7], QoS Medium Access Control (MAC) [8], and QoS routing [9–11]. In this paper, we will survey only QoS routing problem in Ad-hoc networks. The QoS routing searches for a path with enough resources for QoS requirements. The QoS metrics could be additive, concave or productive. It is proved that if QoS contains at least two additive metrics, then the QoS routing is a NP complete problem [11]. Therefore, searching for the shortest path with minimal cost and finding delay constrained least-cost paths are NP-complete problems. For this reason, approximated solutions and heuristic algorithms should be developed for multi-path constraints QoS routing. Also, to cope with changing of Ad-hoc networks topology, routing methods should be adaptive, flexible, and intelligent.

Use of intelligent algorithms based on Fuzzy Logic (FL), Neural Networks (NNs) and Genetic Algorithms (GAs) can prove to be efficient for traffic control in telecommunication networks [12,?,?,?]. In difference from non-linear programming methods, GA, FL and NNs are heuristic methods which use explicit rules to find feasible routes. The GA uses the genetic operators for optimization. By finding good values of crossover probability, mutation probability, and population size, the response time of the GA can be improved. However, there is a trade-off between the selected route and the response time.

The GA tries to find a route with a good fitness value, but based on genetic operators probability this may be not the global optimum. In the case of mobile Ad-hoc networks, it is better to find a route very fast in order to have a good response time to the speed of topology change, than to search for the best route but without meaning, because the network topology is changed and this route does not exist any-more.

In this paper, we use GAs and multi-objective optimization for QoS routing in Ad-hoc Networks. In order to reduce the search space of GA, we implemented a Search Space Reduction Algorithm (SSRA), which reduces the search space of GAMAN. We evaluate the performance of GAMAN by computer simulations and show that GAMAN has better behaviour than a previous method called Genetic Load Balancing Routing (GLBR).

The paper is organized as follows. In Section 2, we discuss QoS Support in Ad-hoc Networks and QoS metrics. In Section 3, we deal with related work on QoS routing. In Section 4, we present GA-based QoS routing. Our proposed multi-objective optimization method is presented in Section 5. The simulation results are presented in Section 6. Finally, conclusions are given in Section 7.

2. QoS support in Ad-hoc networks and QoS metrics

2.1. QoS support in Ad-hoc networks: Issues and difficulties

The QoS is usually defined as a set of services requirements that needs to be met by the network while transporting a packet stream from a source to a destination. The network needs are governed by the service requirements specified by the end user applications. The network is expected to guarantee a set of measurable pre-specified service attributes to the users in terms of end-to-end performance metrics, such as delay, bandwidth, transmission success ratio, packet loss, and jitter. The power consumption and service coverage area are two other QoS metrics that are more specific to Ad-hoc networks [16].

The Ad-hoc networks differ from traditional wire-line networks. The difference introduce unique issues and difficulties for supporting QoS in Ad-hoc networks environment, which are summarized in following.

- **Unpredictable Link Properties**
  Wireless media is very unpredictable. Packet collision is intrinsic to wireless network. Signal propagation faces difficulties such as signal fading, interference, and multi-path cancellation. All these properties make the measure of bandwidth and delay of a wireless link unpredictable.

- **Hidden-Exposed Terminal Problem**
  Multi-hop packet relaying introduces the hidden-exposed terminal problem. This problem happens when signals of two nodes which are out of transmission range of each other collide at the common receiver. Because of the local nature of transmissions, hidden and exposed stations abound in Ad-hoc networks. A hidden station is a host that is within the range of the receiver but not the transmitter, while an exposed station is within the range of the transmitter but not the receiver.

- **Node Mobility**
  Mobility of the nodes create a dynamic network topology. Links will be dynamically formed when two nodes come into transmission range of each other and be torn down when they move out of the range.

- **Route Maintenance**
  The dynamic nature of the network topology and the changing behaviour of the communication medium make the precise maintenance of network information very difficult. Thus, the routing algorithms in Ad-hoc networks have to operate with inherently imprecise information. Furthermore, in Ad-hoc networks environments, node can join and leave any-time. The established routing paths may be broken even during the process of data transfer. Thus arises the need for maintenance and reconstruction of routing paths with minimal overhead and delay.

- **Limited Battery Life**
  Mobile devices generally are dependent on finite battery sources. The resource allocation for QoS provisioning must consider the residual battery power and the rate of battery consumption corresponding to the resource utilization. Thus, all the techniques for QoS provisioning should be power-aware and power-efficient.

- **Security**
  Security can be considered as a QoS attribute. Without adequate security, unauthorized access and usages may violate the QoS negotiations. The nature of broadcasts in wireless networks potentially results in more security exposures. The physical medium of communications is inherently insecure. So, we need to design security-aware routing algorithms for Ad-hoc networks.

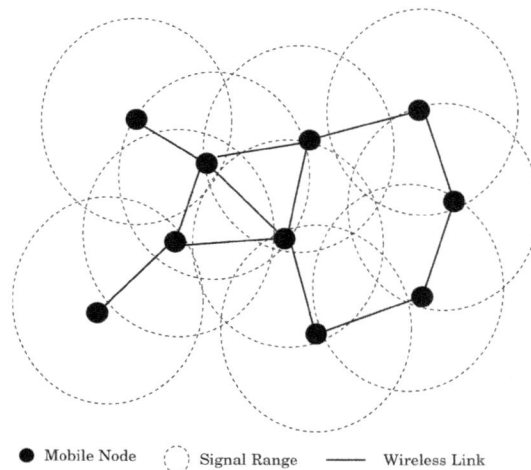

Fig. 1. An Ad-hoc network.

2.2. QoS routing metrics

In [17], the authors present the following metrics for QoS route discovery and selection.

– Minimum Required Throughput or Capacity (bps)
  This is the desired application data throughput.
– Maximum Tolerable Delay (s)
  Usually defined as the maximum tolerable end-to-end (source to destination) delay for data packets.
– Maximum Tolerable Delay Jitter
  One widely accepted definition of this metric is the difference between the upper bound on end-to-end delay and the absolute minimum delay. The former incorporates the queuing delay at each node and the latter is determined by the propagation delay and the transmission time of a packet. This metric can also be expressed as delay variance.
– Maximum Tolerable Packet Loss Ratio (PLR)
  The acceptable percentage of total packets sent, which are not received by the transport or higher layer agent at the packet's final destination node.

An application may typically request a particular quality of service by specifying its requirements in terms of one or more of the above metrics.

3. Related work on QoS routing

In this section, we will review the routing schemes that can support QoS in Ad-hoc networks. In Fig. 1 is shown an Ad-hoc network. The wireless topology derived from Fig. 1 is shown in Fig. 2. The mobile nodes are labelled as A, B, C, ..., K. The numbers beside each edge represent the available bandwidths of the wireless links. Suppose we want to find a route from Source Node (SN) A to a Destination Node (DN) G. For conventional routing using shortest path (in terms of the number of hops) as metric, the route "A-B-H-G" will be chosen. It is quite different in QoS route selection. Suppose we consider bandwidth as QoS metric and desire to find a route from A to G with a minimum bandwidth of 4. Now, the feasible

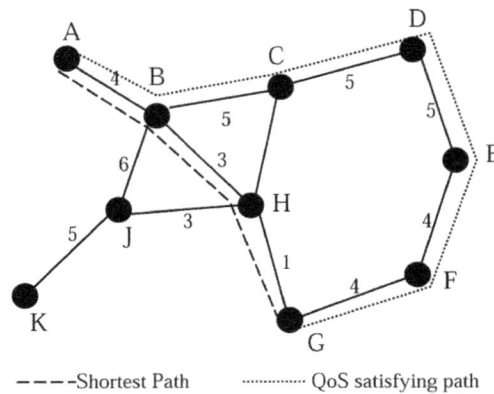

Fig. 2. An example of QoS in Ad-hoc networks.

route will be "A-B-C-D-E-F-G". The shortest path route "A-B-H-G" will not be adequate for providing the required bandwidth.

The primary goal of the QoS routing protocol is to determine a path from a SN to a DN that satisfies the needs of the desired QoS. The QoS path is determined within the constraints of minimal search, distance and traffic conditions. We discuss some QoS routing algorithms in following [17].

3.1. Ticket-based probing algorithm

A ticket-based probing algorithm with imprecise model was proposed by Chen and Nahrstedt [9]. While discovering a QoS-aware routing path, this algorithm tries to limit the amount of flooding (routing) messages by issuing a certain amount of logical tickets. Each probing message must contain at last on ticket. When a probing message arrives at a node, it may be split into multiple probes and forwarded to different next-hops. Each child probe will contain a subset of tickets from their parent. Obviously, a probe with a single ticket cannot be split any more. When one or more probe(s) arrive(s) at the destination, the hop-by-hop path is known and delay/bandwidth information can be used to perform resource reservation for the QoS-satisfying path.

In wire-line networks, a probability distribution can be calculated for a path based on the delay and bandwidth information. In a mobile Ad-hoc network, however, building such a probability distribution is not suitable, because wireless links are subject to breakage and state information is imprecise in nature. Hence a simple imprecise model was proposed for ticket-based probing algorithm. It uses history and current (estimated) delay variations and a smoothing formula to calculate the current delay, which is represented as a range of [delay $- \delta$, delay $+\delta$]. To adapt to the dynamic topology of mobile Ad-hoc networks, this algorithm allows different level of route redundancy. It also uses re-routing and path-repairing techniques for route maintenance. When a node detects a broken path, it will notify the SN, which will reroute the connection to a new feasible path, and notify the intermediate nodes along the old path to release the corresponding resources. Unlike the re-routing technique, path-repairing technique does not find a completely new path. Instead, it tries to repair the path using local reconstructions.

3.2. CEDAR

CEDAR, a Core Extraction Distributed Ad-hoc Routing is proposed as a QoS routing scheme for small and medium scale mobile Ad-hoc networks [10]. It dynamically establishes the core of the network, and then incrementally propagates the link states of stable high-bandwidth links to the core nodes. The route computation is on-demand basis, and is performed by the core nodes using only local state. CEDAR has three key components has follows.

Core Extraction: A set of nodes elected to form the core that maintains the local topology of the nodes in its domain, and also perform route computation. The core nodes are elected by approximating a minimum dominating set of the mobile Ad-hoc Networks.

Link State Propagation: QoS routing in CEDAR is achieved by propagating the bandwidth availability information of stable links to all core nodes. The basic idea is that the information about stable high-bandwidth links can be made known to nodes far away in the network, while information about the dynamic or low bandwidth links remains within the local area.

Route Computation: Route computation first establishes a core path from domain of the SN to the domain of DN. Using the directional information provided by core path, CEDAR iteratively tries to find a partial route from the source to the domain of the furthest possible node in the core path satisfying the requested bandwidth. This node then becomes the source of the next iteration. In CEDAR approach, the core provides an efficient and low-overhead infrastructure to perform routing, while the state propagation mechanism ensures the availability of the link-state information at the core nodes without incurring high overheads.

3.3. Interference-aware QoS routing

In [18], the authors consider throughput-constrained QoS routing based on knowledge of the interference between links. So-called clique graphs are established, which reflect which links interfere with each other, thereby preventing simultaneous transmission. The proposed solution operates by first recording the channel usage (bps) of each existing data session on each link. It is noted that the total channel usage of the sessions occupying the links within the same clique must not exceed the channel capacity. A link's residual capacity is then calculated by subtracting the channel usage of all sessions on links in the same clique from the link's nominal capacity. This link capacity information may then be used in any known distributed Ad-hoc routing protocol to solve the throughput-constrained routing problem.

3.4. Cross-layer multi-constraint QoS routing

In [19], the author focuses on performing multi-constraint QoS routing with three metrics: delay, link reliability and throughput. The author reiterates the fact that the multi-constraint QoS routing problem is NP-complete when a combination of additive and multiplicative metrics is considered. Among the above metrics, delay is additive, link reliability is multiplicative and achievable throughput is concave. However, methods have been proposed for reducing this NP-complete problem to one that can be solved in polynomial time. In one such method, all QoS metrics, except one, take bounded integer values. Then, the task of finding a path to satisfy all constraints can be performed by a modified Dijkstra's algorithm.

In [19], the multiplicative metric is reduced to an additive one by taking the logarithm of the reliability percentage of a link. Also, the delay metric is reduced such that each link is represented by the percentage of the allowable total delay it introduces. The resulting problem in the new metric space can be solved in polynomial time. Then, a modified Bellman-Ford or Dijkstra's algorithm with the new reliability metric for link weights can be used to find an approximation to the optimal path.

An obvious advantage of this approach is the concurrent consideration of several important QoS metrics in path selection. However, the QoS state for all paths must be discovered and kept fresh. This incurs extra overhead and the details of this mechanism are not discussed in the paper.

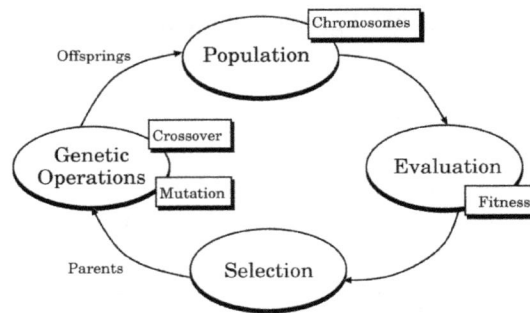

Fig. 3. GA cycle.

4. GA-based QoS routing

4.1. GA cycle

The GA cycle is shown in Fig. 3. At the beginning, an initial population of potential solutions is created as a starting point for the search. In the next stage, the performance (fitness) of each individual is evaluated with respect to the constraints imposed by the problem. Based on each individual's fitness, a selection mechanism chooses "parents" for the crossover and mutation operators.

The crossover operator takes two chromosomes and swaps part of their genetic information to produce new chromosomes. The mutation operator introduces new genetic structures in the population by randomly modifying some of genes, helping the search algorithm to escape from local optimum.

The offspring produced by the genetic manipulation process are the next population to be evaluated. GA can replace either a whole population or just its less fit members. The creation-evaluation-selection-manipulation cycle repeats until a satisfactory solution to the problem is found, or some other termination criteria are met [20].

4.2. GAMAN QoS routing algorithm

In [21], we proposed a GA-based QoS routing algorithm for mobile Ad-hoc networks. We will explain in following GAMAN algorithm.

4.2.1. Network model assumptions

We assume that all the hosts communicate on the same shared wireless channel. Each node has a unique identifier and has at least one transmitter and one receiver. Assume that the effective transmission distance of every node is equal. Two nodes are neighbours and have a link between them if they are in the transmission range of each other.

We assume that there exists a neighbour discovering protocol. Each node periodically transmits a BEACON packet identifying itself, so that every node knows the set of neighbour nodes. We assume the existence of a MAC protocol, which resolves the media contention, supports resource reservation and ensures that among the neighbours in the local broadcast range only the intended receiver keeps the message and the other neighbours discard the message.

We assume small to medium networks. For larger networks some cluster based algorithms or distributed algorithms can be used. We assume that the change in network topology is frequent, but not frequent enough to render any sort of route computation useless. Specifically, we assume that topology changes are typically followed by at least a short period of stability. Note that we only care about the relative mobility of hosts not the absolute mobility of the hosts.

In particular, even if a platoon of solders or cars is moving, the mobile Ad-hoc Networks would be considered to be stable as long as the neighbourhood of each solder or car does not change. The links between the stationary or slow moving nodes are likely to exist continuously. Such links are considered as stationary links. The links between fast moving nodes are likely to exist only for a short period of time. Such links are considered as transient links. A routing path should use the stationary links whenever possible in order to reduce the probability of the path breaking when the network topology changes.

4.2.2. GAMAN goals

We consider a type of mobile Ad-hoc network whose topologies are not changing that fast to make the QoS routing meaningless. We want to emphasize that GAMAN supports soft QoS without hard guarantees. The soft QoS means that there may exist transient time periods when the required QoS is not guaranteed due path breaking or network partition. However, the required QoS should be ensured when the established paths remain unbroken. Many multimedia applications accept soft QoS and use adaptation techniques to reduce the level of QoS disruption.

In CEDAR, the bandwidth is used as the only QoS parameter for routing. Also, in ticked-based routing method the delay and bandwidth are used for QoS routing but not together. They are implemented as different algorithms. In this version, the GAMAN algorithm uses two parameters: delay and transmission success rate to decide the QoS path. To our best knowledge, this is the first work for MANET QoS routing to consider two QoS parameters.

The routing in mobile Ad-hoc networks has the following goals: a) the routing computation for small networks should be made at the SN in order to avoid the computation at intermediate nodes; b) as few nodes as possible must be involved in route computation; c) each node must only care about the routes corresponding to its destination, and must not be involved in frequent topology updates for parts of the network to which it has no traffic; d) broadcast must be avoided as far as possible; e) it is desirable to have a backup route when the primary route has become stale and is being recomputed; f) messages should be not cycled around loops.

Our GAMAN algorithm can satisfy these requirements: a) the GAMAN is a source-based routing algorithm; b) by using a SSRA and small population size few nodes are involved in route computation; c) by taking a sub-population, the nodes in this sub-population care only about the routes in this sub-population; d) the broadcast is avoided because the information is transmitted only for the nodes in a population; e) the GA search different routes and they are sorted by ranking them. So the first one is the best route, but other ranked routes can be used as backup routes; f) by using a tree based GA method, the loops can be avoided.

In summary, our goal is to compute good routes quickly, and react to the dynamics of the network very fast. As a results, we sacrifice optimality of routes. Robustness, rather than optimality is the key requirement.

4.2.3. SSRA

By SSRA a network with many nodes will be reduced in a network with a small number of nodes. Thus, GAMAN is able to cope with increase of number of nodes in Ad-hoc networks.

The flowchart of SSRA is shown in Fig. 4. The key element of SSRA is topology extraction. In order to extract the topology, the node information and QoS requirement are needed. In this work, we use TSR (Th), DT (Th), CC (Th) to specify the QoS requirements. If DT < DT (Th), TSR > TSR (Th), CC

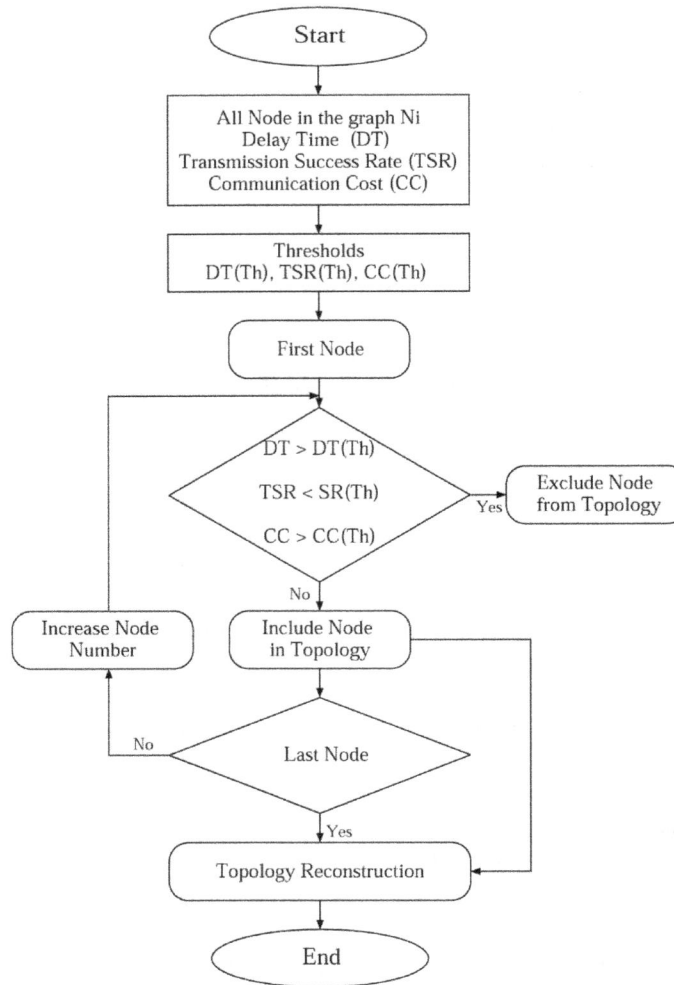

Fig. 4. SSRA Flowchart.

< CC (Th), then the node is included in the extracted topology, otherwise the node is excluded from the extracted topology. The procedure is repeated until all nodes are finished. After all nodes are checked, the network topology is constructed and the complete procedure is finished.

4.2.4. Gene coding

To explain this procedure, we use a small mobile Ad-hoc network with 8 nodes as shown in Fig. 5. Let consider node A as a SN and node H as a DN. All routes are expressed by the network tree model shown in Fig. 6. The shaded areas show the same routes from node C to H. In order to decrease the chromosome gene number, the network tree model of Fig. 6 is reduced as shown in Fig. 7. In the reduced network tree model, each tree junction is considered as a gene and the path is represented by the chromosome.

The most important factor to achieve efficient genetic operations is gene coding, because it has influence on the efficiency of genetic operations.

In [22], the GLBR method which uses GA has been proposed. The adaptive routing mechanism has a load balancing system among alternative paths. In the GLBR method, the genes are put in a chromosome

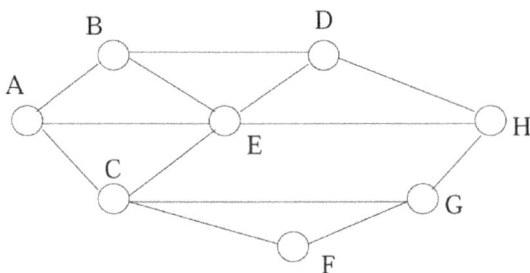

Fig. 5. A mobile Ad-hoc network with 8 nodes.

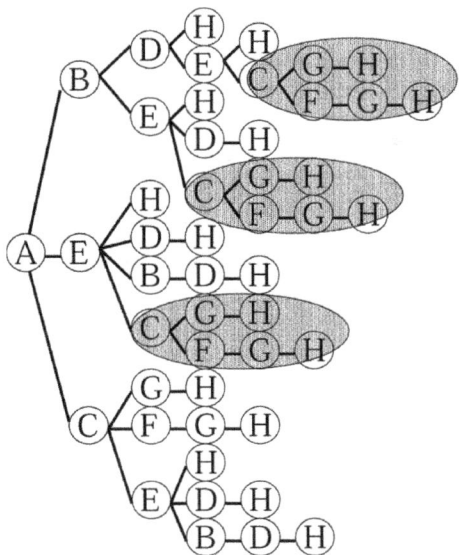

Fig. 6. Network tree model.

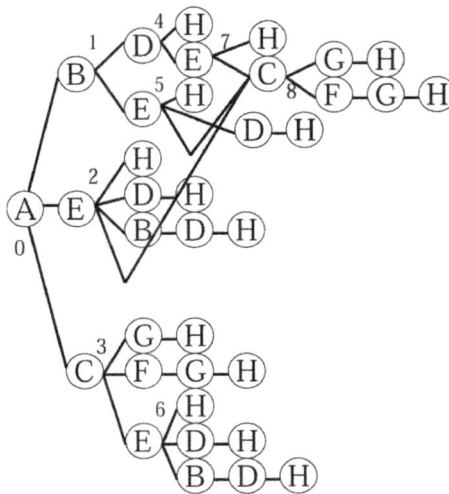

Fig. 7. Reduced network tree model.

in the same order the nodes form a route as shown in Fig. 8(a). Therefore, the chromosomes have different sizes. If genetic operations are chosen randomly, the new off-springs of a population may be unsuitable individual populations. Also, because the individuals of a population have different sizes, the crossover operations are complicated.

In order to simplify the genetic operations of GLBR method, in GAMAN, the network is changed in a tree, then the tree network is reduced by grouping together the same routes. After, the tree reduction, the tree junctions are expressed by genes as shown in Fig. 8(b). In Fig. 8(b) is shown a chromosome example for the route "A-B-D-E-C-F-G-H" of network example. Each chromosome expresses only one route. The genes contain the information of the adjacent nodes. As is shown in Fig. 8(b), the chromosomes have the same length. Therefore, the crossover operation becomes very easy. The genes in a chromosome have two states "active" and "inactive". A gene is called active if the junction is in the route, otherwise the gene is in "inactive" state.

In GLBR method, the interaction between the adjacent genes in a chromosome is necessary. While, in GAMAN this interaction is not necessary. So, the mutation operation also becomes easy. Furthermore, the GAMAN has a good gene coding method, which is able to create various individuals, which result in a fast evolution.

0	1	2	3	4	5	6	7
A	B	D	E	C	F	G	H

(a) GLBR gene coding.

0	1	2	3	4	5	6	7	8
BEC	DE	HD BC	GFE	HE	HDC	HDB	HC	GF

(b) GAMAN gene coding.

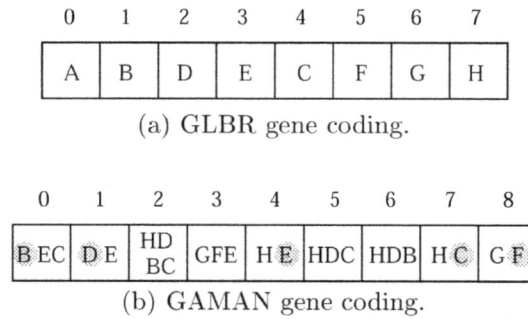

Fig. 8. Chromosome examples for the route "A-B-D-E-C-F-G-H".

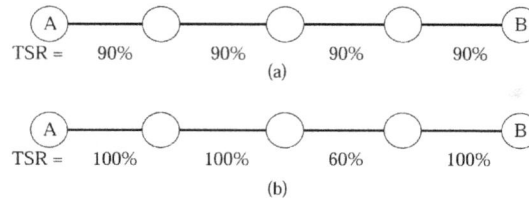

Fig. 9. An example of TSR calculation.

In GAMAN, the selection operation uses both the ranking and elitist models. As the crossover method is used the single point crossover. In the mutation operation, the genes are chosen randomly in the range from zero up to mutation probability $p\_mutation \leqslant \frac{1}{l}$, where l is the chromosome length.

4.2.5. QoS routing parameters

The GAMAN algorithm can use many parameters for QoS routing. In following we consider the case for two QoS parameters: Delay Time (DT) and Transmission Success Rate (TSR). The DT means the time it takes a packet to go from one node to another one. The TSR shows the rate of correctly transmitted packets (without loss).

Let consider a wireless network with wireless nodes and links as shown in Fig. 9. The node A is the source node and node B is the destination node. Let node A sends to node B 10 packets. The total TSR value for Figs 9(a) and 9(b) is calculated by Eqs (1) and (2), respectively.

$$10 \times 0.9 \times 0.9 \times 0.9 \times 0.9 = 6.561 \tag{1}$$

$$10 \times 1.0 \times 1.0 \times 0.6 \times 1.0 = 6.000 \tag{2}$$

The best route in this case is that of Fig. 8(a), because the total TSR is higher compared with that of Fig. 8(b).

Let consider another example, when the values of DT and TSR are considered as shown in Fig. 10. The value of T parameter is decided as follows.

$$T = \frac{\sum_{i=1}^{n} DT_i}{\prod_{i=1}^{n} TSR_i} \tag{3}$$

where n is the number of wireless links in a path.

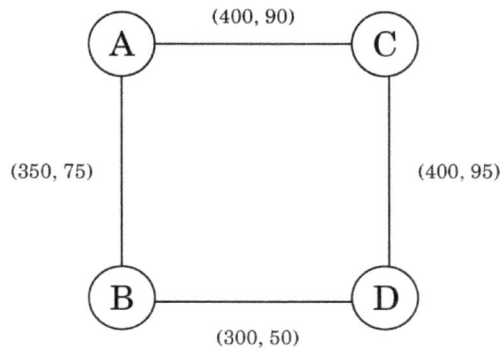

Fig. 10. A network example for mobile Ad-hoc network QoS routing.

When node A wants to communicate with node D, there are two possible routes: "A-B-D" and "A-C-D". The T value for these routes is calculated by Eqs (4) and (5), respectively.

$$T_{A-B-D} = \frac{350 + 300}{75 \times 50} = \frac{650}{3750} = 0.1733 \tag{4}$$

$$T_{A-C-D} = \frac{400 + 400}{90 \times 95} = \frac{800}{8550} = 0.0468 \tag{5}$$

The delay time of "A-B-D" route is lower than "A-C-D" route, but the T value of "A-C-D" route is lower than "A-B-D", so "A-C-D" route is the better one. This shows that other good candidate routes can be found when two QoS parameters are used for routing in mobile Ad-hoc networks.

4.2.6. GAMAN operation

The GAMAN is a source-based routing mechanism and uses two QoS parameters for routing. When a node of mobile Ad-hoc network wants to transmit information to a DN, this node becomes the SN. The network first is transformed in a tree network with the SN as the root of tree. After that, the tree network is reduced in the parts where are the same routes. By reducing the tree network, the chromosome length is shorten so the genetic operations become simple.

After the reduction of the tree network, the tree junctions are coded as genes. The genes in a chromosome have the information of the adjacent nodes. Because, the individual and chromosome are the same, the route is represented by the chromosome and the population is a collection of wireless links.

After the gene coding, GAMAN starts the genetic operations. First, an initial population is selected. In the selected population, the ranking selection model is used to select individuals in order to carry out the genetic operations. The ranking model ranks each individual by their fitness. The rank is decided based on the fitness and the probability is decided based on the rank. The individual fitness is based on T value. When T value is small, the individual fitness is high.

The genetic operations are the crossover and mutation. The GAMAN algorithm uses the single point crossover, because simple operations are needed to get a fast response. In the mutation operation, the genes are chosen randomly in the range from zero up to mutation probability $p\_mutation \leqslant \frac{1}{\ell}$, where l is the chromosome length.

After the crossover and mutation, the elitist model is used. Based on the elitist model the individual which has the highest fitness value in a population is left intact in the next generation. Therefore, the best value is always kept and the routing algorithm can converge fast. The off-springs produced by the

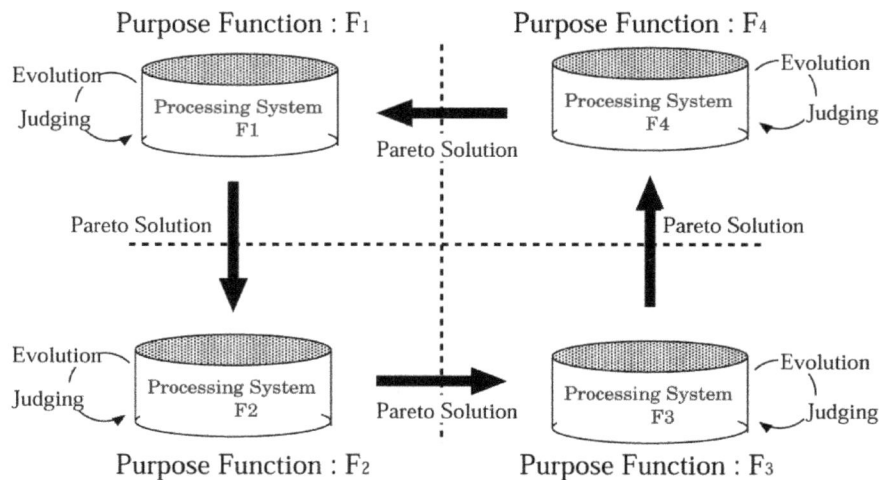

Fig. 11. Multiple-objective optimization.

genetic operations are the next population to be evaluated. The genetic operations are repeated until the initialized generation size is achieved or a route with a minimum T value is found.

The route selection in GAMAN is based on T value, which is the ratio of DT with TSR. T is used as a fitness function to evaluate the selected individuals (routes). By minimizing the T value, the DT value is minimized and the TSR value is maximized. This means that a packet from node SN to node DN is transmitted with a small delay and a high transmission success rate.

5. Proposed multi-objective optimization method

In this section, we propose a new method for Ad-hoc networks considering GA and multi-objective optimization. The proposed method uses the multi-division group model for multi-objective optimization. When a function can be divided in different objective functions, the global domain can be divided in different domains and each individual can evolve in its domain.

The procedure is shown in Fig. 11, where four different objective functions are independent from each other and each process is operating independently in each domain. In Fig. 12 is shown an example of Delay Time (DT) and Communication Cost (CC). The vertical axis shows the DT and horizontal axis the CC. The DT and CC have trade-off relations. The points in the figure show the individuals (routes). The individuals which are in the left-upper part of the figure have the lowest CC. On the other hand, the individuals which are in the right-lower part of the figure have the lowest DT values. The individuals which are in the shaded area have good values for both DT and CC. The shaded area is called "pareto solution". The individuals near pareto solution can be found by exchange the solutions of different domains.

5.1. QoS routing search engine

The structure of QoS Routing Search Engine (QRSE) is shown in Fig. 13. It includes Cache Search Engine (CSE) and GAMAN. CSE and GAMAN operate independently, but they cooperate together to update the route information. When the RSE receives a request for QoS routing from a client, it forwards

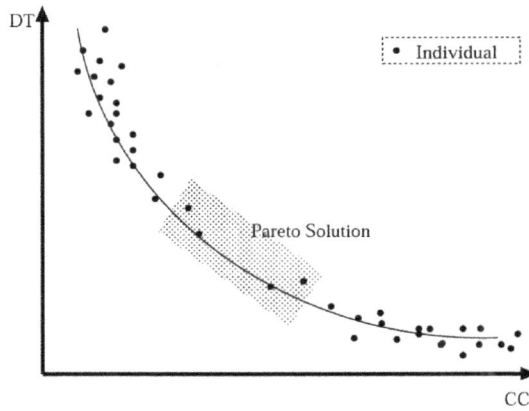

Fig. 12. Pareto solution for DT and CC.

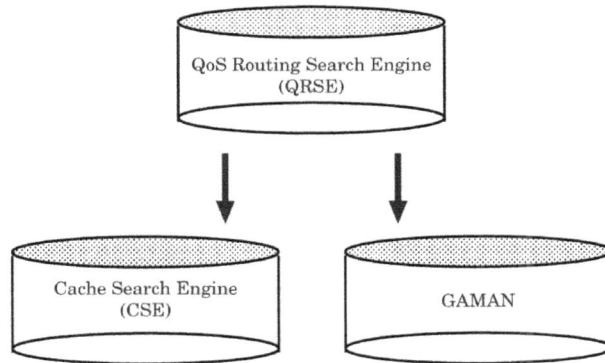

Fig. 13. RSE structure.

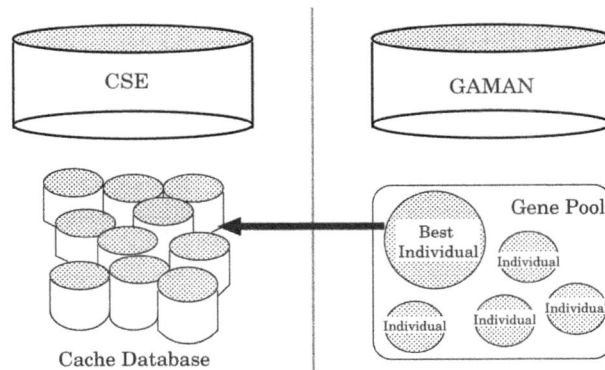

Fig. 14. CSE and GAMAN cooperation1.

the request in parallel to CSE and GAMAN. Then, the CSE and GAMAN search in parallel to find a route satisfying the required QoS. The CSE searches for a route in the cache database (in the cache database, the destination and route information is saved as a database item). If the found route by CSE

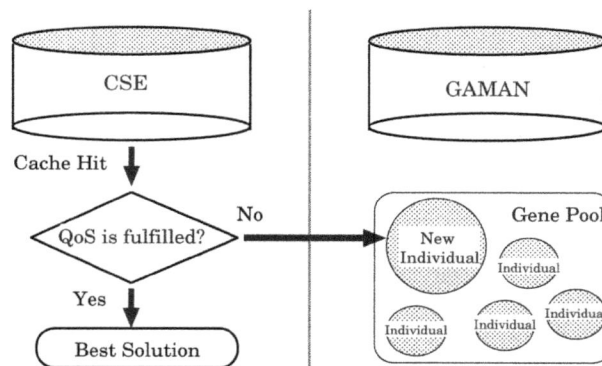

Fig. 15. CSE and GAMAN cooperation2.

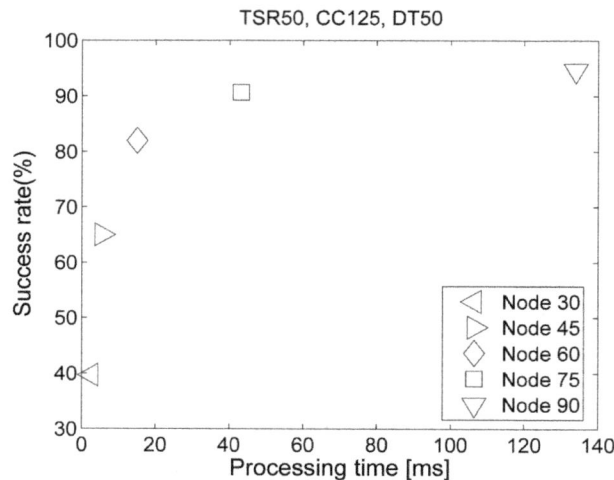

Fig. 16. Algorithm success rate versus processing time.

satisfies the required QoS, this route information is sent to RSE, otherwise the route is put in the gene pool as new individual. If a QoS route can not be found by CSE, the route found by GAMAN is sent to RSE.

5.2. CSE-GAMAN cooperation and database updating

The database information should be updated because the network traffic and the network state change dynamically. In order to update the database, the CSE and GAMAN cooperate together. When the GAMAN finds a good QoS routing, it puts this route information in the cache database as shown in Fig. 14. The route information which is used frequently is given high priority, thus this route can be searched very fast by CSE. In the case when the CSE finds a route in the database, but this route does not satisfy the required QoS by client, this route information is put as a new individual in the genepool of GAMAN as shown in Fig. 15.

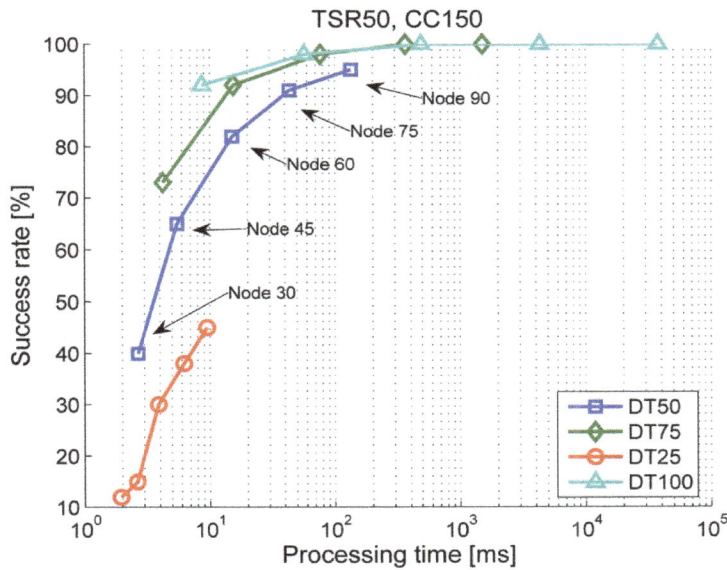

Fig. 17. Algorithm success rate versus processing time for different DTs.

6. Simulation results

We show in Figs 16 and 17 the simulation results for SSRA. By reducing the search space, the GAMAN can find faster a new route. We use as QoS parameters: Transmission Success Rate (TSR), DT and CC. As the simulation environment was used desk-top computer (OS: Windows XP Pro; IDE: Visual C++6.0; CPU: Athlon64 +3500, Memory: 1024 MB). The QoS parameters were set up in a random way. The DT for each link was set between 1 and 100, TSR was set between 1 and 100% and CC between 1 and 200 units. In Fig. 16, we set the threshold parameters $TSR(Th) = 50$, $CC(Th) = 125$, $DT(Th) = 50$. Then, we changed the number of nodes from 30 to 45, 60, 75 and 90 and measure the algorithm success rate. As shown in these figures, with the increase of the number of nodes the success rate to find a feasible route increases. In Fig. 17, we consider $TSR(Th) = 50$, $CC(Th) = 150$ and change $DT(Th)$ from 25 to 100. As shown in the figure with the increase of the DT threshold the success rate is increased. However, when the threshold is very small the success rate is less than 20%. Therefore, there are trade-offs for deciding the threshold parameters.

In Table 1, we show the simulation results of the proposed method for population size 10 and for different number of generations, crossover rates and mutation rates.

We can see that with increase of mutation rate, the average number of generations and average processing time is decreased. But, with increase of crossover rate, the average number of generations and average processing time is increased. However, when the mutation rate is 8%, the performance of proposed method is better than 10%. Also for crossover rate 70% and mutation rate 8% we have the best performance. In Table 2 are shown the simulation results of GLRB. Comparing Tables 1 and 2, GAMAN has a better performance than GLBR.

The simulation results of GAMAN for the time needed for one generation are shown in Table 3. If there are few individuals in the population, the Generation Number (GN) which shows the number of generations needed to find a solution becomes large. On the other hand, when the number of individuals is high, the GN to find a solution becomes small. However, when the number of individuals is 12 and

Table 1
GAMAN Simulation results for population size 10

Mutation rate [%]	Crossover rate [%]	Average no. of generations	Average processing time [ms]
1	70	15.7	468.2
	80	16.7	487.5
	90	21.2	590.7
	100	19.3	536.7
5	70	11.3	213.3
	80	9.8	185.9
	90	10.7	203.9
	100	14.1	275.2
8	70	7.4	120.6
	80	7.8	137.5
	90	7.9	133.3
	100	9.1	157.3
10	70	9.1	149.3
	80	8.9	143.8
	90	8.2	135.5
	100	7.4	120.6

Table 2
GLBR Simulation results for population size 10

Mutation rate [%]	Crossover rate [%]	Average no. of generations	Average processing time [ms]
1	70	20.7	2017.1
	80	23.6	1652.8
	90	22.3	1842.9
	100	21.5	1872.0
5	70	15.4	497.3
	80	17.5	647.4
	90	17.7	668.0
	100	22.1	589.9
8	70	12.4	404.6
	80	10.0	389.9
	90	24.4	538.8
	100	21.4	494.4
10	70	14.2	384.1
	80	11.5	442.1
	90	17.2	468.2
	100	17.3	368.8

16, the difference is very small. This is because some individuals become the same in the genepool. Considering the exchange of individuals between domains, it can be seen that when the exchange interval is short the solution can be found very fast. This shows that by exchanging the individuals the algorithm can approach very quickly to the pareto solution.

In Fig. 18 are shown the simulation results for number of generations and the processing time. In Fig. 18(a), we show the Generation Size (GS) versus the number of individuals. While, in Fig. 18(b) is shown the processing time versus the number of individuals. We consider 4 cases when the number of exchanging individuals is 3, 5, 7, and 10, respectively. If there are few individuals in the population, the GS which shows the number of generations needed to find a solution becomes large.

Table 3
Time needed for one generation (ms)

Number of individuals	GN exchange			
	3	5	7	10
4	44.43	50.45	46.19	55.59
8	26.83	28.01	40.26	31.17
12	23.55	26.49	26.04	26.71
16	22.22	22.23	23.25	24.04

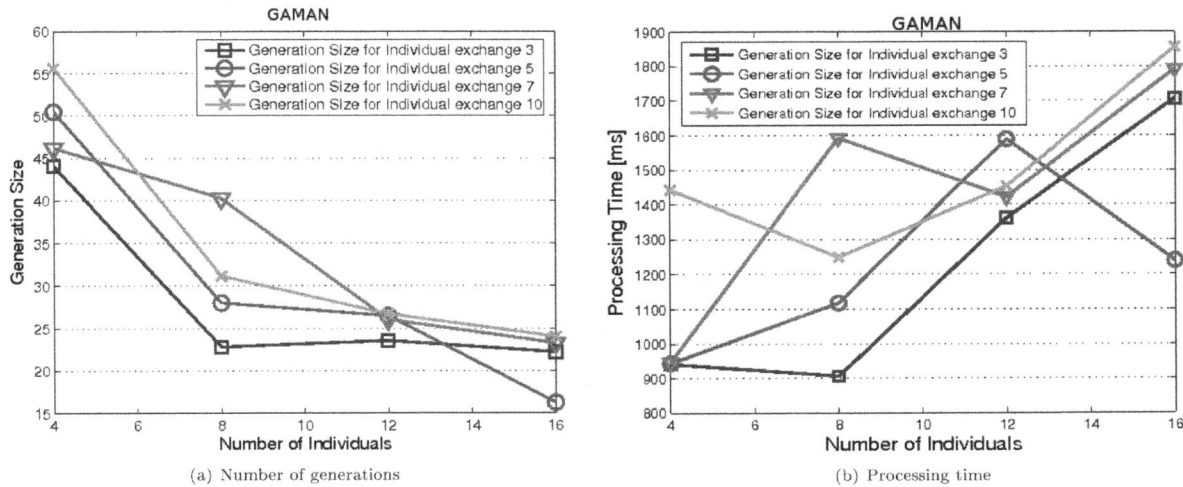

(a) Number of generations

(b) Processing time

Fig. 18. Relation between generation size and number of individuals. (Colours are visible in the online version of the article; http://dx.doi.org/10.3233/MIS-2011-0116)

7. Conclusions

In this paper we propose a new method for Ad-hoc networks considering GA and multi-objective optimization. The proposed method uses the multi-division group model for multi-objective optimization.

We evaluated SSRA and GAMAN by computer simulations. From the simulation results, we found the following results.

– The SSRA can reduce the search space of GAMAN, so GAMAN can find faster a new route.
– With increase of mutation rate, the average number of generations and average processing time is decreased. But, with increase of crossover rate, the average number of generations and average processing time is increased. However, when the mutation rate is 8%, the performance of proposed method is better than 10%. Our proposed method has the best performance for crossover rate 70% and mutation rate 8%.
– If there are few individuals in the population, the GS which shows the number of generations needed to find a solution becomes large.
– GAMAN has better performance than GLBR.

We would like to consider in the future a policy-based routing using FL. We want to put the policies in the Fuzzy Rule Base (FRB) and use this policies for QoS-routing. We also want to investigate an integrated approach of GA-based routing and policy-based routing. We also want to consider the swarm intelligence approach combined with GA for QoS routing in Ad-hoc networks. Population of the agents (ants) can be adapted according to the problem size. In swarm intelligence the scalability can be promoted by local and distributed agent interactions. Swarm intelligent processes do not rely on a centralized control mechanism. Therefore the loss of a few agents does not result in catastrophic failure, but rather leads to graceful, scalable degradation. Agents can change, die or reproduce, according to system changes.

References

[1]   C.E. Perkins, *Ad-hoc Networking*, Addison-Wesley, 2001.

[2]   E.M. Royer and C.K. Toh, A Review of Current Routing Protocols for Ad-hoc Mobile Wireless Networks, *IEEE Personal Communications* (April 1999), 46–55.

[3]   S. Chen and K. Nahrstedt, An Overview of Quality of Service Routing for Next-Generation High-Speed Networks: Problems and Solutions, *IEEE Network*, Special Issue on Transmission and Distribution of Digital Video, Vol. 12, No. 6, pp. 64–79, 1998.

[4]   W.C. Lee, M.G. Hluchyj and P.A. Humblet, Routing Subject to Quality of Services Constraints in Integrated Communications Networks, *IEEE Network* **9**(4) (1995), 46–55.

[5]   H. Xiao, W.K.G. Seah, A. Lo and K.C. Chua, A Flexible Quality of Service Model for Mobile Ad-hoc Networks, *Proc. of IEEE VTC-2000*, Tokyo, May 2000.

[6]   G.S. Ahn, A.T. Campbell, A. Veres and L.H. Sun, Supporting Service Differentiation for Real-Time and Best-Effort Traffic in Stateless Wireless Ad-hoc Networks (SWAN), *IEEE Transactions on Mobile Computing* **1**(3) (2002), 192–207.

[7]   S.B. Lee, G.S. Ahn, X. Zhang and A.T. Campbell, INSIGIA: An IP-Based Quality of Service Framework for Mobile Ad-hoc Networks, *Journal of Parallel and Distributed Computing* **60** (2000), 374–406.

[8]   S. Chen, S.H. Shah and K. Nahrstedt, Cross-Layer Design for Data Accessibility in Mobile Ad-hoc Networks, *Journal of Wireless Personal Communications* **21** (2001), Special Issue on Multimedia Network Protocols and Enabling Radio Technologies, Kluwer Academic Publishers, 49–75.

[9]   S. Chen and K. Nahrstedt, Distributed Quality-of-Service Routing in Ad-hoc Networks, *IEEE Journal of Selected Areas in Communications* **17**(8) (1999), 1–18.

[10] R. Sivakumar, P. Sinha and V. Bharghavan, CEDAR: a Core-Extraction Distributed Ad-hoc Routing Algorithm, *IEEE Journal of Selected Areas in Communications* **17**(8) (1999), 1454–1465.

[11] K. Wu and J. Harms, QoS Support in Mobile Ad-hoc Networks, *Crossing Boundaries – An Interdisciplinary Journal* **1**(1) (Fall 2001), 92–106.

[12] Special Issue on Computational Intelligence in Telecommunication Networks, *Computer Communications* **25**(16) (2002).

[13] L. Barolli, A Speed-Aware Handover System for Wireless Cellular Networks Based on Fuzzy Logic, *Mobile Information Systems (MIS)* **4**(1) (2008), 1–12.

[14] J. Anno, L. Barolli, A. Durresi, F. Xhafa and A. Koyama, Performance Evaluation of Two-Fuzzy based Cluster Head Selection Systems for Wireless Sensor Networks, *Mobile Information Systems (MIS)* **4**(4) (December 2008), 297–312.

[15] Gj. Mino, L. Barolli, F. Xhafa, A. Durresi and A. Koyama, Implementation and Performance Evaluation of Two Fuzzy-based Handover Systems for Wireless Cellular Networks, *Mobile Information Systems (MIS)* **5**(4) (2009), 339–361.

[16] E. Kulla, M. Hiyama, M. Ikeda, L. Barolli, V. Kolici and R. Miho, MANET Performance for Source and Destination Moving Scenarios Considering OLSR and AODV Protocols, *Mobile Information Systems (MIS)* **6**(4) (December 2010), 325–339.

[17] L. Hanzo (II.) and R. Tafazolli, A Survey of QoS Routing Solutions for Mobile Ad-hoc Networks, *IEEE Communications Surveys and Tutorials* **9**(2) (2nd Quarter 2007), 50–70.

[18] R. Gupta, Z. Jia, T. Tung and J. Walrand, Interference-aware QoS Routing (IQRouting) for Ad-hoc Networks, *Proc. of Global Telecommunications Conference* **5** (November 2005), 2599–2604.

[19] Z. Fan, QoS Routing Using Lower Layer Information in Ad-hoc Networks, *Proc. of Personal, Indoor and Mobile Radio Communications Conference*, pp. 135–139, September 2004.

[20] D.E. Goldberg, *Genetic Algorithms in Search, Optimization, and Machine Learning*, Addison–Wesley, 1989.

[21] L. Barolli, A. Koyama, T. Suganuma and N. Shiratori, GAMAN: A GA Based QoS Routing Method for Mobile Ad-hoc Networks, *Journal of Interconnection Networks (JOIN)* **4**(3) (September 2003), 251–270.

[22] M. Munetomo, Y. Takai and Y. Sato, An Adaptive Routing Algorithm with Load Balancing by a Genetic Algorithm, *Trans of IPSJ* **39**(2) (1998), 219–227.

[23] J. Anno, L. Barolli, A. Durresi, F. Xhafa and A. Koyama, Performance Evaluation of Two Fuzzy-based Cluster Head Selection Systems for Wireless Sensor Networks, *Mobile Information Systems* **4**(4) (December 2008), 297–312.

[24] L. Barolli, M. Ikeda, G. De Marco, A. Durresi, A. Koyama and J. Iwashige, A Search Space Reduction Algorithm for Improving the Performance of a GA-based Routing Method in Ad-hoc Network, *International Journal of Distributed Sensor Networks (IJDSN)* **3**(1) (January 2007), 41–57.

[25] S. Zhao and D. Raychaudhuri, Policy-based Adaptive Routing in Mobile Ad-hoc Wireless Networks, *Proc. IEEE Sarnoff-2006 Symposium*, Available on line at http://www.winlab. rutgers.edu/~sulizhao/szhao_sarnoff06.pdf, March 2006.

[26] E. Bonabeau, M. Dorigo and G. Theraulaz, *Swarm Intelligence: from Natural to Artificial Systems*, Oxford University Press, 1999.

[27] M. Dorigo, V. Maniezzo and A. Colorni, The Ant System: Optimization by a Colony of Cooperating Agents, *IEEE Transactions on Systems, Man, and Cybernetics, Part B* **26**(1) (1996), 29–41.

[28] A. Koyama, L. Barolli, G. Capi, B.O. Apduhan, J. Arai and A. Durresi, An Efficient Multi-Purpose Optimization Method for QoS Routing Using Genetic Algorithm, *Journal of Interconnection Networks (JOIN)* **5**(4) (December 2004), 409–428.

[29] Y.C. Hu and D.B. Johnson, Caching Strategies in On-demand Routing Protocols for Wireless Ad-hoc Networks, *Proc. of IEEE/ACM MobiCom-2000*, 2000.

Admir Barolli was graduated from Agricultural University of Tirana, Albania. He got his Diploma Degree in April 2008. From October 2009 to June 2010, he was a Visiting Researcher at Curtin University of Technology, Australia. Presently, he is a Visiting Researcher at the Department of Computer and Information Science, Seikei University, Japan and working toward his PhD Degree. He has been working for many years in Agriculture Sectors in Albania. His research interest are in genetics, genetic algorithms, agricultural engineering, intelligent algorithms, climate change, global warming, computer networks and P2P systems.

Evjola Spaho received her B.S and M.S degrees at Faculty of Information Technology, Polytechnic University of Tirana (PUT) in 2008 and 2010, respectively. Presently, she is a Ph.D Student at Graduate School of Engineering, Fukuoka Institute of Technology (FIT), Japan. Her research interests include P2P networks, vehicular networks, ad-hoc networks and robot control.

Leonard Barolli is a Professor at the Department of Information and Communication Engineering, Fukuoka Institute of Technology (FIT), Japan. He received BE and PhD Degrees from Tirana University and Yamagata University in 1989 and 1997, respectively. He has published about 300 papers in Journals, Books and International Conference. He has served as a Guest Editor for many Journals. He was PC Chair of IEEE AINA-2004 and ICPADS-2005. He was General Co-Chair of IEEE AINA-2006, AINA-2008 and AINA-2010. He is Steering Committee Chair of CISIS International Conference. His research interests include, P2P, intelligent algorithms, ad-hoc and sensor networks. He is a member of IEEE, IEEE Computer Society, IPSJ and SOFT.

Fatos Xhafa joined the Department of Languages and Informatics Systems of the Technical University of Catalonia as an Assistant Professor in 1996 and is currently an Associate Professor and member of the ALB- COM Research Group of this department. His current research interests include parallel algorithms, combinatorial optimisation, approximation and meta-heuristics, distributed programming, grid and P2P computing. He has published in leading international journals and conferences and has served in the organising committees of many conferences and workshops. He is also a member of the editorial board of several international journals.

Makoto Takizawa is a Professor at the Department of Computer and Information Science, Seikei University. He was a Professor and the Dean of the Graduate School of Science and Engineering, Tokyo Denki University. He was a Visiting Professor at GMD-IPSI, Keele University, and Xidian University. He was on the Board of Governors and a Golden Core member of IEEE CS and is a fellow of IPSJ. He received his DE in Computer Science from Tohoku University. He chaired many international conferences like IEEE ICDCS, ICPADS, and DEXA. He founded IEEE AINA. His research interests include distributed systems and computer networks.

3

Sensor relocation for emergent data acquisition in sparse mobile sensor networks

Wei Wu[a,*], Xiaohui Li[a], Shili Xiang[a], Hock Beng Lim[b] and Kian-Lee Tan[a]

[a]*School of Computing, National University of Singapore, Singapore*
[b]*Intelligent Systems Centre, Nanyang Technological University, Singapore*

Abstract. In this paper, we study the problem of sensor relocation for emergent data acquisition (initiated by a base station) in sparse mobile sensor networks. We propose a distributed scheme called BRIDGE that relocates mobile sensors to fulfill an emergent data acquisition task with the objective to minimize the task completion time. BRIDGE gradually finds a sensor that is close to the task location and relocates that sensor to the task location, and at the same time relocates some other sensors to connect that sensor to the base station. BRIDGE exploits the encountered sensors during relocation, and handles the challenges caused by intermittent connections. Our extensive performance study shows the effectiveness of our proposed scheme.

Keywords: Sensor network, data acquisition, sensor relocation

1. Introduction

Recently an increasing number of research activities are being carried out for Mobile Sensor Networks (MSNs) [1,5,8,12,15,20] where the sensors are capable of moving. The mobility of the sensors increases the sensors' coverage [13], and makes them possible to adapt to the environments, because the sensors are not constrained by the initial deployment and can be relocated to desirable locations when necessary.

We in this paper consider a class of MSNs applications which involve a base station and a number of mobile sensors. The sensors' basic task is to explore a large area and send collected information to the base station. The base station sometimes may have an emergent data acquisition task that requires one sensor to sense a specific location and send the data back to the base station as soon as possible.

For the sake of concreteness, let us look at an **application example**. In battlefield, a command center (CC) dispatches a small number of UAVs (Unmanned Aerial Vehicle) to scan a big area. These UAVs scan the area by taking pictures of the region that they fly over. They send back the pictures to the CC when they have wireless connection to it. Based on the pictures received or some other intelligence sources, the experts at the CC may find a region suspicious. When this happens, the CC issues an emergent task about that region and wants the UAVs to collect detailed information about that region and send the data back as soon as possible.

---

*Corresponding author: Wei Wu, School of Computing, National University of Singapore, Singapore 117417. E-mail: wuw@nus.edu.sg.

We study the problem of relocating mobile sensors to carry out the emergent data acquisition tasks in *sparse* MSNs where the connections among the nodes (sensors and the base station) are intermittent. The aim is to make the sensors complete an emergent task as early as possible.

We study this problem in the context of sparse MSNs because of two reasons. First, MSNs are likely to be sparse. Mobile sensors are more expensive than stationary sensors, so it may not be feasible to deploy a large number of them. Moreover, dense MSNs can become sparse due to node failures caused by environmental hazards or even intentional demages (e.g. by adversaries in the battlefield). Second, solutions designed for sparse MSNs are more robust and versatile, because they will also work well in dense MSNs.

Although several works have studied the mobile sensor relocation problem [5,11,22,23,25], our problem is different in relocation objectives and in system settings and therefore existing approaches cannot be used to solve our problem. Existing works investigate the problem of relocating certain number of sensors to a region so that the region is covered by the sensors with a certain density. We are interested in relocating some sensors so that an emergent data acquisition task can be fulfilled in a short period of time. Existing works assume a system where all the sensors are connected and reachable. We look at a system where the sensor network is sparse and the connections in the system are intermittent.

Relocating sensors to fulfill an emergent data acquisition task in a sparse MSNs is challenging. If all the sensors are connected, then the following straightforward solution would be good enough: find a sensor near the task location, let it move to the task location to collect data, and use connected nodes to relay the data back to the base station. However, in sparse networks, this straightforward solution will not work well, because: 1) when the base station issues the task, most sensors are not reachable, therefore the base station does not know which sensor is near the task location and have no way to contact that sensor because the network is partitioned; 2) the base station therefore has to select a sensor that is connected to it to carry out the task; it may take much time for the sensor to move to the task location; 3) after collecting the data from the task location, the sensor may have no connection to the base station so it cannot send the acquired data directly back to the base station; it has to move towards the base station; again, it may need to move a long way before it is connected to the base station.

We propose a distributed sensor relocation scheme, called BRIDGE, by which the mobile sensors relocate themselves to carry out an emergent data acquisition task cooperatively. The main idea is to gradually find a sensor that is close to the task location and at the same time relocate some sensors to build a connection between the sensor at the task location and the base station. BRIDGE exploits the connected and encountered sensors to minimize the task completion time, by relocating proper sensors to proper locations.

The contributions of this paper are:

1. We identify an interesting problem: relocation of mobile sensors for emergent data acquisition in sparse mobile sensor networks.
2. We propose a distributed relocation scheme in which the mobile sensors relocate themselves to fulfill an emergent data acquisition task through collaboration. In the scheme we deal with the various problems caused by intermittent connections.
3. We show through an extensive simulation study that the proposed scheme is effective.

The rest of the paper is organized as follows. In Section 2 we briefly survey the related works. In Section 3 we describe the system model and the emergent data acquisition task, and define the associated sensor relocation problem. We present our distributed relocation scheme in Sections 4. Results of experimental study are shown in Section 5. We finally conclude the paper in Section 6.

2. Related works

In static wireless sensor networks, various techniques [3,7,16,26] have been proposed to organize sensors into logical structures to facilitate data collection and query processing. Unfortunately, these techniques cannot be applied to mobile sensor networks where the topology of the network is very dynamic.

Task execution using mobile sensors (e.g., UAVs) has attracted much research attention. In [21], multiple mobile UAVs cooperate to facilitate probabilistic information fusion to achieve high-accuracy environment perception and target tracking. The objective is to determine the actions a UAV should carry out so as to maximize the belief of the current information. This is different from our objective, which is to minimize the time for fulfilling an emergent data acquisition task. There are also works on multi-task allocation and path planning for cooperating UAVs, to minimize the task completion time, using market based approach [9] or mixed-integer linear programming [2]. However, these works did not consider the opportunity that we exploit in this paper, that is, some mobile sensors could be relocated to certain locations to relay information for a task, to further reduce the task completion time.

[6,14,18,24,28] propose methods for data collection using mobile elements in wireless sensor networks or mobile Ad-Hoc networks. The basic idea is to use mobile elements as message carriers. They collect information from sensors when they are in close range to the sensors, buffer the information when they move around, and pass the information to the base station when they become near to the base station. In these works, the time a mobile element takes to deliver the information is not critical. Our work differs from them in that we want to reduce the time the sensors take to fulfill an emergent data acquisition task.

The authors of [27] propose a data acquisition framework called SenseSwarm for mobile sensor networks. SenseSwarm partitions the sensors into perimeter and core nodes. Perimeter nodes are responsible for data acquisition while core nodes take care of storage and replication. The aim of the SenseSwarm framework is to improve data availability.

[5,11,22,23,25] study the mobile sensor relocation problem [23,25] focus on fine-grained relocation to deal with a coverage hole caused by a sensor failure [5,22] investigate event-based relocation where the sensor locations and density are adapted to properly sense and control a large event area [11] proposes self-deployment algorithms for sensors to achieve a focused coverage around a Point of Interest. In all these studies, the sensor networks under consideration are assumed to be fully connected, and the aim of relocation or self-deployment is to meet a certain coverage requirement. As mentioned in Section 1, the differences between our work and these existing studies are twofold: we look at the sensor relocation problem in sparse sensor networks; the objective of relocation in our work is to fulfill an emergent data acquisition task as soon as possible.

3. System model and problem definition

3.1. System model

The system consists of a stationary base station BS and n mobile sensors (s_1, s_2, ..., s_n) that are sparsely distributed in an area A. Each mobile sensor knows the BS's location and its own location. The mobile sensors and the BS use wireless technology (such as Wi-Fi) for communication and there is no direct long-range communication. Two sensors (or the BS and a sensor) can communicate *directly* only if the distance between them is smaller than the wireless technology's communication range r. The sensors and the BS form a mobile ad-hoc network (MANET) where one can communicate with another if they are connected either directly or through other sensors. Since the sensors' communication range is limited and the sensors are sparse, the network formed by the sensors is not fully connected, and can even be severely partitioned. The topology of the network changes with time as the sensors move.

The general task of the mobile sensors is to explore (sense) the area A by moving in it following a certain mobility pattern. The mobile sensors' move speed is v. Each sensor senses the region that it passes by, carries the sensed data, and forwards the data to the BS when it is connected to the BS.

3.1.1. Emergent data acquisition task

An emergent data acquisition task (emergent task for short) $ET(L)$ specifies a location L (within A). Given an emergent task $ET(L)$, the mobile sensors shall carry out the task by having one sensor going to L, sensing for a period of time T_s, and sending the sensed data back to the BS. T_s models the time a sensor needs to collect enough information around L, and the length of T_s is determined by the applications. The BS would like the time from the moment an emergent task is issued to the moment relevant data is received to be as short as possible.

Note that when an emergent task arises, it is possible that only a small number of sensors are connected to the BS.

3.1.2. Assumptions

Since the focus of this work is on sensor relocation scheme and our optimization metric is the time the sensors take to fulfill an emergent task, for simplicity we in this paper do not consider energy consumption. In the class of applications that we are considering, all the sensors are moving to collect information. The energy spent on moving will be much more than the energy spent on communication. [17] shows that when a 0.5 kg UAV flies horizontally at a 10–12 m/s speed its minimum energy consumption is 10–25 J(joule). It is reported in [4] that a normal (Lucent) IEEE802.11 wireless network card consumes about 1.5 J per second when in active transmission mode. Although more powerful wireless transmitters may be used, they will be equipped only on larger UAVs. Clearly, larger UAVs will need much more energy for flying, or for simply staying in the air. It is also mentioned in [19,23] that a mobile sensor on ground spends much more energy on motion than on wireless communication when moving around.

Since all the sensors are moving most of the time, they will spend similar amount of energy no matter how they communicate with each other and how they move during a relocation. For this reason, we believe that: 1) trying to save energy by controlling the communication between the sensors will not be very helpful; 2) although controlling the move distances of the sensors during the relocations may help save some energy, the save will be marginal because the relocation only happens during the ad-hoc emergent tasks.

We will also neglect data transmission time, because we believe that in a sparse MSN, the relocation time will be much longer than the data transmission time so it does not affect our design of the relocation scheme.

3.2. Problem definition

We define the problem of sensor relocation for emergent data acquisition tasks in the system model described in Section 3.1 as the following problem: relocating mobile sensors to proper locations so that a given emergent data acquisition task is fulfilled in a minimum period of time, with the constraint that a sensor (or the BS) can communicate with another sensor only if they are connected in the sparse MANET formed by the BS and the sensors.

Let us call the sensor that is assigned to sense the task location L as the Scanner. The time for fulfilling the task $ET(L)$ can be divided into three parts:

– T_{go}: from the time the task is issued to the moment the Scanner arrives at L;
– T_s: the time for sensing the task location;
– T_{return}: from the Scanner finishes sensing the task location to the moment the BS receives the sensed data.

The goal is to minimize $T_{go} + T_s + T_{return}$. Since T_s is fixed for a given task (T_s is determined by the application), we would like to minimize $T_{go} + T_{return}$. Both T_{go} and T_{return} can be significant in sparse MSNs because: the Scanner could be far away from L; after sensing the task location the Scanner may be disconnected from the BS so it has to move to find a connection to the BS. We shall reduce both T_{go} and T_{return}.

4. BRIDGE: A distributed relocation scheme

We propose a distributed relocation scheme called BRIDGE for sparse mobile sensors to carry out emergent tasks. The main ideas are as follows:

1. relocate the sensor that is the nearest to the task location to sense the task location so that it can arrive at the task location as early as possible. We call the sensor relocated to the task location as the *Scanner*. This is to reduce T_{go}.
2. relocate some sensors to help connect the Scanner to the BS so that the Scanner can send the acquired data to the BS without moving towards the BS. We refer to these sensors as the *Connectors*. This is to reduce the T_{return} component of the task execution time.
3. adjust the relocation with the sensors encountered during relocation, see whether they can be better Scanner or better Connectors. This is to exploit the encountered sensors to reduce T_{go} and T_{return}.

In BRIDGE, we store the information about the sensors involved in the relocation for an emergent task in a *relocation plan*. It basically tells which sensors are involved and what are their relocation destinations. Details about relocation plan is presented in Section 4.3.

In Fig. 1 we use the processing of an emergent task as an example to illustrate these ideas. In the figures, the rectangle marked with L is the task location, the circles marked with numbers are the sensors, the arrows on the sensors indicate the sensors' moving directions, the sensor in dark (e.g. s_3 in (b)) is the Scanner, the sensors in light gray (e.g. s_1, s_2 in (b)) are the Connectors, and the ones in white (e.g. s_4, s_5, s_6 in (b)) are not involved in the relocation. A line between two sensors means that they are connected.

Figure 1(a) depicts the scenario when the emergent task is issued but before the initial relocation plan is executed. Figure 1(b) illustrates the initial relocation plan where s_3 is assigned as the Scanner and s_1 and s_2 are assigned as the Connectors. s_3 is assigned as the Scanner because it is the nearest to the task location among the sensors that are connected to the BS. s_1 and s_2 are relocated to the locations between the task location and the BS so that they will help connect the Scanner with the BS. Figure 1(c) depicts the event where s_3 encounters s_4 and s_5. s_3 adjusts the initial relocation plan to a new relocation plan that is depicted in Fig. 1(d). In the new relocation plan, s_5 is assigned as the Scanner, s_1, s_2, s_3 and s_4 work as Connectors. s_5 is selected as the new Scanner because it is nearer to the task location than the existing Scanner (s_3) is. s_3 and s_4 also work as Connectors because the existing Connectors (s_1 and s_2) are not enough to connect the Scanner to the BS. In Fig. 1(e), the Scanner is sensing the task location and the Connectors connect the Scanner with the BS. After the Scanner finishes sensing the task location and sending the data to the BS, the emergent task is done. Then all the participating sensors can move freely to continue their basic exploration task, as shown in Fig. 1(f).

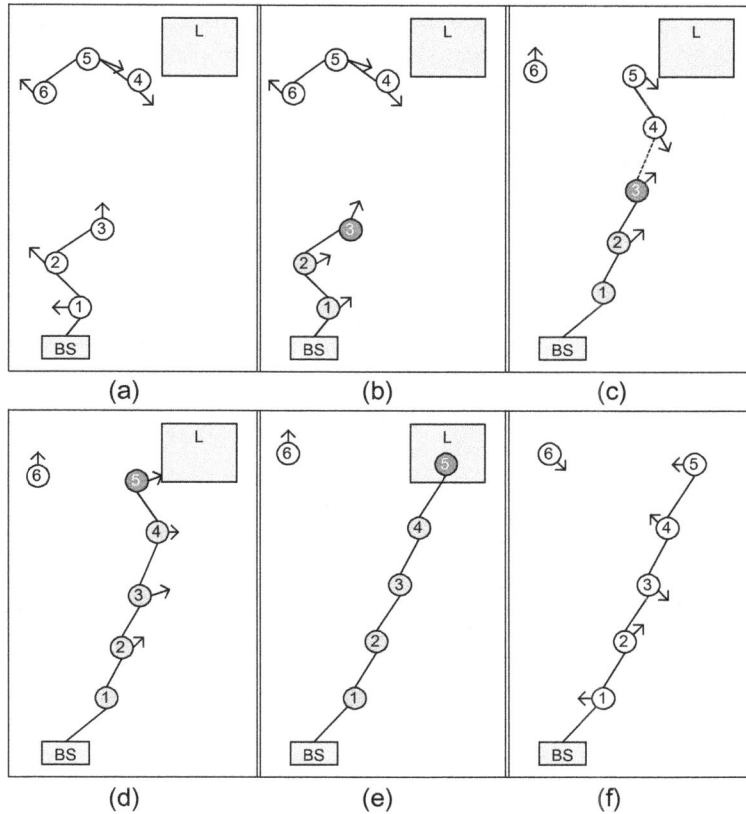

Fig. 1. A simple example illustrating the BRIDGE.

To realize the ideas of BRIDGE in a sparse MSNs, we have to deal with the problems caused by the intermittent connections.

First, when the BS has an emergent task, it is likely only few sensors are connected with it. BRIDGE must enable the BS to initiate the processing of the task by generating an initial relocation plan as long as at least one sensor is connected with the BS. This problem is solved in BRIDGE with an algorithm called `Init`.

Second, during the execution of a relocation plan, involved sensors may encounter sensors that were not connected. BRIDGE shall make use of the newly connected sensors to generate better relocation plans. This problem is solved in BRIDGE with an algorithm called `Adjust`.

Third, sensors involved in a same relocation plan can get disconnected during the relocation. They may encounter new neighbors. BRIDGE shall let them have enough information to generate better relocation plans even when being disconnected from other involved sensors. This problem is solved in BRIDGE by keeping enough information of the emergent task and involved sensors in a relocation plan and letting each involved sensor have a copy of the relocation plan.

Fourth, since involved sensors may generate different new relocation plans when they are disconnected, BRIDGE has to help them merge different relocation plans when they get connected again. This problem is solved in BRIDGE with an algorithm called `Merge`.

Fifth, after an emergent task is fulfilled, BRIDGE needs to make sure that all the sensors involved in the relocation finally know that the emergent task is done so that they can return to work on their general task. This problem is solved in BRIDGE by letting the sensors keep the information about the emergent tasks (e.g. whether an emergent task is done) and synchronize the information when being connected.

Before presenting the algorithms, we will first describe the sensor's states (Section 4.1), the idea and definition of the *ConnectorPoints* (Section 4.2), what makes a relocation plan (Section 4.3), and the execution of a relocation plan (Section 4.4).

4.1. Sensor states

In BRIDGE, a mobile sensor is always in one of the following states: Free, Scanner, Connector, and Returner.

A sensor is in the Free state if it is not involved in the fulfillment of an emergent task. A Free sensor works on the general task of the application, e.g. explore the area.

A sensor is in the Scanner state if it is assigned to go to sense the task location. After being assigned as a Scanner, the sensor moves towards the task location.

A sensor is in the Connector state if its task is to help connect the Scanner with the BS. A Connector sensor will be given an ID (independent from its sensor ID) called the Connector-ID. A Connector sensor given a Connector-ID j will be called Connector-j. The Connector-ID tells the Connector where it shall move to. We will discuss this in Section 4.2.

A sensor is in the Returner state if it is carrying the acquired data of the emergent task. The Returner is responsible for sending the data to the BS. A Scanner becomes a Returner when it finishes sensing the task location.

The state of a sensor may change during the relocation because the sensors may adjust the relocation plan when they encounter new neighbors.

4.2. ConnectorPoints, Connector-ID

To connect the Scanner (at task location) and the BS with a minimum number of Connectors, we relocate the Connectors onto the line between the BS and the task location, and let them maintain a distance that is shorter than the sensors' communication range r from its neighboring Connectors. In this way, the number of Connectors, denoted as N_c, is minimized. N_c is computed as follows:

$$N_c = \lceil (Distance(BS, L)/r) \rceil - 1 \tag{1}$$

Here L is the task location, r is the sensors' communication range, $Distance(BS, L)$ means the distance between the BS and the task location.

These N_c Connector sensors can build a link on the line between the task location and the BS. For example, if the distance between the task location and the BS is 2400 meters and the sensors' communication range is 500 meters, we will need 4 sensors to work as Connectors. Figure 1 (e) shows the example.

We define the locations that the Connectors shall move to as the *ConnectorPoints* of the emergent task. Let $Line(BS, L)$ be the line segment between the BS and the task location L. There are N_c ConnectorPoints on the $Line(BS, L)$. We define $ConnectorPoint_j$ as the point on $Line(BS, L)$ whose distance to the BS is

$$j * Distance(BS, L)/(N_c + 1) \tag{2}$$

A Connector sensor given a Connector-ID j moves to the $ConnectorPoint_j$.

4.3. Relocation plan

A Relocation Plan of an emergent task $ET(L)$ is an assignment of a set of sensors to their roles in the processing of the task. The roles tell the sensors where they shall move to.

A relocation plan RP contains the following information:

- $ET(L)$: the emergent task, which includes the task location information.
- Sensor ID → Scanner: this specifies which sensor shall work as the Scanner. Later we will use RP.Scanner to refer to it.
- {Sensor ID → Connector-j}, $1 \leqslant j \leqslant N_c$: a map of sensor IDs to the Connector-IDs. This specifies which sensors shall be Connectors. We will use RP.Connector-j to refer to an entry in it. There could be fewer than N_c Connectors. If no sensor is assigned as Connector-j, RP.Connector-j is *null*.
- The time when the relocation plan is generated.
- The locations of the involved sensors (the Scanner and the Connectors) when the relocation plan is generated.

Note that in the relocation plan there is information about all the involved sensors, including their locations at the time the relocation plan is generated. By such, each involved sensor will have enough information to generate a better relocation plan when it encounters new neighbors as long as it has a copy of the current relocation plan. This design is important for the involved sensors to deal with the problems caused by intermittent connections.

4.4. Execution of a relocation plan

To execute a relocation plan, the sensor that computes the relocation plan simply disseminates the relocation plan to all the connected sensors. Upon receiving a relocation plan, a sensor checks whether it has a role in the plan. If yes, it saves a copy of the relocation plan and sets its state according to its role in the plan; otherwise it sets itself to the Free state. The Scanner moves towards the task location. The Connectors compute their ConnectorPoints based on their Connector-IDs, and then move to their ConnectorPoints.

If a new relocation plan is generated during the execution of a relocation plan (e.g. when new sensors are encountered), the new one will be executed in the connected sensors so that all the connected sensors follow the new plan.

4.5. Initiate a relocation for an emergent task

When the BS has a new emergent task, it uses the `Init` algorithm presented here to generate an initial relocation plan based on the Free sensors that are currently connected to it. `Init` assigns the Free sensor that is the nearest to the task location as the Scanner and assigns at most N_c Free sensors as the Connectors based on their vicinity to the ConnectorPoints.

The pseudocode of `Init` is listed in Algorithm 1. In the pseudocode, $Nearest(S, location)$ finds the sensor in set S that is the nearest to the *location*. $RP.Scanner$ denotes the information in the relocation plan while $Scanner$ denotes the corresponding sensor. This also applies to $RP.Connector_j$ and $Connector_j$. CP_j denotes $ConnectorPoint_j$.

Let $Frees$ be the set of Free sensors that are connected to the BS. If $Frees$ is empty, the BS has to wait until there is at least one sensor in Frees. The BS chooses the sensor that is the nearest to the task

---

Algorithm 1: `Init`

   Input: an emergent task $ET(L)$
   Output: a relocation plan RP for $ET(L)$
1 $Frees \leftarrow$ the Free sensors that are connected to BS;
2 $RP.Scanner \leftarrow Nearest(Frees, L)$;
3 $Frees \leftarrow Frees - \{Scanner\}$;
4 $N_c \leftarrow \lceil Distance(BS, L)/r \rceil - 1$;
5 $m \leftarrow Min(N_c, |Frees|)$;
6 **for** $j \leftarrow 1$ **to** m **do**
7     $RP.Connector_j \leftarrow Nearest(Frees, CP_j)$;
8     $Frees \leftarrow Frees - \{Connector_j\}$
9 **return** RP

---

location as the Scanner of the relocation plan (lines 2–3). Among the remaining Free sensors, the BS selects Connector sensors (lines 4–8) as follows. Let $m = Min(N_c, |Frees|)$. Here N_c is the number of Connectors that the relocation plan wants, and $|Frees|$ is the number of remaining Free sensors. m will be the number of Connectors in the initial relocation plan. For each $ConnectorPoint_j$ ($1 \leqslant j \leqslant m$), the BS finds the sensor in $Frees$ that is the nearest to it, and sets the sensor as the Connector-j of the relocation plan.

The time complexity of the `Init` algorithm is $O(n * N_c)$ where n is the number of Free sensors that are connected to the BS and N_c is the number of Connectors needed in the relocation plan. N_c typically is a small number. Since n will also be a small number in a sparse mobile sensor network, the initial plan of an emergent task can be computed efficiently.

Once the initial relocation plan for an emergent task is generated, it is executed among the connected sensors.

4.6. Adjust a relocation on meeting sensors

In BRIDGE, the participants of a relocation plan always try to find better relocation plans when they encounter new neighbors during the relocation. They do it using the `Adjust` algorithm listed in Algorithm 2. In this algorithm s_i denotes the involved sensor that encounters a set of new neighbors, and RP denotes the relocation plan that s_i currently has.

Algorithm 2: `Adjust`

   Input: a relocation plan RP
   Output: a new relocation plan
1 $Frees \leftarrow$ the Free sensors that are connected to s_i;
2 $Scanner' \leftarrow Nearest(Frees, L)$;
3 **if** $Scanner'$ can reach L earlier than $RP.Scanner$ **then**
4   **if** $RP.Scanner$ is connected **then**
5      $Frees \leftarrow Frees + \{Scanner\}$
6    $RP.Scanner \leftarrow Scanner'$;
7    $Frees \leftarrow Frees - \{Scanner'\}$;
8 $RP \leftarrow AdjustConnectors(RP, Frees)$;
9 **return** RP

On encountering new neighbors, s_i finds out all the Free sensors that it now can reach (line 1). It first checks whether it can find a better Scanner among the Free sensors (lines 2–7). If there is a better Scanner, s_i puts the current Scanner into the $Frees$ set if it is connected, and then updates the relocation plan with the new Scanner.

After that, s_i adjusts the Connectors with the remaining Free sensors using the algorithm AdjustConnectors which is listed in Algorithm 3. s_i divides the ConnectorPoints into two sets (lines 1–2): the ones for which no Connectors are assigned, and the ones with corresponding Connectors in current relocation plan. It first assigns Free sensors to the ConnectorPoints that have no Connectors (lines 3–6). Then it tries to find better Connectors for existing Connectors (lines 7–12).

Algorithm 3: AdjustConnectors

    Input: a relocation plan RP
    Input: a set of Free sensors $Frees$
    Output: a new relocation plan
1   $EmptyCPs \quad \{j|j \leqslant N_c \wedge RP.Connector_j = null\}$;
2   $ExisingCPs \quad \{j|RP.Connector_j \neq null\}$;
3   **for** j in EmptyCPs **do**
4   **if** Frees is not empty **then**
5       $RP.Connector_j \quad Nearest(Frees, CP_j)$;
6       $Frees \quad Frees - \{Connector_j\}$;
7   **for** j in ExisingCPs **do**
8   **if** $Frees$ is not empty **then**
9       $Connector'_j := Nearest(Frees, CP_j)$;
10   **if** $Connector'_j$ can reach CP_j earlier than $RP.Connector_j$ **then**
11         $RP.Connector_j \leftarrow Connector'_j$;
12         $Frees \leftarrow Frees - \{Connector_j\}$;
13 **return** RP

The complexity of both Adjust and AdjustConnectors is $O(n * N_c)$ where n is the number of Free sensors that are connected to s_i (the sensor that adjusts the relocation plan) and N_c is the number of Connectors needed in the relocation plan.

Note that when s_i encounters new neighbors, it is possible that some other participants of the relocation plan are disconnected from s_i. For example, when there are fewer than N_c Connectors, the Scanner will get disconnected from the Connectors when it moves to the task location. This kind of disconnection between the participants of a same relocation plan has two influences on the design of BRIDGE.

First, at the time a participating sensor meets new neighbors it may not be able to connect to other participating sensors and get up-to-date information from them. However, to find out whether the new neighbors can be helpful it needs information about current participants. For example, it needs the location information of the current Scanner to determine whether a new neighbor is nearer to the task location (as in line 3 of the Adjust algorithm). To resolve this problem, as described in Section 4.3, in the relocation plan we keep information about each participant's role and location at the time the relocation plan was generated. In this way, each participating sensor will have enough information to compute other participating sensors' locations using the following information: their starting locations, their relocation destinations (determined by roles), the elapsed time, and moving speed (all sensors have the same speed).

Second, disconnected participants will not receive the new relocation plans that are generated by other participants. In this case, some will have the old relocation plan while the others have a new one. Furthermore, partitioned groups of participants may independently adjust the relocation plan to new ones. Therefore, there can be more than one relocation plan for an emergent task being executed. In such cases, different relocation plans have to be merged when sensors having different relocation plans get connected again.

Table 1
Relocation Plans

Relocation Plan	Scanner	Connectors
RP1	s_2	$s_1{:}1$
RP2	s_4	$s_1{:}1$; $s_2{:}2$
RP3	s_2	$s_1{:}1$; $s_3{:}2$; $s_5{:}3$
RP4	s_4	$s_1{:}1$; $s_3{:}2$; $s_2{:}3$

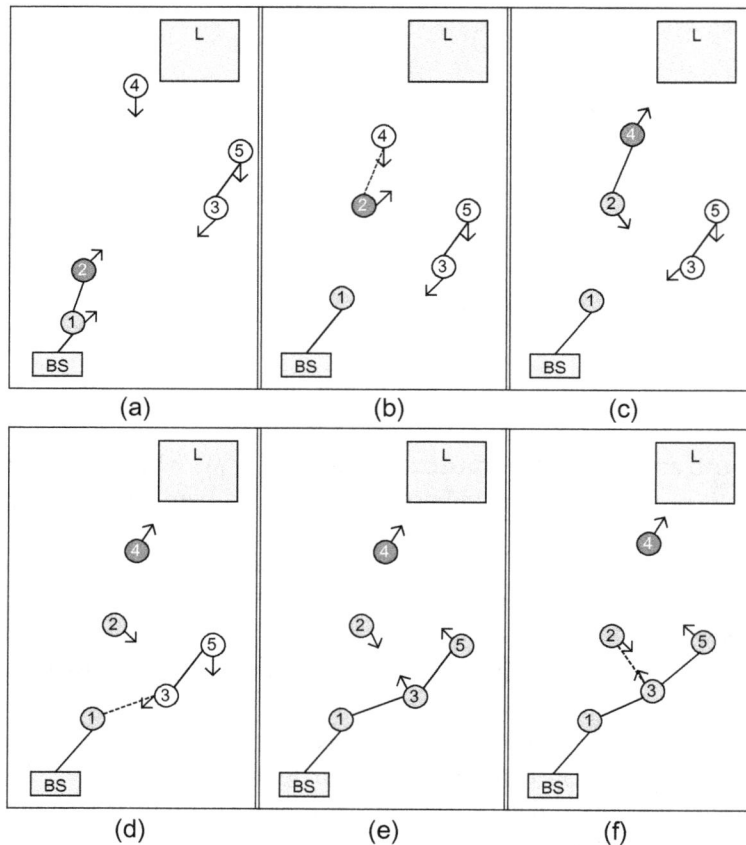

Fig. 2. Example for Merge.

4.7. Merge relocation plans

Figure 2 shows an example where participants of a relocation plan get disconnected, generate new relocation plans, and the participants having different relocation plans meet. The relocation plans are sketched in Table 1. In the table, "$s_i : j$" means that sensor s_i works as Connector-j.

- Figure 2(a) shows the scene after the initial relocation plan RP1 is executed. As the Scanner (s_2) moves towards the task location, it gets disconnected from s_1.
- When the Scanner (s_2) encounters s_4 (a Free sensor) as shown in Fig. 2(b), s_2 adjusts the initial relocation plan to a new relocation plan RP2.

- Figure 2(c) shows the scene after s_2 executes RP2. At this time s_2 and s_4 are disconnected from s_1 therefore they cannot update s_1 about the new relocation plan.
- As s_3 and s_5 (two Free sensor) move, they encounter s_1 as shown in Fig. 2(d). s_1 adjusts its relocation plan RP1 to a new relocation plan RP3.
- Figure 2(e) shows the scene after s_1 executes RP3. Note that when s_1 meets s_3 and s_5 it has the initial relocation plan RP1. So s_1 thinks that s_2 is still the best Scanner because it is not aware of s_4 and RP2.
- As the sensors move, s_2 encounters s_3. Figure 2(f) depicts this event. s_2 and s_3 have two different relocation plans. RP4 in Table 1 is the resultant relocation plan after s_2 and s_3 merge their relocation plans.

In BRIDGE, when two sensors having different relocation plans meet, one of them will merge the relocation plans to a new relocation plan and execute the new one among the connected sensors. The algorithm for merging two relocation plans is called `Merge`. Algorithm 4 lists its pseudocode.

Algorithm 4: `Merge`

Input: two relocation plans $RP1$ and $RP2$
Output: a relocation plan
1   **if** $RP1.Scanner \neq RP2.Scanner$ **then**
2     $Scanners \quad \{RP1.Scanner, RP2.Scanner\}$;
3     $RP.Scanner \quad Nearest(Scanners, R)$;
4   **if** *the other Scanner is connected* **then**
5   put it to $Frees$;
6   **else**
7     $RP.Scanner = RP1.Scanner$;
8   **for** $Connector_j$ in *RP1* and *RP2* **do**
9   **if** $Connector_j$ *is connected* **then**
10     $Frees \quad Frees + \{Connector_j\}$;
11   for **For** $j \quad 1$ to N_c **do**
12   **if** $RP1.Connector_j \neq null \parallel RP2.Connector_j \neq null$ **then**
13       $Connectors \quad \{RP1.Connector_j, RP2.Connector_j\}$;
14       $RP.Connector_j \quad Nearest(Connectors, CP_j)$;
15   $RP \quad AdjustConnectors(RP, Frees)$;
16   **return** RP

Let s_i be the sensor that merges two relocation plans. In `Merge`, s_i first checks whether the two relocation plans have the same Scanner. If not, s_i chooses the one that can arrive at the task location earlier as the Scanner in the new plan, and put the other Scanner into the $Frees$ set if it is connected. Then for all the Connectors in the two relocation plans, if a Connector is connected s_i puts it into the $Frees$ set and clears the corresponding information in the relocation plan (lines 8–10). After this, all the available sensors will be in the $Frees$ set, and the two input relocation plans only contain information about Connectors that are currently disconnected. For each Connector slot in the new relocation plan, if the input relocation plans have disconnected Connector(s) for it, s_i picks the one that is nearer to the corresponding ConnectorPoint (lines 11–14). s_i finally uses the `AdjustConnectors` algorithm to: (1) assign Free sensors to empty Connector slots; (2) find better Connectors for disconnected existing Connectors.

The complexity of the `Merge` algorithm is $O(n * N_c)$ where n is the number of sensors that are connected to s_i (the sensor that merges the relocation plans) and N_c is the number of Connectors needed in the relocation plan.

4.8. Relocation after acquiring data

A Scanner becomes a Returner when it finishes sensing the task location. The Returner behaves according to the following rules.

- If the Returner is connected to the BS through the Connectors, it sends the data to the BS and the task is done. The Returner and all the connected Connectors are set to the Free state.
- Otherwise, if the Returner is connected to a Connector, it passes the data to that Connector and changes itself to the Free state. The Connector that receives the data becomes a Returner.
- Otherwise, the Returner moves towards the BS.

4.9. Handle obstacles

In the previous sections, we have implicitly assumed that there are no obstacles in the area. Here we discuss how BRIDGE handles obstacles.

Obstacles only affects how BRIDGE computes the ConnectorPoints between the BS and the task location L. If an obstacle is not on the line between BS and L, ConnectorPoints are computed on $Line(BS, L)$ as described in Section 4.2. If an obstacle is on $Line(BS, L)$, however, then we cannot relocate sensors onto $Line(BS, L)$ because the obstacle can be a region where the sensors cannot move into or a region where wireless communication is jammed by adversaries. Figure 3(a) shows an example where an obstacle intersects the line between BS and L.

The basic idea for handling an obstacle (that intersects $Line(BS, L)$) is to relocate the Connectors onto the shortest simple polyline between BS and L that does not intersect with the obstacle. Such a polyline is computed by first computing the convex hull using BS, L and the vertices of the obstacle, and then taking the shorter path from BS to L on the convex hull perimeter. Figure 3(b) shows the convex hull computed for the scenario shown in Fig. 3(a). The polyline BS-v_2-L is shorter than the other polyline between BS and L on the convex hull perimeter, so it is taken as the polyline on which the ConnectorPoints will be computed.

To make sure that two Connectors at consecutive ConnectorPoints can communicate with each other, a line of sight between them is necessary (in particular when the obstacle blocks the wireless communication). This requirement is satisfied as follows. BRIDGE first computes the ConnectorPoints on the selected polyline (shown in Fig. 3(c)). If the line between any two consecutive ConnectorPoints intersects the obstacle, the polyline is adjusted by adding a line segment that does not intersect the obstacle, and then the ConnectorPoints are re-computed on the adjusted polyline. Figures 3(d-f) show how the polyline is adjusted. In this example (see Figures 3(d)), the line between CP_2 and CP_3 intersects the obstacle. Because of this, a line segment near vertex v_2 is added and the polyline is adjusted by incorporating the new line segment (shown in Fig. 3(e)). The length of the new line segment is no longer than the communication range of the mobile sensors so that the two Connectors relocated on to the segment can communicate with each other. The ConnectorPoints are re-computed on the adjusted polyline. Figure 3(f) shows the final polyline and the ConnectorPoints on it.

Note that the obstacles only affect the locations of the ConnectorPoints. The polyline and the ConnectorPoints are computed (on BS) in the initial relocation plan, and they are fixed during the relocation. Therefore only the `Init` algorithm needs to be slightly modified to handle the obstacles in the field.

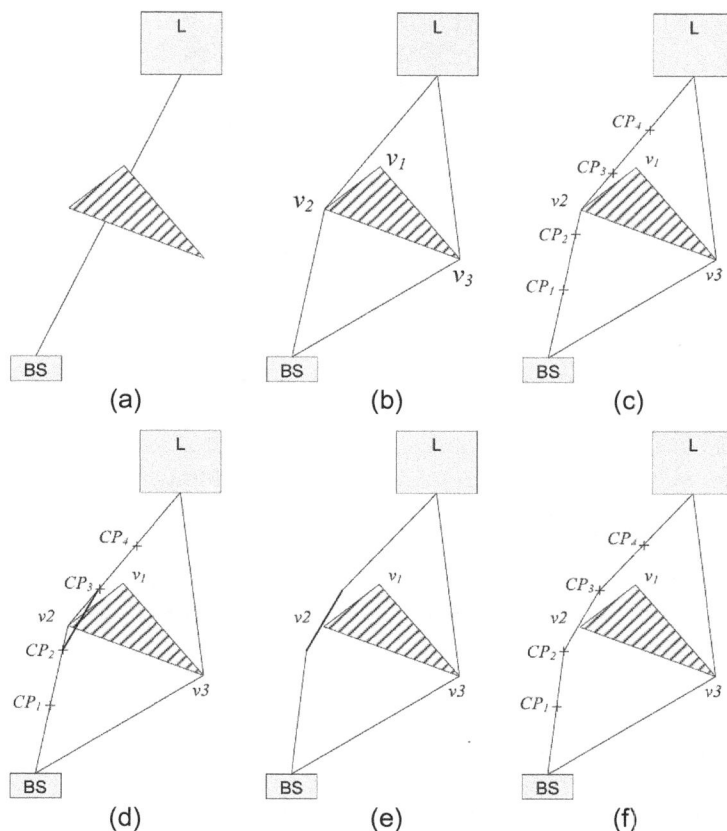

Fig. 3. Example for handling an obstacle.

5. Experimental study

We use simulation to study the performance of our proposed BRIDGE scheme. Since this is the first work on relocation of mobile sensors for emergent data acquisition tasks, there is no existing solution to compare with. We compare the proposed scheme to two variants: a simpler version (Simple-BRIDGE) and a motion-aware version (Motion-Aware-BRIDGE).

In the Simple-BRIDGE scheme, we set the number of Connectors (N_c) to 0. By this, a relocation plan only contains a Scanner and no sensors will be relocated to help connect the Scanner with the BS. The goal of comparing BRIDGE with this simpler version is to study the effect of relocating some sensors to connect the Scanner to the BS.

The Motion-Aware-BRIDGE solution is an ideal scheme with perfect information. We assume that the Scanner know all other sensors' planned trajectories. Rather than moving directly to the task location, the Scanner in this version moves to meet a currently disconnected sensor that can reach the task location earlier. The Scanner finds such a sensor by computing the sensor that minimizes the expression ($T_{meet} + T_{move}$) where T_{meet} is the minimal time the Scanner needs to move to meet that sensor and T_{move} is the time that sensor needs to move to the task location. The goal of comparing BRIDGE with this version is to see how much the BRIDGE algorithm will improve if all sensors' future trajectories are given.

Table 2
Parameter Settings

Parameter	Default value	Value range
$FieldWidth$	30 km	10–50 km
$FieldHeight$	30 km	
Number of Sensors n	20	5–50
Sensor speed v	0.05 km/s	0.02–0.2 km/s
Communication range r	5 km	1–10 km
T_s	40 (s)	20–100 (s)

The simulation model follows the System Model that we describe in Section 3. The system parameters and their values are listed in Table 2. The parameter values are chosen based on the setting used in [10]. This simulates a system where a set of UAVs are carrying out a reconnaissance task and the base station sometimes issues emergent data acquisition tasks to the UAVs.

In the experiments the mobile sensors are placed randomly in the simulated area and they follow the Random WayPoint mobility model. Emergent tasks are also placed randomly in the simulated area. When we generate an emergent task we always generate a new scenario (placement of the sensors) so that all the solutions have the same start point.

Because obstacles only affect the number of sensors needed to connect BS and a task location, its effect is the same as shortening the communication range of the mobile sensors. For this reason, we do not generate obstacles in the experiments.

The performance metric is the average completion time of a large number of emergent data acquisition tasks.

In the figures that show the experimental results, "Simple", "BRIDGE", and "MA" ("S", "G", "M" in some bar figures) denote the Simple-BRIDGE, the BRIDGE, and the Motion-Aware-BRIDGE schemes respectively. In the figures that show the breakdown of task completion times, the number under each group of three bars denotes the value of the parameter under investigation. Each bar has three parts: Go, Scan, and Return. They correspond to the T_{go}, T_s, and T_{return} times defined in Section 3.

5.1. Basic performance study

Here we study the solutions' performance under the default parameters setting. Figure 4 shows the breakdowns of the solutions' average processing times. We see that BRIDGE and MA have similar performance and they perform much better than Simple. In particular, BRIDGE and MA spend much less time in the Return phase. MA performs a little better than BRIDGE because MA spends a little less time in the Go phase than BRIDGE does. However, the performance difference between BRIDGE and MA is minor.

We learn two things from this basic performance study. 1) The idea of relocating some sensors to connect the Scanner with the base station is very effective in reducing task completion time. 2) A more complex solution (MA) that makes use of sensors' motion information does not improve BRIDGE much. They show the merits of the BRIDGE scheme: it is effective and widely applicable (since it does not make any assumption of the sensors' mobility pattern).

5.2. Effect of sensors density

We use the average number of sensors that are within each sensor's communication range as the measurement of the sensor density. Three parameters affect the sensor density: the number of sensors n,

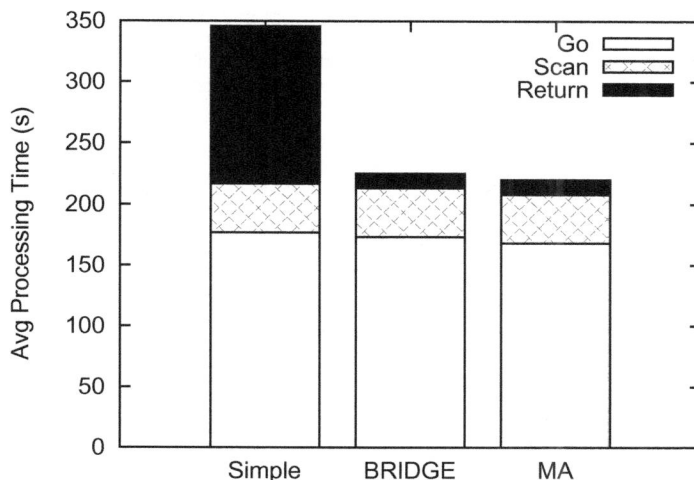

Fig. 4. Default setting.

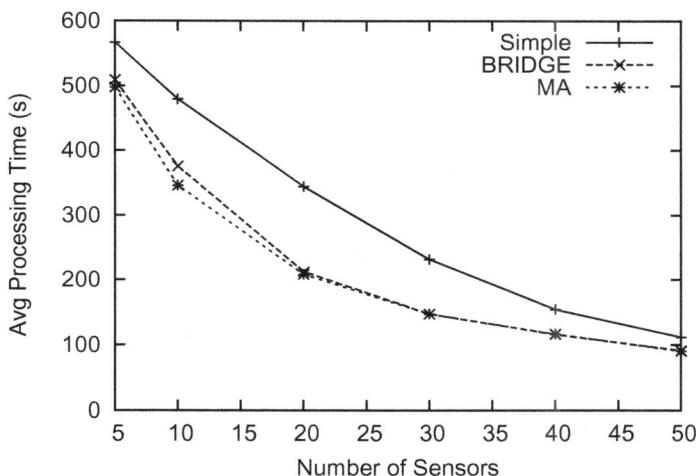

Fig. 5. Effect of n.

$FieldWidth$ that controls the field size, and the sensors' communication range r. As n or r increases, the sensor density increases; as $FieldWidth$ increases, the sensor density decreases. Figures 5, 6, and 8 show the effect of the three parameters on the solutions' performance. We have several observations here.

1) As the sensor density increases, all schemes perform better. This is because when sensor density increases, more sensors are connected, then it becomes easier to find a sensor that is near to the task location, and it is more likely that the acquired data can be sent back to the BS directly.

2) BRIDGE's performance is very close to MA's. Only in very sparse networks, MA performs a little better than BRIDGE does. In very sparse networks, the chance of encountering new neighbors during the relocation is small. In this circumstance, having the motion information of other sensors (in MA) helps the Scanner to meet more sensors.

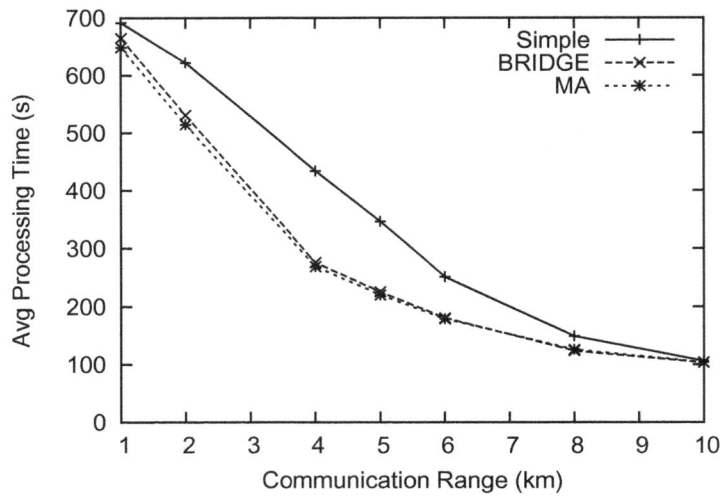

Fig. 6. Effect of r.

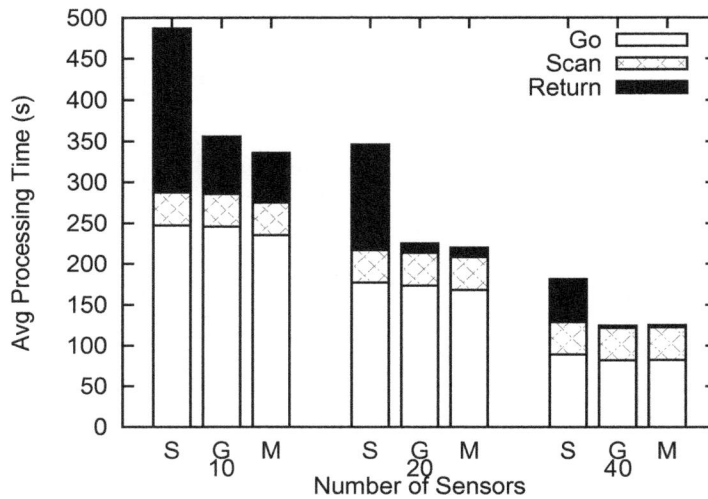

Fig. 7. Breakdowns, n.

3) When the number of sensors or the communication range increases from small to large (sensor density increases from very sparse to very dense) in Figs 5 and 6, the performance gap between Simple and BRIDGE first increases and then decreases. This is explained by Fig. 7 which shows the breakdowns of the task completion times when the number of sensors is 10, 20 and 40. Please notice the difference between Simple and BRIDGE's Return times first increases and then decreases. The gap increases first because as sensor density increases from very sparse to medium, BRIDGE can find more sensors to work as Connectors, and this reduces the time of the Return phase greatly. Then, when the sensor density increases from medium to dense, it is more likely that the Scanner at the task location is connected to the BS so that it is less necessary to relocate some sensors to work as Connectors.

4) When the field size increases (in Fig. 8), the gap between the Simple and BRIDGE always increases.

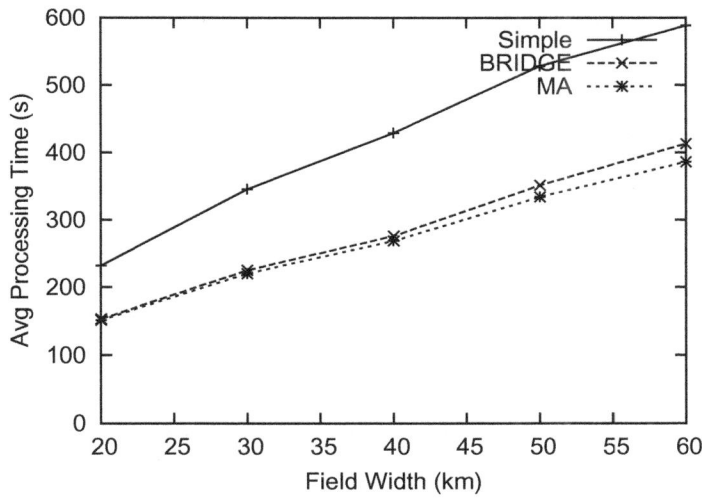

Fig. 8. Effect of field size.

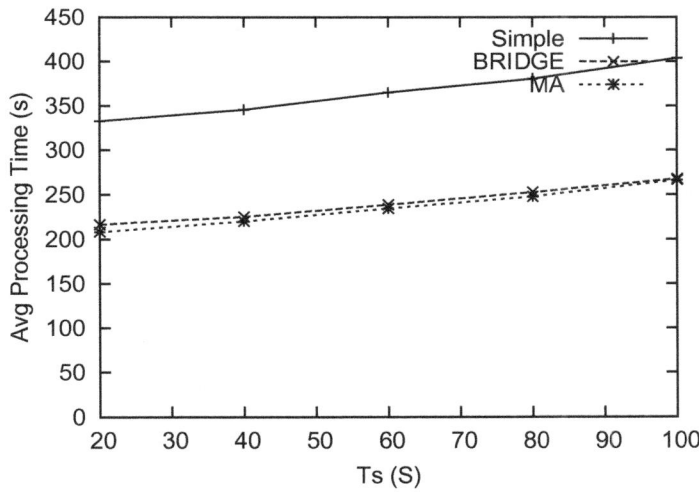

Fig. 9. Effect of T_s.

This is expected as the increase in field size not only makes the sensor network sparser, but also makes the task location farther from the base station (in the experiments the emergent task locations are randomly generated in the area). When an emergent task is farther from the BS, it is more important to have some sensors working as Connectors.

5.3. Other sensitivity study

5.3.1. Effect of T_s

Figure 9 shows the effect of T_s on the schemes' performance. Figure 10 shows the breakdowns of processing times when T_s is 40 and 80 (s). From the figures, we see that the task completion times increase as T_s increases.

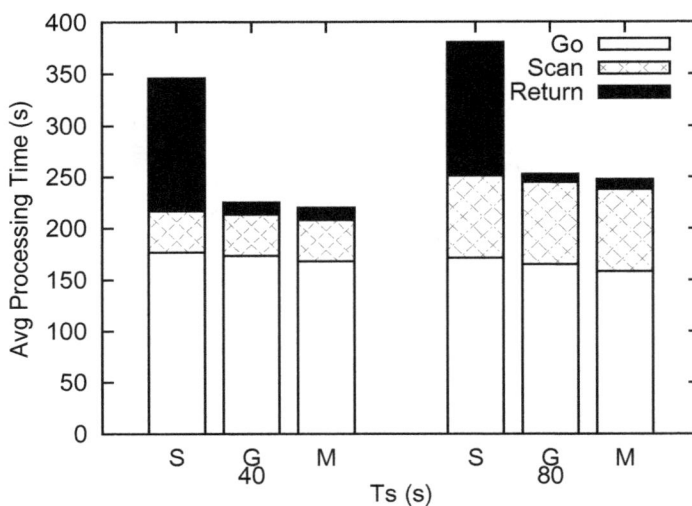

Fig. 10. Breakdowns, T_s.

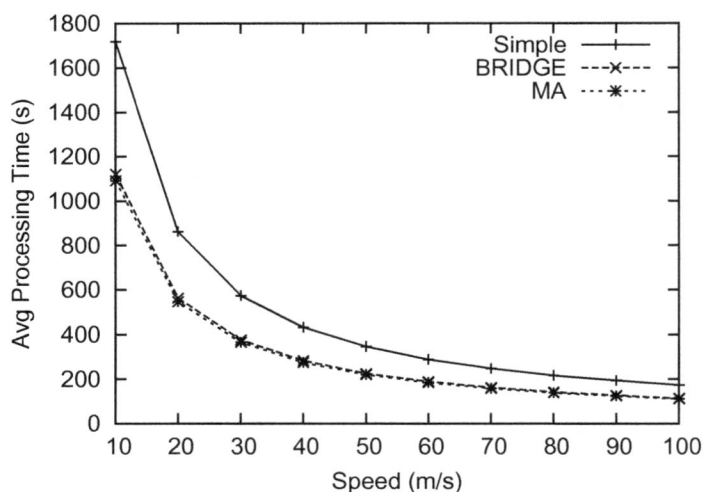

Fig. 11. Effect of move speed.

5.3.2. Effect of sensor speed

The sensors' move speed affect the time for a sensor to relocate to a location and the dynamism of the sensor network's topology. When sensors move faster, it takes shorter time for a sensor to meet new neighbors. Figures 11 and 12 show the effect of sensors' move speed on the solutions' performance. It is clear, and as expected, that as sensors' move speed increases, the processing times.

6. Conclusion and future work

In this paper, we have identified the problem of sensor relocation for emergent data acquisition in sparse mobile sensor networks. Mobile sensors are tasked to go to a given location to gather information

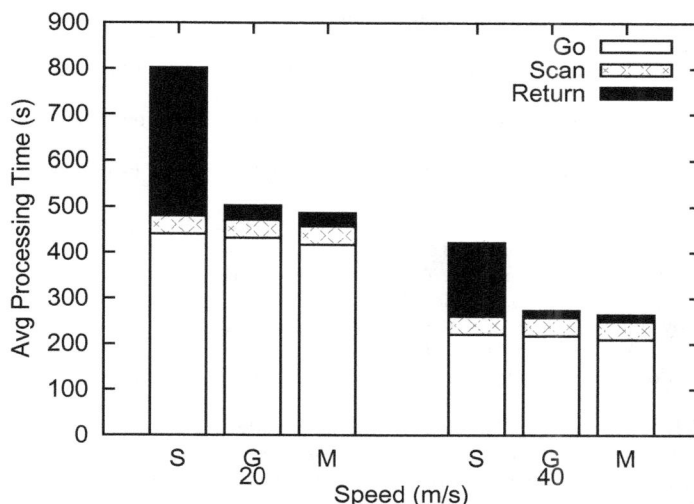

Fig. 12. Breakdowns, v.

and to return the sensed data to the base station as early as possible. We proposed a BRIDGE scheme in which the sensors collaborate to fulfill an emergent task by gradually finding a sensor that is near the task location and relocating some sensors to connect the base station to the sensor that is responsible for sensing the task location. Our extensive performance evaluation showed the effectiveness of the proposed solution.

There are a number of directions in which we would like to expand this research. We are interested, for example, in how the BRIDGE scheme can be extended to handle multiple emergent data acquisition tasks, and how to adapt BRIDGE to deal with obstacles that are not known in advance. For multiple emergent data acquisition tasks, the challenge is to compute a relocation plan so that the average completion time of several concurrent emergent data acquisition tasks is minimized. In scenarios where not all the obstacles are known in advance, the requirement for the relocation is dynamic and the relocation plan may have to be adjusted dramatically during relocation. For example, not only the involved sensors, but also the locations of the ConnectorPoints will be different after encountering an un-anticipated obstacle. The challenge here is to coordinate the mobile sensors (whose connections are intermittent) so that they will be informed when a relocation plan has been changed significantly.

Acknowledgements

This project is partially supported by a research grant R-252-000-352-232 (TDSI/08-001/1A) from the Temasek Defense Systems Institute.

References

[1]   J. Allred, A.B. Hasan, S. Panichsakul, W. Pisano, P. Gray, J. Huang, R. Han, D. Lawrence and K. Mohseni, Sensorflock: an airborne wireless sensor network of micro-air vehicles, in: *SenSys*, New York, NY, USA, ACM, 2007, pp. 117–129.
[2]   J. Bellingham, M. Tillerson, A. Richards and J.P. How, Multi-task allocation and path planning for cooperating uavs, in: *Conference on Cooperative Control and Optimization*, 2001.

[3] J.Y. Chen, G. Pandurangan and D. Xu, Robust computation of aggregates in wireless sensor networks: Distributed randomized algorithms and analysis, *Transactions on Parallel and Distributed Systems* **17**(9) (2006), 987–1000.

[4] L.M. Feeney and M. Nilsson, Investigating the energy consumption of a wireless network interface in an ad hoc networking environment, in: *INFOCOM*, 2001.

[5] M. Garetto, M. Gribaudo, C.F. Chiasserini and E. Leonardi, A distributed sensor relocatlon scheme for environmental control, in: *MASS*, 8–11 Oct. 2007, pages 1–10.

[6] Y.Y. Gu, D. Bozdağ, R.W. Brewer and E. Ekici, Data harvesting with mobile elements in wireless sensor networks, *Comput Netw* **50**(17) (2006), 3449–3465.

[7] H. Gupta, Z.H. Zhou, S.R. Das and Q. Gu, Connected sensor cover: self-organization of sensor networks for efficient query execution, *IEEE/ACM Transactions on Networking* **14**(1) (Feb 2006), 55–67.

[8] B. Hull, V. Bychkovsky, Y. Zhang, K. Chen, M. Goraczko, A. Miu, E. Shih, H. Balakrishnan and S. Madden, Cartel: a distributed mobile sensor computing system, in: *SenSys*, New York, NY, USA, ACM, 2006, pages 125–138.

[9] Y. Jin, A.A. Minai and M.M. Polycarpou, Cooperative real-time search and task allocation in uav teams, in: *IEEE Conference on Decision and Control*, (Vol. 1), 9–12 Dec. 2003, pp. 7–12.

[10] E. Kuiper and S. Nadjm-Tehrani, Mobility models for uav group reconnaissance applications, in: *ICWMC*, 29–31 July 2006, pages 33–33.

[11] X. Li, H. Frey, N. Santoro and I. Stojmenovic, Focused-coverage by mobile sensor networks, in: *Proc IEEE 6th International Conference on Mobile Adhoc and Sensor Systems MASS '09*, October 12–15, 2009, pp. 466–475.

[12] X. Li and N. Santoro, An integrated self-deployment and coverage maintenance scheme for mobile sensor networks, in: *Mobile Ad-hoc and Sensor Networks, Second International Conference*, 2006, pp. 847–860.

[13] B.Y. Liu, P. Brass, O. Dousse, P. Nain and D. Towsley, Mobility improves coverage of sensor networks, in: *MobiHoc*, New York, NY, USA, 2005. ACM, pp. 300–308.

[14] W. Liu, J.P. Wang, G.L. Xing and L.S. Huang, Throughput capacity of mobility-assisted data collection in wireless sensor networks, in: *Proc IEEE 6th International Conference on Mobile Adhoc and Sensor Systems MASS '09*, October 12–15, 2009, pp. 70–79.

[15] J. Luo, D. Wang and Q. Zhang, Double mobility: Coverage of the sea surface with mobile sensor networks, in: *Proc INFOCOM 2009. The 28th Conference on Computer Communications. IEEE*, April 19–25, 2009, pp. 118–126.

[16] S. Madden, M.J. Franklin, J.M. Hellerstein and W. Hong, Tinydb: An acquisitional query processing system for sensor networks, *ACM TODS* **30**(1) (November 2005).

[17] E.D. Margerie, Jean baptiste Mouret, Stphane Doncieux, Jean arcady Meyer, Thomas Ravasi, Pascal Martinelli, and Christophe Gr. Flapping-wing flight in bird-sized uavs for the robur project: from an evolutionary optimization to a real flapping-wing mechanism, in: *3rd US-European Competition and Workshop on Micro Air Vehicle Systems (MAV07)*, 2007.

[18] R.C. Shah, S. Roy, S. Jain and W. Brunette, Data mules: modeling a three-tier architecture for sparse sensor networks, in: *IEEE SNPA Workshop*, pages 30–41, 11 May 2003.

[19] G.T. Sibley, M.H. Rahimi and G.S. Sukhatme, Robomote: a tiny mobile robot platform for large-scale ad-hoc sensor networks, in: *Proc. IEEE International Conference on Robotics and Automation ICRA '02*, (Vol. 2), 11–15 May 2002, pp. 1143–1148.

[20] W. Wang, V. Srinivasan and K.-C. Chua, Trade-offs between mobility and density for coverage in wireless sensor networks, in: *MobiCom*, New York, NY, USA, 2007. ACM, pp. 39–50.

[21] S. Sukkarieh, E. Nettleton, J.H. Kim, A. Goktogan and H. Durrant-Whyte, The anser project: data fusion across multiple uninhabited air vehicles, *The International Journal of Robotics Research* **22**(7–8) (2003), 505–539.

[22] G. Trajcevski, P. Scheuermann and H. Brönnimann, Mission-critical management of mobile sensors: or, how to guide a flock of sensors, in: *DMSN*, New York, NY, USA, 2004. ACM, pp. 111–118.

[23] G. Wang, G. Cao, T.L. Porta and W. Zhang, Sensor relocation in mobile sensor networks, in: *INFOCOM*, (Vol. 4), 13–17 March 2005, pp. 2302–2312.

[24] G.L. Xing, T. Wang, W.J. Jia and M.M. Li, Rendezvous design algorithms for wireless sensor networks with a mobile base station, in: *MobiHoc '08: Proceedings of the 9th ACM international symposium on Mobile ad hoc networking and computing*, New York, NY, USA, 2008. ACM, pp. 231–240.

[25] N. Santoro, X. Li and I. Stojmenovic, Mesh-based sensor relocation for coverage maintenance in mobile sensor networks, in: *Ubiquitous Intelligence and Computing*, 2007.

[26] O. Younis and S. Fahmy, Heed: A hybrid, energy-efficient, distributed clustering approach for ad hoc sensor networks, *IEEE Transactions On Mobile Computing* **3**(4) (December 2004), 366–379.

[27] D. Zeinalipour-Yazti, P. Andreou, P.K. Chrysanthis and G. Samaras, Senseswarm: a perimeter-based data acquisition framework for mobile sensor networks, in: *DMSN*, New York, NY, USA, 2007. ACM, pp. 13–18.

[28] W. Zhao, M. Ammar and E. Zegura, A message ferrying approach for data delivery in sparse mobile ad hoc networks, in: *MobiHoc*, New York, NY, USA, 2004. ACM, pp. 187–198.

Wei Wu is a research fellow at the School of Computing, National University of Singapore (NUS). He completed his PhD in computer science in 2009 at Singapore-MIT Alliance, NUS, and received his Bachelor and Master from Nanjing University in 2002 and 2005 respectively. His research interests include data management, mobile computing, and sensor networks.

Xiaohui Li is currently a PHD candidate in School of Computing (SoC), National University of Singapore (NUS). His current research interest is moving objects data management. He received a B.Comp (Hons.) and a B.S. from NUS. Prior to joining SoC, he worked as IT Analyst in Center for Life Science (CeLS).

Shili Xiang is a Ph.D. student in Computer Science department at National University of Singapore, under supervision of Professor Kian-Lee Tan. Her research interests include query processing in sensor networks, streaming processing and pervasive computing.

Hock Beng Lim is program director of the Intelligent Systems Center at Nanyang Technological University, Singapore. He received his BS in Computer Engineering, MS in Electrical Engineering, and PhD in Electrical and Computer Engineering from the University of Illinois at Urbana-Champaign, and his MS in Management Science and Engineering from Stanford University. His research interests include sensor networks and sensor grids, cyber-physical systems, cloud computing, parallel and distributed computing, wireless and mobile networks, computer architecture, embedded systems, performance evaluation, e-Science and high-performance computing.

Kian-Lee Tan is a Professor of Computer Science at the School of Computing, National University of Singapore (NUS). He received his Ph.D. in computer science in 1994 from NUS. His current research interests include multimedia information retrieval, query processing and optimization in multiprocessor and distributed systems, database performance, and database security. He has published numerous papers in conferences such as SIGMOD, VLDB, ICDE and EDBT, and journals such as TODS, TKDE, and VLDBJ. Kian-Lee is a member of ACM.

4

An enhanced MPR-based solution for flooding of broadcast messages in OLSR wireless ad hoc networks

Tzu-Hua Lin[a], Han-Chieh Chao[a,b,*] and Isaac Woungang[c]

[a]*Department of Electrical Engineering, National Dong Hwa University, Hualien, Taiwan*
[b]*Institute of Computer Science & Information Engineering, and Department of Electronic Engineering, National Ilan University, I-Lan, Taiwan*
[c]*Department of Computer Science, Ryerson University, Toronto, Canada*

Abstract. In an Optimized Link State Routing (OLSR)-based mobile wireless network, optimizing the flooding of broadcast messages is a challenging task due to node's mobility and bandwidth resource consumption. To complement existing solutions to this problem, the Multi-Point Relays (MPR) selection has recently been advocated as a promising technique that has an additional feature of reducing the number of redundant re-transmission occurring in the network. This paper continuous on the investigation of an existing MPR-based solution, arguing that by considering a cost factor as an additional decision parameter in selecting the MPR nodes, the enhanced MPR selection algorithm leads to less packet loss in the network. Simulation experiments are presented to validate the stated goal, using the average packet loss ratio as the performance metric.

Keywords: Mobile wireless ad hoc network, link stability, OLSR, proactive routing, flooding, MPR selection

1. Introduction

With the widespread availability and rapid evolution of wireless local area network technologies such as 802.11 [8], Bluetooth [6], to name a few, the use of MANETs is growing fast [9–11]. In a MANET, there is no need for an infrastructure or a centralized administration for nodes to communicate with each other. Nodes cooperate to provide connectivity and services. However, this flexibility is accompanied by several challenges, in particular, from a routing prospective. In MANETs, the bandwidth, power, speed of mobile nodes, density of the topology, distribution and location of mobile nodes, are among the various factors that can influence the routing process [3,12], hereby the establishment of a routing path between a node pair. These factors, if not well controlled, can lead to an increasing packet loss and a decreasing network performance and link stability. As far as routing is concerned, one of the primary challenges in a MANET is to determine methods for reducing the waste of bandwidth and power consumption while ensuring a stable transmission of information, and quickly responding to network

*Corresponding author: Han-Chieh Chao, Institute of Computer Science and Information Engineering, National I-lan University, Sec. 1, Shen-Lung Rd., I-Lan, Taiwan, Republic of China.
E-mail: hcc@niu.edu.tw.

changes. Of course, these methods depend on the capability of optimizing broadcast messages in the network.

To cope with this issue, one of the recent proactive protocols known as OLSR [13] has been advocated as a promising routing protocol for multi-hop wireless ad-hoc networks. In OLSR, only a subset of pre-selected nodes referred to as MPR nodes have specific functionalities of performing topological advertisements as well as broadcasting and forwarding of control messages, with the goal to reduce the impact of message flooding and control overhead. The set of MPR nodes is chosen in such a way that a minimum of one-hop symmetric neighbors is able to reach all the symmetric two-hop neighbors. To calculate the MPR set, the node must possess the knowledge of the link state information about all one-hop and two-hop neighbors. The objective of the MPR technique is to reduce the number of redundant retransmissions, while ensuring reliable delivery of broadcast messages. In a MPR-based OLSR wireless ad hoc network, every node needs to issue periodic Hello message to update the 1-hop neighbors set and 2-hop neighbors set tables. Every node must dynamically select a neighbor node as MPR node at a time. After this selection is completed, messages are broadcasted, and control messages as well as traffic data that all packets should transmit through MPR nodes are identified.

This paper continues on the investigation of a heuristic introduced in [4] (our so–called MPR-based OLSR heuristic) for the selection of MPR nodes. We consider a cost factor [4] as an additional decision parameter within the MPR-based OLSR heuristic, and we demonstrate by simulation that the resulting heuristic (so-called Enhanced MPR-based OLSR heuristic – here denoted as EMPR-based OLSR) has superior performance over MPR-OLSR in terms of average packet loss ratio in the network. Our primary objective is to investigate the impact of packet loss during transmission when the coverage of a node's selected MPR increases due to the introduction of the above-mentioned additional cost factor in the computation of the MPR selection node.

The paper is organized as follows. Section 2 presents the related works. Section 3 overviews the MPR selection concept. Section 4 describes the proposed EMPR-based OLSR heuristic and contrasts it against the MPR-based OLSR approach. In Section 5, simulation results comparing the proposed EMPR-based heuristic against the MPR-based heuristic are presented. To this effect, OPNET [14] is used as the simulation tool, and the packet loss is considered as the performance metric. Finally, in Section 6, some concluding remarks are given.

2. Related work

The selection of the MPR set of nodes is a fundamental operation in the OLSR protocol. Most solutions to this problem are in the form of heuristics. Representative ones deal with MPR-based flooding techniques [2,4,15,16], as well as reducing the number of collisions or maximizing the bandwidth [7, 17]. With respect to link stability, the authors in [18] studied the link lifetime using various mobility models. They derived a formula to calculate the expected value of each node in order to select high expected value of MPR. Similarly, in [5], some statistical techniques based on link stability metrics were used for selecting MPR nodes while enhancing the route reliability and decreasing the packet loss. Other OLSR enhancements deal with quality of service routing based MPR selection, by combining the MPR selection mechanism with the path determination algorithm [19–21]. In almost all these representative solutions, the goal has been to determine the impact the selection of MPRs has on the performance of OLSR

This paper introduces an additional parameter in the MPR selection method proposed in [4], namely the cost values calculation by all 1-hop and 2-hop neighbor nodes. Our goal is to improve the process of MPR

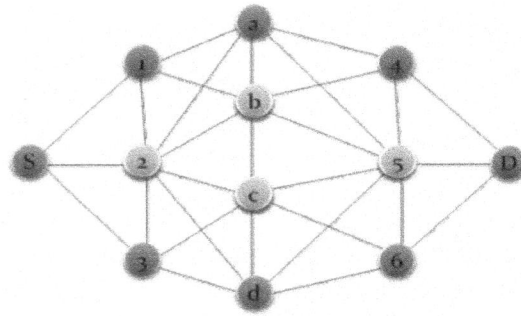

Fig. 1. Example network.

through reducing the total packet loss during transmission, thereby, achieving a superior performance than that of the MPR-based OLSR heuristic.

3. Overview of MPR selection

The OLSR protocol has been extensively investigated in the literature. Rather than re-describing its basic functionality, this Section is meant to setup the context of the study carried in this paper. OLSR is an optimization of the pure link state-based routing protocol [13]. Two important concepts were introduced in the RFC3626 OLSR draft [23] that can enable OLSR extensions, namely, neighbor discovery and Multipoint-Relay selection (MPR). This paper focuses on using MPR. The later is particularly important since it has an immediate effect on the routing protocol's performance. Indeed, the overhead of control traffic generated by the OLSR protocol as well as the flooding efficiency depends on the MPR selection, which itself, is known as a NP-hard problem [17]. In the MPR-based OLSR protocol, each node selects a subset of its one-hop neighbors as its MPR list, based on periodic HELLO messages received by these neighborhood nodes (one-hop neighbors and two-hop neighbors). The compilation of MPR lists then builds a database of nodes that are used in the routing process to compute the shortest paths to all possible destinations in the network. Compared to classical link-state algorithms, MPR-based OLSR algorithms have been proved to significantly reduce the number of message re-transmissions and control traffic [1, 2,4,7,16,18,19,22–24], to name a few.

This paper continues the investigation of the MPR-based OLSR heuristic proposed in [4]. One of the drawbacks of this heuristic is that in the employed MPR technique, only a selected number of nodes propagate the message, which might limit the chances that the message reaches all the nodes in the network. The situation is even worse when the bit error rate in the radio transmission is high [4,13], leading to dramatic packets loss in the network. The MPR-based OLSR heuristic [4] proceeds in three main steps (as detailed in the next section) among which the second one is devoted to the optimization of the MPR set. An analysis of this MPR-based OLSR heuristic has shown that it is within a log n factor from optimality [4].

Our enhancement to the above MPR-based OLSR (so-called EMPR-based OLSR) consists of keeping the core of the MPR-based OLSR heuristic, but introducing a cost factor (in the above-mentioned Step 2) as an additional decision parameter when designing the optimized MPR set.

The EMPR-based OLSR heuristic produces a larger cover range for the MPR set compared to that of the MPR-based OLSR heuristic. As an example, the MPR-based OLSR heuristic applied to the example network shown in Fig. 1 produces {2, b, c, 5} as optimized MPR set with a cover range as depicted in

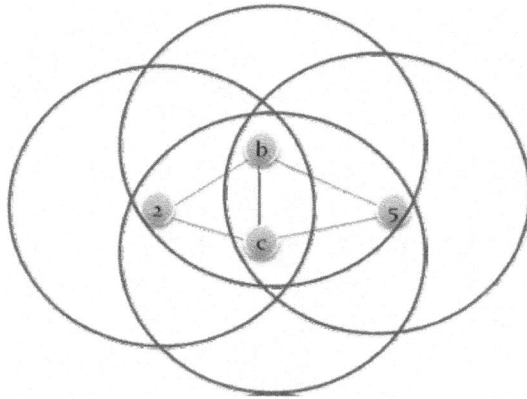

Fig. 2. MPR set cover range when using the MPR-based OLSR heuristic.

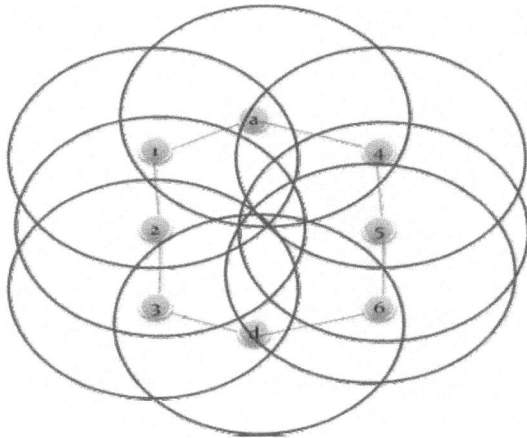

Fig. 3. MPR set cover range when using the EMPR-based OLSR heuristic.

Fig. 2, whereas for the same network, our EMPR-based OLSR heuristic yields $\{1, 2, 3, a, d, 4, 5, 6\}$ as optimized MPR set with a cover range as depicted in Fig. 3.

Our goal is to demonstrate that the EMPR-based OLSR heuristic outperforms the MPR-based OLSR heuristic in terms of packets loss reduction in the network. In the sequel, we adopt the same notations that were used in [4] and we describe the MPR-based OLSR and the EMPR-based OLSR heuristics.

4. Enhanced MPR-based OLSR

The network is represented as a graph G (i.e. a set V of nodes and a set of neighbors for each node). For each element x in V, $M(x)$ is the neighborhood of x, $N(x)$ is the set of one-hop neighbors of x (i.e. x covers any element of $N(x)$), $N^2(x)$ is the set of two-hop neighbors of x, $D(x)$ is the degree of x (i.e. the number of links connected to x), and $MPR(x)$ is the selected MPR set of node x.

The MPR-based OLSR heuristic as introduced in [9] can be summarized as follows:

– Step 1: Start with an empty MPR set $MPR(x)$

– Step 2: Calculate $N(x)$, $N^2(x)$ and $D(x)$ for all x in $N(x)$
– Step 3: Select as MPR nodes those nodes in $N(x)$ which are the only neighbor of some node in $N^2(x)$
– Step 4: As long as there exist some node in $N^2(x)$ which is not covered by the set $MPR(x)$, do the following:

  * For each node z in $N(x)$ which is not element of $MPR(x)$, determine the number of nodes that z covers among uncovered nodes in $N^2(x)$.
  * Add in $MPR(x)$ the node z for which this number is maximum

Let S_1 be the set of nodes selected in Step 3 and $(x_1, \ldots, x_k)$ be the set of nodes selected in the above-mentioned Step 3 and Step 4 of MPR-based OLSR heuristic respectively. Let N_1^2 be the set of nodes in $N^2(s)$ that are neighbors of some nodes s in S_1. Let $N^{2'} = N^2(s) - N_1^2$. A unit cost Cy, associated with each node y in $N^{2'}$, is defined as follows [8]:

$$c_y = \frac{1}{|N'(x_j) - \cup_{j=1}^{j-1} N'(x_j)|} \tag{1}$$

where $|P|$ denotes the cardinality of the set P. As pointed out in [4], for each x_t chosen by the MPR-based OLSR algorithm, this unit cost is equally distributed among nodes that are newly covered in N^2. Our proposed EMPR-based OLSR heuristic takes advantage of this feature. Its stepwise description follows.

– Step 1: Same as in the MPR-based OLSR heuristic
– Step 2: Calculate $N(x)$, $N^2(x)$ and $D(x)$ for all x in $N(x)$. Calculate the cost value C_z associated to each node z in $N(x)$ using Eq. (1)
– Step 3: Select as MPR nodes those nodes in $N(x)$ which are the only neighbor of some node in $N^2(x)$. Enlarge the $MPR(x)$ set by adding those nodes in $N(x)$ that have the lowest cost values (there might be only one such node or many such nodes)
– Step 4: As long as there exists some node in $N^2(x)$ which is not covered by the set $MPR(x)$, do the following:

  * For each node z in $N(x)$ which is not element of $MPR(x)$, determine the number of nodes that z covers among uncovered nodes in $N^2(x)$.
  * Add in $MPR(x)$ the node z for which this number is maximum or add in $MPR(x)$ the node z with lower cost value.

The distinction between the MPR-based OLSR heuristic and the EMPR-based heuristic relies in the introduction of the aforementioned cost value C_z associated with each node z in $N(x)$, which contributes to the construction of the optimized $MPR(x)$ set.

It should be noticed that the introduction of the cost value in the MPR selection calculation has resulted to the following impacts. The first one is an increase of the MPR count. In this case, Step 4 of the proposed EMPR heuristic will attempt to control this side effect by allowing each node to periodically update its neighbor table using "HELLO Messages", which would result in allowing each MPR selected node to flood TC message to all other MPR nodes in the network. The second one is an increase of the coverage of a node's selected MPRs, which in turn will increase the redundancy of TC-messages forwarding. In this case, the MPR-forwarding strategy described in Step 4 of our EMPR heuristic would handle the duplicate re-transmissions directly at the design level when selecting the MPR node set. Indeed, when a mobile node n1 will receive a packet, each node n2 would be required to determine whether node n1 has been selected as MPR node – note that this is realized through computing the node

Table 1
Simulation Environment and Parameters setup

Parameter	Meaning
Number of nodes	50
Communication range	200
Mobility model	Random waypoint
Wireless LAN	802.11g
IP protocol	IPv6
Transmission type	G723 voice bps
Data rate	11 Mbps
Maximum received lifetime	0.5 sec
Simulation start time	10 sec
Channel match criterion	Strict match
Traffic duration	3600 sec
Traffic mix	All explicit
Traffic start time	60 sec
Simulation area	$2*2 \text{ km}^2$

n2's neighbor's table, the node n2's two-hop neighbor table and the number of neighbors of node n2. If the answer is yes, then node n1 will be sending the packets to all other designated neighbor nodes. Otherwise, node n1 will not send the packets, but will remain active.

For both heuristics, experimenting the processes of MPR selection within a specific node using OPNET revealed that this additional decision parameter helps the EMPR-based OLSR heuristic producing a larger cover range for the MPR set compared to that generated by the MPR-based OLSR heuristic. This feature helps reducing the number of packets loss in the network during the transmission period (as illustrated through the simulation results shown in Section 5).

5. Performance evaluation

In this Section, we evaluate the performance of our EMPR-based OLSR heuristic against that of the MPR-based heuristic via simulations. The performance comparison metric is the average packet loss ratio, representing the ratio of packets lost to the total packets generated in a certain time period. The simulation tool is OPNET [14] and the location of the nodes is random. The simulation environment and parameters setup are captured in Table 1.

We have used a single seed value and have ran several simulations. For the same scenario, the simulation time was set to 1 hour and the node's speed was varied from 10 m/s to 90 m/s.

Figure 4 depicts the speed of the mobile node (x axis) versus the amount of traffic it has received (y axis) in terms of number of bits. Figure 5 illustrates the MPR count time average versus the mobile node's speed when running both heuristics. Finally, Fig. 6 depicts the average packet loss ratio versus the mobile node's speed when running both heuristics. It is observed that the EMPR-based heuristic outperforms the MPR-based OLSR heuristic in terms of total number of packet loss during transmission.

6. Concluding remarks

In this paper, we have proposed an enhanced version (called EMPR-based OLSR heuristic) of an existing MPR-based OLSR selection method in MANETs (referred to as MPR-based OLSR heuristic). Our proposal considers the cost value as an additional factor of MPR selection. Our simulation results

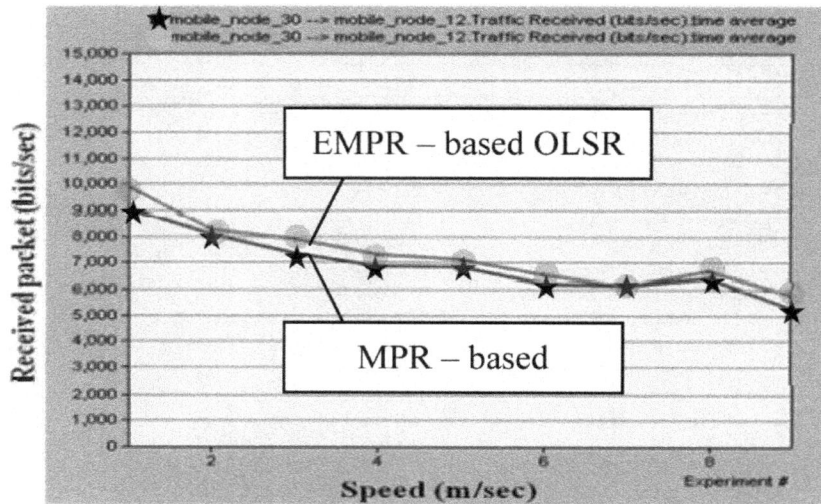

Fig. 4. Mobile node's speed versus the amount of traffic received.

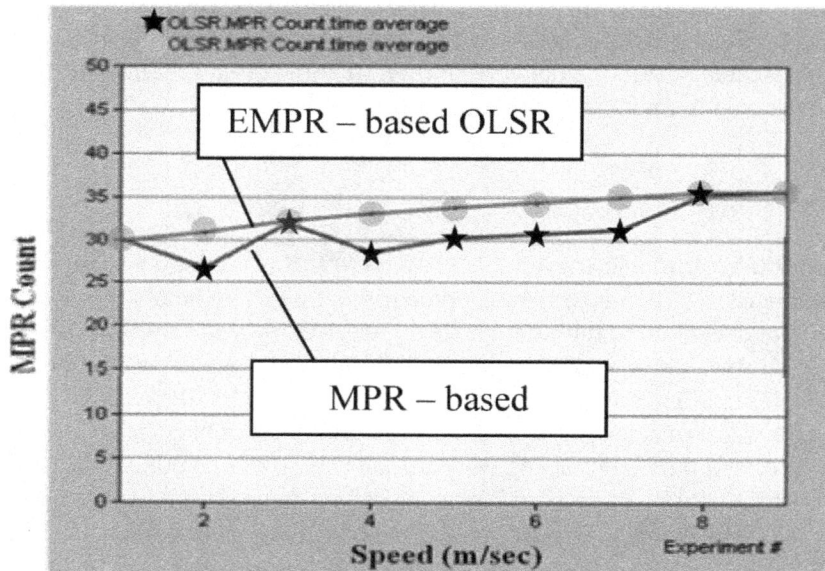

Fig. 5. MPR count versus mobile node's speed.

have shown that the EMPR-based OLSR heuristic yields a better average packet loss ratio compared to the MPR-based OLSR heuristic. As future work, comparing the proposed EMPR-based OLSR heuristic against few other existing MPR-based selection methods for link stability in MANETs would be an interesting work. Also, we have not simulated the change in MPR coverage, nor investigating the NS_MPR forwarding TC message strategy.

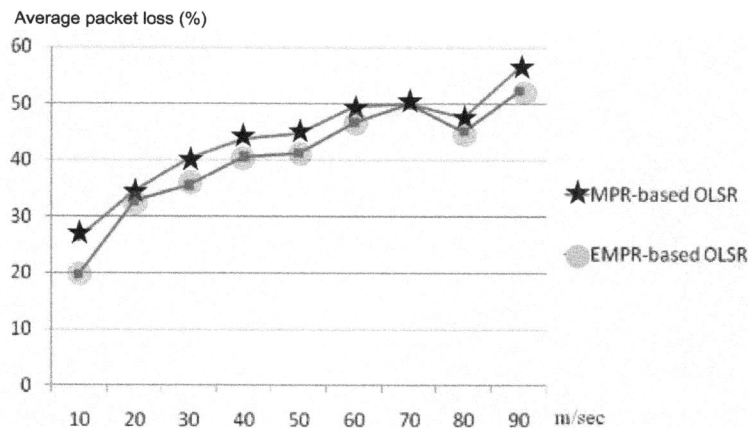

Fig. 6. Average packet loss versus the mobile node's speed.

Acknowledgements

This work is partially supported by National Science Council of Taiwan, under grant number NSC97-2219-E-197-001 and NSC97-2219-E-197-002.

References

[1] A. Benslimane, R. El Khoury, R. El Azouzi and S. Pierre, *Energy Power-Aware Routing in OLSR Protocol*, In Proc. 1st International Mobile Computing and Wireless Communication Conference, Sept. 2006, pp. 14–19.

[2] A. Laouiti, P. Muhlethaler, A. Najid and E. Plakoo, *Simulation Results of the OLSR Routing Protocol for Wireless Network*, MedHoc-Net, Sardegna, Italy, 2002.

[3] A.M. Hanashi, I. Awan and M.E. Woodward, Performance Evaluation with Different Mobility Models for Dynamic Probabilistic Flooding in MANETs, *Mobile Information Systems* 5(1) (2009), 65–80.

[4] A. Qayyum, L. Viennot and A. Laouiti, *Multipoint Relaying for Flooding Broadcast Messages in Mobile Ad Hoc Networks*, Proc. of 35th International Conference on System Sciences, Hawaii, USA, 2002.

[5] B. Mans and N. Shrestha, *Performance Evaluation of Approximation Algorithms for Multipoint Relay Selection*, Med-Hoc-Net04, Bodrum, Turkey, June 27–30, 2004

[6] Bluetooth Specifications, http://www.bluetooth.org/ (Last visited June 3, 2010).

[7] D. Gantsou, P. Sondi and S. Hanafi, Revisiting Multipoint Relay Selection in the Optimized Link State Routing Protocol, International Journal of Communication Networks and Distributed Systems, *Inderscience, UK* 2(1) (2009), 4–15.

[8] IEEE Standards, Standards.ieee.org/getieee802/802.11.html (Last visited June 3, 2010).

[9] S. Corson and J. Macker, *Mobile Ad hoc Networking (MANET): Routing Protocol Performance Issues and Evaluation Considerations*, RFC2501, Jan. 1999, http://www.faqs.org/rfcs/rfc2501.html (Last visited June 3, 2010).

[10] H. Zhang, H. Zhang, J. Guan and Y. Qin, A New Architecture of Multicast Interworking between MANET and Fixed Network, *Journal of Internet Technology* 8(1) (2007), 141–150.

[11] J. Chen, AMNP: ad hoc multichannel negotiation protocol with broadcast solutions for multi-hop mobile wireless networks, *IET Communications* 4(5) (2010), 521–531.

[12] V. Pham, E. Larsen, ØKure and P. Engelstad, Routing of internal MANET traffic over external networks, *Mobile Information Systems* 5(3) (2009), 291–311.

[13] T. Clausen, P. Jacquet, A. Laouiti, P. Muhlethaler, A. Qayyum and L. Viennot, *Optimized Link State Routing Protocol*, IEEE INMIC Pakistan, 2001.

[14] OPNET Simulator, http://www.opnet.com/ (Considerations, RFC2501, Jan. 1999, http://www.faqs.org/rfcs/rfc2501.html), (Last visited June 3, 2010).

[15] T. H. Clausen, G. Hansen, L. Christensen and G. Behrmann, *The Optimized Link State Routing Protocol, Evaluation through Experiments and Simulation*, IEEE Symposium on Wireless Personal Mobile Communications, Sept. 2001.

[16] P. Jacquet, A. Laouiti, P. Minet and L. Viennot, *Performance of Multipoint Relaying in Ad Hoc Mobile Routing Protocols*, Networking 2002, Pise, Italy, 2002.

[17] P. Jacquet, A. Laouiti, P. Minet and L. Viennot, *Performance Analysis of OLSR Multipoint Relay Flooding in Two Ad Hoc Wireless Network Models*, Research Report-4260, INRIA, Sept. 2001, RSRCP Journal, Special Issue on Mobility and Internet.

[18] E. Baccelli and P. Jacquet, Flooding Techniques in Mobile Ad-Hoc Networks, Research Report RR-5002, INRIA, 2003.

[19] M. Gerharz, C. Waal, de, M. Frank and P. Martini, *Link Stability in Mobile Wireless Ad Hoc Networks*, In Proc. of the 27th Annual IEEE Conference on Local Computer Networks (LCN), Nov 6–8, Tampa, Florida, USA, pp. 30–39, 2002.

[20] S. Obilisetty, A. Jasti and R. Pendse, *Link Stability Based Enhancements to OLSR (LS-OLSR)*, In Proc. of 62nd IEEE Vehicular Technology Conference (VTC-2005), Dallas, USA, Sept. 2005.

[21] N. Ghanem, S. Boumerdassi and E. Renault, *New Energy Saving Mechanisms for Mobile Ad Hoc Networks using OLSR*, In Proc. of the 2nd ACM Intl. Workshop on Performance Evaluation of Wireless Ad Hoc, Sensor, and Ubiquitous Networks, Oct. 2005, pp. 273–274.

[22] Z. Guo and B. Malakooti, *Energy Aware Proactive MANET Routing with Prediction on Energy Consumption*, In Proc. of the Intl. Conference on Wireless Algorithms, Systems and Applications, Aug. 2007, pp. 287–293.

[23] RFC3626 Draft, http://www.ietf.org/rfc/rfc3626.txt (Last visited June 3, 2010).

[24] M. Ikeda, L. Barolli, G. De Marco, T. Yang, A. Durresi and F. Xhafa, Tools for Performance Assessment of OLSR Protocol, *Mobile Information Systems* 5(2) (2009), 165–176.

Lin Tzu-Hua was born in Kaohsiung, Taiwan in 1984. He received his B.S. and M.S. degrees from the Department Electrical Engineering, Da-Yeh University and National Dong Hwa University in 2003 and 2009 respectively. Since then, he has been with the National Cheng Kung University, Tainan, Taiwan, working towards his Ph. D degree. His research interests include Mobile communication systems, Vehicle wireless communication, Cooperative networking, and Network coding.

Han-Chieh Chao is a joint appointed Full Professor of the Department of Electronic Engineering and Institute of Computer Science & Information Engineering. He also serves as the Dean of the College of Electrical Engineering & Computer Science for National Ilan University, I-Lan, Taiwan, R.O.C. He has been appointed as the Director of the Computer Center for Ministry of Education starting from September 2009 as well. His research interests include High Speed Networks, Wireless Networks, IPv6 based Networks, Digital Creative Arts and Digital Divide. He received his MS and Ph.D. degrees in Electrical Engineering from Purdue University in 1989 and 1993 respectively. He has authored or co-authored 4 books and has published about 200 refereed professional research papers. He has completed 80 MSEE thesis students and 2 PhD students. Dr. Chao has received many research awards, including Purdue University SRC awards, and NSC research awards (National Science Council of Taiwan). He also received many funded research grants from NSC, Ministry of Education (MOE), RDEC, Industrial Technology of Research Institute, Institute of Information Industry and FarEasTone Telecommunications Lab. Dr. Chao has been invited frequently to give talks at national and international conferences and research organizations. Dr. Chao is also serving as an IPv6 Steering Committee member and co-chair of R&D division of the NICI (National Information and Communication Initiative, a ministry level government agency which aims to integrate domestic IT and Telecom projects of Taiwan), Co-chair of the Technical Area for IPv6 Forum Taiwan, the executive editor of the Journal of Internet Technology and the Editor-in-Chief for International Journal of Internet Protocol Technology and International Journal of Ad Hoc and Ubiquitous Computing. Dr. Chao has served as the guest editors for Mobile Networking and Applications (ACM MONET), IEEE JSAC, IEEE Communications Magazine, Computer Communications, IEE Proceedings Communications, the Computer Journal, Telecommunication Systems, Wireless Personal Communications, and Wireless Communications & Mobile Computing. Dr. Chao is an IEEE senior member and a Fellow of IET (IEE). He is a Chartered Fellow of British Computer Society.

Isaac Woungang received his M.S. and Ph. D degrees in Mathematics from the Universit¤f de la M¤fditerran¤fe-Aix Marseille II, France, and the Universit¤f du Sud, Toulon et Var, France, in 1990 and 1994 respectively. In 1999, he received a M.S from the INRS-Materials and Telecommunications, University of Quebec, Montreal, Canada. From 1999 to 2002, he worked as a software engineer at Nortel Networks. Since 2002, he has been with Ryerson University, where he is an Associate professor of Computer Science. In 2004, he founded the Distributed Applications and Broadband NEtworks Laboratory (DABNEL) R&D group. His research interest includes Network security, Computer communication networks, Mobile communication systems, computational intelligence applications in telecommunications, and Coding theory. Dr. Woungang serves as Editor-in-Chief of the International Journal of Communication Networks & Distributed Systems (IJCNDS), Inderscience, U.K, and the International Journal of Information and Coding Theory (IJICoT), Inderscience, U.K, as Associate Editor of the International Journal of Communication Systems (IJCS), Wiley, and Associate Editor of Computer and Electrical Engineering, An International Journal (C&EE), Elsevier. Dr. Woungang edited several books in the areas of wireless ad hoc networks, wireless sensor networks, wireless mesh networks, communication networks & distributed systems, and information & coding theory, published by reputed publishers such as Springer, Elsevier, and World Scientific.

5

Energy efficient resource management in mobile grid

Chunlin Li* and Layuan Li

Department of Computer Science, Wuhan University of Technology, Wuhan, P.R. China

Abstract. Energy efficient computing has recently become hot research area. Many works have been carried out on conserving energy, but considering energy efficiency in grid computing is few. This paper proposes energy efficient resource management in mobile grid. The objective of energy efficient resource management in mobile grid is to maximize the utility of the mobile grid which is denoted as the sum of grid application utility. The utility function models benefits of application and system. By using nonlinear optimization theory, energy efficient resource management in mobile grid can be formulated as multi objective optimization problem. In order to derive a distributed algorithm to solve global optimization problem in mobile grid, we decompose the problem into the sub problems. The proposed energy efficient resource management algorithm decomposes the optimization problem via iterative method. To test the performance of the proposed algorithm, the simulations are conducted to compare proposed energy efficient resource management algorithm with other energy aware scheduling algorithm.

Keywords: Mobile grid, resource management, energy efficient

1. Introduction

Mobile grid means that movable wireless devices are integrated into traditional wired grid through wireless channel to share grid resources (CPU power, storage capacity, instrument, devices, data, software, etc.), meanwhile, mobile devices can provide service or resource to grid users, such as PDAs, cellular phones, handsets or wearable computers, laptops with GPS service, mobile service, etc. [1]. Mobile grid can provide end users with grid services such as music, medical, accounting, message alert service, function estimation, data mining, assurance, and so on. Mobile grid may be constructed on current network infrastructure, integrate continually developing wireless network technologies, enrich network contents and software platform function. Mobile grid includes various kinds of mobile devices, and then leads to the grid system more complicated than wired grid system due to the mobile grid node dynamically behavior in the grid system. Mobile grid requires dynamic management of distributed resources, and such management needs to meet application quality requirements and prolong application lifetimes. Mobile grid application' lifetime is determined by available energy on the mobile devices. Mobile devices are battery-driven, and hence operate on an extremely frugal energy budget. Considering energy efficiency in mobile grid is an important challenge.

This paper proposes energy efficient resource management in mobile grid. The objective of energy efficient resource management in mobile grid is to maximize the utility of the mobile grid which is denoted as the sum of grid application utility. The utility function models benefits of application and

---

*Corresponding author. E-mail: chunlin74@whut.edu.cn; chunlin74@yahoo.com.cn.

system. By using nonlinear optimization theory, energy efficient resource management in mobile grid can be formulated as multi objective optimization problem. In order to derive a distributed algorithm to solve global optimization problem in mobile grid, we decompose the problem into the sub problems. The proposed energy efficient resource management algorithm decomposes the optimization problem via iterative method. To test the performance of the proposed algorithm, the simulations are conducted to compare proposed energy efficient resource management algorithm with other energy aware algorithm.

The rest of the paper is structured as followings. Section 2 discusses the related works. Section 3 presents energy efficient resource management in mobile grid. Section 4 presents energy efficient resource management algorithm in mobile grid. In Section 5 the simulations are conducted. Section 6 gives the conclusions to the paper.

2. Related work

Energy efficient computing has recently become hot research area. Many works have been carried out on conserving energy, but considering energy in grid computing is few. Tao Xie et al. [2] address the issue of allocating tasks of parallel applications in heterogeneous embedded systems with an objective of energy-saving and latency-reducing. They proposed BEATA (Balanced Energy-Aware Task Allocation), a task allocation scheme considering both energy consumption and schedule length, is developed to solve the energy-latency dilemma. Y. Huang et al. [3] present techniques for exploiting intermittently available resources in grid infrastructures to support QoS-based multimedia applications on mobile devices. They integrate power aware admission control, grid resource discovery, dynamic load-balancing and energy adaptation techniques to enable power deficient devices to run distributed multimedia applications. Ziliang Zong et al. [4] design energy efficient scheduling algorithms for parallel applications running on clusters, they propose a scheduling strategy called energy efficient task duplication schedule, which can significantly conserve power by judiciously shrinking communication energy cost when allocating parallel tasks to heterogeneous computing nodes. Tarek A. AlEnawy et al. [5] propose to minimize the number of dynamic failures while remaining within the energy budget. They propose techniques to statically compute the speed of the CPU in order to meet the (m,k)-firm deadline constraints. Kyong Hoon Kim et al. [6] provide power-aware scheduling algorithms for bag of tasks applications with deadline constraints on DVS enabled cluster systems in order to minimize power consumption as well as to meet the deadlines specified by application users. Eunjeong Park et al. [8] designed an entire process of multimedia service composition for mobile computing. Their approach adapts the composition graph and the use of service routing for the context of mobile devices with the support of monitoring components. The works [9–13] mainly deal with resource allocation, QoS optimization in the computational grid and don't consider energy consumption for mobile grid. Narottam Chand et al. [17] propose a utility based cache replacement policy, least utility value (LUV), to improve the data availability and reduce the local cache miss ratio. Markus Aleksy et al. [18] present a generic architecture, which can be used for the development of context-sensitive mobile applications. Vincent Reinhard et al. [19] propose a mechanism for the introduction of application data in the network with respect to CARRIOCAS architecture. P. Hellinckx et al. [18] introduced a new multipurpose Lightweight Grid (LWG) system under the acronym Computational Basic Reprogrammable Adaptive (CoBRA) grid. The methods and contributions of this paper are different from above works. This paper is targeted to manage energy consumption without compromising system's performance in mobile grid.

3. Energy efficient resource management in mobile grid

3.1. Energy efficient resource management mathematic model

In mobile grid, energy resources distribution and computation workloads are not balanced within mobile devices. It is important for the mobile grid system to design energy efficient resource management. Firstly, let consider the notations used in the following sections. e_i^l is the energy obtained by grid application i from energy resource l. Ce_l is the capacity of energy resource l. Cn_k is the capacity of network resource k. Cc_j is the capacity of computing power j. t_i^n is the time taken by the grid application i to complete nth job. Pc_i^j is the payments of the grid application i to the computing power j. Pn_i^k is the payments of the grid application i to the network resource k. Pe_i^l is the payments of the grid application i to energy resource l. B_i is the expense budget of grid application i. er_l is the energy consumption rate of energy resource l. e_i^n is energy dissipation caused by grid application i's nth job. cq_i^n is the computation task of ith grid application's nth job. bq_i^n is the transmission task of ith grid application's nth job. eq_i^n is the energy storage task of ith grid application's nth job. np_k the price of network resource k. cp_j is the price of computing resource j. ep_l is the price of energy resource l. T_i is the deadline given by the grid application i to complete its all jobs. y_i^k is the network allocation obtained by grid application i from the network resource k. x_i^j is the computing power obtained by grid application i from computing power j.

We assume that the mobile grid has heterogeneous nodes with different system performance rates and network conditions. This means that the energy consumption of the mobile device can vary with the response time of the application and the network bandwidth. We denote e_i^l is the consumed energy fraction of the energy l (e.g. a battery) by grid application i. Total consumed energy of all grid applications $\sum_{i=1}^{I} e_i^l$ does not exceed the total capacity Ce_l of energy l. Energy consumption rate of each node in the system is measured by Joule per unit time. Let e_i^n be an energy dissipation caused by grid application i's nth job, t_i^n is the execution time of job n of grid application i on the grid node. er is the energy consumption rate of energy resource l. If the energy consumption is proportional to execution time of job n, as is the case with battery energy. We define the energy consumption of each application A_i as the sum of the energy consumed by N grid jobs $\sum_{n=1}^{N} e_i^n$. The energy consumption of all grid jobs of each application A_i should be less than the available resources of e_i^l which is limited energy budget of grid user application i.

Now, we formulate the problem of energy efficient resource management in mobile grid as constraint optimization problem, the utility of the mobile grid $U_{\text{mobilegrid}}$ is defined as the sum of grid application utilities $\sum_{i=1}^{I} U_i$. The objective of energy efficient resource management in mobile grid is to maximize the utility of the mobile grid $U_{\text{mobilegrid}}$. By using nonlinear optimization theory, energy efficient resource management in mobile grid can be formulated as follows.

$$Max U_{\text{mobilegrid}}$$

$$\text{s.t.} \; T_i \geqslant \sum_{n=1}^{N} t_i^n, Cn_k \geqslant \sum_{i=1}^{I} y_i^k, \sum_{n=1}^{N} e_i^n \leqslant e_i^l, Cc_j \geqslant \sum_{i=1}^{I} x_i^j, \sum_{i=1}^{I} e_i^l \leqslant Ce_l \qquad (1)$$

$$B_i \geqslant \sum_{l=1}^{L} Pe_i^l + \sum_{j=1}^{J} Pc_i^j + \sum_{k=1}^{K} Pn_i^k$$

Where e_i^l is the energy obtained by grid application i from the energy l. x_i^j is CPU allocation obtained by grid application i from the computing resource provider j. y_i^k is bandwidth allocation obtained by grid application i from the network resource provider k. The utility function for application A_i depends on allocated resources x_i^j, y_i^k and consumed energy e_i^l. In the problem Eq. (1), the first type of constraints is related with different resource capacity. The QoS constraint implies that the aggregate network resource units $\sum_{i=1}^{I} y_i^k$ do not exceed the total capacity Cn_k of network resource provider k, aggregate consumed energy of all grid application $\sum_{i=1}^{I} e_i^l$ does not exceed the total Ce_l of energy l, aggregate computing power $\sum_{i=1}^{I} x_i^j$ does not exceed the total resource Cc_j of the computing resource provider j. The second type of constraints is related with grid application expense budget. Grid application needs to complete a sequence of jobs in a specified amount of time, T_i, while the payment overhead accrued cannot exceed B_i, which is the expense budget of grid application i. Pe_i^l, Pc_i^j, Pn_i^k are the payments of the grid application i to the energy storage provider l, computing resource provider j and network resource provider k. The total payments of the grid application

$$i \sum_{l=1}^{L} Pe_i^l + \sum_{j=1}^{J} Pc_i^j + \sum_{k=1}^{K} Pn_i^k$$

dose not exceed B_i. The total energy consumed by all jobs of grid application i $\sum_{n=1}^{N} e_i^n$ cannot exceed the energy budget e_i^l which is the available energy obtained by grid application i from the energy storage l.

3.2. Solutions to mathematic model

We apply the Lagrangian method to solve such a problem Eq. (1). The Lagrangian approach can be used to solve constrained optimization problems. Let us consider the Lagrangian form of energy efficient resource management in mobile grid:

$$L = \sum_{i=1}^{I} U_i - \lambda_i \left(\sum_{i=1}^{I} e_i^l - Ce_l \right) - \beta_i \left(\sum_{i=1}^{I} x_i^j - Cc_j \right) - \varphi_i \left(\sum_{i=1}^{I} y_i^k - Cn_k \right)$$
$$- \gamma_i \left(\sum_{l=1}^{L} Pe_i^l + \sum_{j=1}^{J} Pc_i^j + \sum_{k=1}^{K} Pn_i^k - B_i \right) - \mu_i \left(\sum_{n=1}^{N} e_i^n - e_i^l \right) - \alpha_i \left(\sum_{n=1}^{N} t_i^n - T_i \right)$$

where λ_i, β_i and φ_i are the Lagrange multipliers of grid application with their interpretation of energy price, computing resource capacity price, and network resource capacity price, respectively. Since the Lagrangian is separable, this maximization of Lagrangian over (x_i^j, y_i^k, e_i^l) can be conducted in parallel at each application A_i. In problem Eq. (1), though the allocated resources x_i^j, y_i^k and consumed energy e_i^l are coupled in their constraints, respectively, but they are separable. Given that the grid knows the utility functions U of all the grid applications, this optimization problem can be mathematically tractable. However, in practice, it is not likely to know each application's utility, and it is also infeasible for mobile grid environment to compute and allocate resources in a centralized fashion. In order to derive a distributed algorithm to solve problem Eq. (1), we decompose the problem into the sub problems.

The mobile grid utility denoted as the sum of grid application utility can be defined as follows Eq. (2):

$$
\begin{aligned}
U_{\text{mobilegrid}} = {} & \left(B_i - \sum_{l=1}^{L} Pe_i^l - \sum_{j=1}^{J} Pc_i^j - \sum_{k=1}^{K} Pn_i^k \right) + \left(T_i - \sum_{n=1}^{N} t_i^n \right) \\
& + \sum_{i=1}^{I} \left(Pe_i^l \log e_i^l + Pc_i^j \log x_i^j + Pn_i^k \log y_i^k \right) + \left(e_i^l - \sum_{n=1}^{N} e_i^n \right)
\end{aligned}
\tag{2}
$$

Mobile grid utility function is maximally optimized with specific constraints. In Eq. (2), $Pe_i^l \log e_i^l + Pc_i^j \log x_i^j + Pn_i^k \log y_i^k$ present the revenue of energy storage resource, computing power and network resource provider. We could have chosen any other form for the utility that increases with x_i^j, y_i^k, e_i^l. But we chose the *log* function because the benefit increases quickly from zero as the total allocated resource increases from zero and then increases slowly. Moreover, *log* function is analytically convenient, increasing, strictly concave and continuously differentiable. The benefits of grid resource provider are affected by payments of grid applications and allocated resources. It means that the revenue increases with increasing allocated resources and increasing payment.

The Lagrangian form of the problem Eq. (1) can be reformulated as follows Eq. (3):

$$
\begin{aligned}
L = {} & \left(B_i - \sum_{l=1}^{L} Pe_i^l - \sum_{j=1}^{J} Pc_i^j - \sum_{k=1}^{K} Pn_i^k \right) + \sum_{i=1}^{I} \left(Pe_i^l \log e_i^l + Pc_i^j \log x_i^j + Pn_i^k \log y_i^k \right) \\
& + \left(T_i - \sum_{n=1}^{N} t_i^n \right) + \left(e_i^l - \sum_{n=1}^{N} e_i^n \right) - \lambda_i \left(\sum_{i=1}^{I} e_i^l - Ce_l \right) - \beta_i \left(\sum_{i=1}^{I} x_i^j - Cc_j \right) \\
& - \varphi_i \left(\sum_{i=1}^{I} y_i^k - Cn_k \right) - \gamma_i \left(\sum_{l=1}^{L} Pe_i^l + \sum_{j=1}^{J} Pc_i^j + \sum_{k=1}^{K} Pn_i^k - B_i \right) - \mu_i \left(\sum_{n=1}^{N} e_i^n - e_i^l \right) \\
& - \alpha_i \left(\sum_{n=1}^{N} t_i^n - T_i \right)
\end{aligned}
\tag{3}
$$

The system model presented by Eq. (1) is a nonlinear optimization problem with N decision variables. Since the Lagrangian is separable, the maximization of the Lagrangian can be processed in parallel for grid user applications and grid resource providers respectively. From Eq. (3), the resource allocation $\{e_i^l, x_i^j, y_i^k\}$ solves problem Eq. (1) if and only if there exist a set of nonnegative shadow costs $\{\lambda_i, \beta_i, \varphi_i\}$. Generally solving such a problem by typical algorithms such as steepest decent method and gradient projection method is of high computational complexity, which is very time costing and impractical for implementation. In order to reduce the computational complexity, we decompose the utility optimization problem Eq. (1) into two subproblems for grid user applications and grid resource providers so that the computational complexity is reduced. The shadow costs suggest a mechanism to distribute the resource optimization between the grid applications and the grid system. The problem Eq. (1) maximizes the utility of grid applications on the energy price, computing power capacity price, and network resource capacity price, $\sum_{i=1}^{I} U_i(e_i^l, x_i^j, y_i^k)$ is the total utility of mobile grid system, $\beta_i \sum_{i=1}^{I} x_i^j$ is the computing power cost, $\lambda_i \sum_{i=1}^{I} e_i^l$ is the energy cost, $\varphi_i \sum_{i=1}^{I} y_i^k$ is the network resource cost. By decomposing the Kuhn-Tucker conditions into separate roles of consumer and supplier at grid market, the centralized problem Eq. (1) can be transformed into a distributed problem. Grid application's payment is collected by the resource providers. The payments of grid applications paid to resource providers are the payments to resolve the optimality of resource allocation in the grid market. We decompose the problem into the following two subproblems Eq. (4) which is grid users optimization problem and Eq. (5) which is grid resource providers optimization problem, seek a distributed solution where the grid resource provider does not need to know the utility functions of individual grid user application. Equations (4), (5) derived from the distributed approach are identical to the optimal conditions given by centralized energy efficient resource management problem Eq. (1). This means if two subproblems converge to its optimal points, then a globally optimal point is achieved. Total user application benefit of the mobile grid is maximized when the equilibrium prices, obtained through the two subproblems Eqs (4) and (5), equal the Lagrangian multipliers λ_i, β_i and φ_i, where λ_i, β_i and φ_i are the optimal prices charged by resource providers including energy, computing power and network resource to grid applications. Two maximization subproblems correspond to grid application optimization problem as denoted by Eqs (4) and (5).

$$MaxU_{\text{GA}} = \left(T_i - \sum_{n=1}^{N} t_i^n \right) + \left(e_i^l - \sum_{n=1}^{N} e_i^n \right) + \left(B_i - \sum_{l=1}^{L} Pe_i^l - \sum_{j=1}^{J} Pc_i^j - \sum_{k=1}^{K} Pn_i^k \right)$$

$$B_i \geqslant \sum_{l=1}^{L} Pe_i^l + \sum_{j=1}^{J} Pc_i^j + \sum_{k=1}^{K} Pn_i^k \qquad (4)$$

$$\sum_{n=1}^{N} e_i^n \leqslant e_i^l, T_i \geqslant \sum_{n=1}^{N} t_i^n,$$

$$MaxU_{\text{GR}} = \sum_{i=1}^{I} (Pe_i^l \log e_i^l + Pc_i^j \log x_i^j + Pn_i^k \log y_i^k)$$

$$\text{Cc}_j \geqslant \sum_{i=1}^{I} x_i^j, \quad \text{Cn}_k \geqslant \sum_{i=1}^{I} y_i^k, \quad \sum_{i=1}^{I} e_i^l \leqslant Ce_l \qquad (5)$$

In Eq. (4), the grid application gives the unique optimal payment to resource provider under the energy budget, expense budget and the deadline constraint to maximize the user's satisfaction.

$$\left(B_i - \sum_{l=1}^{L} Pe_i^l - \sum_{j=1}^{J} Pc_i^j - \sum_{k=1}^{K} Pn_i^k \right)$$

represents the money surplus of grid application i, which is obtained by expense budgets subtracting the payments to various resource providers. $(T_i - \sum_{n=1}^{N} t_i^n)$ represents the saving times for grid application i, which is gotten by time limit subtracting actual spending time. $(e_i^l - \sum_n e_i^n)$ represents the energy surplus of grid application i which is obtained by the energy budgets subtracting actual energy dissipation. So, the objective of Eq. (4) is to get more surpluses of money and more energy savings, at the same time, complete task for grid user application as soon as possible. In Eq. (5), different resource providers compute optimal resource allocation for maximizing the revenue of their own. Grid application i submits the payment Pe_i^l to the energy resource provider l, Pn_i^k to network resource provider k and Pc_i^j to computing power provider j. $Pe_i^l \log e_i^l$ presents the revenue obtained by energy resource l from grid application i. $Pc_i^j \log x_i^j$ presents the revenue obtained by computing power j from grid application i. $Pn_i^k \log y_i^k$ presents the revenue obtained by network resource k from grid application i. The objective of resource providers is to maximize

$$Pe_i^l \log e_i^l + Pc_i^j \log x_i^j + Pn_i^k \log y_i^k$$

under the constraints of their provided resource amounts. Grid resource providers can't sell the resources to grid applications more than total capacity. In Eq. (4), the grid application adaptively adjusts its payments to computing power, network resource and energy based on the current resource conditions, while in Eq. (5), the grid resource provider adaptively allocates energy, CPU and bandwidth required by the grid application in the Eq. (4). The interaction between two sub-problems is controlled through the use of the price variable λ_i, β_i and φ_i, which is the energy price, computing power price, and network resource price charged from grid applications by grid energy resource, computing power and network resource. The interaction between two sub-problems also coordinates the grid application's payment and the supply of grid resource providers.

Lagrange relaxation and gradient optimization can be applied to decompose such an overall optimization problem into a sequence of two sub-problems, each only involving variables from the grid application and resource providers respectively. In Eq. (4), grid application maximizes its satisfaction and gives the unique optimal payment to resource provider under the energy budget, expense budget and the deadline constraint. We assume that grid application i submits Pe_i^l to energy resource l, Pc_i^j to computing power j and Pn_i^k to network resource k. Then, $Pe_i = [Pe_i^1, \ldots, Pe_i^l]$ represents all payments of grid applications for energy resource l, $Pc_i = [Pc_i^1, \ldots, Pc_i^j]$ represents all payments of grid applications for computing power j, $Pn_i = [Pn_i^1, \ldots, Pn_i^k]$ represents all payments of grid applications for the network resource k. Let

$$m_i = \sum_l Pe_i^l + \sum_j Pc_i^j + \sum_k Pn_i^k,$$

m_i is the total payment of the ith grid application. N grid applications compete for grid resources with finite capacity. The resource is allocated using a market mechanism, where the partitions depend on the relative payments sent by the grid applications. Let ep_l, cp_j, np_k denote the price of the resource unit of energy resource l, the price of the resource unit of computing power j and network resource k respectively. Let the pricing policy, $ep = (ep_1, ep_2, \cdots, ep_l)$, denote the set of resource unit prices of all the energy resources in the grid, $cp = (cp_1, cp_2, \cdots, cp_j)$, denote the set of resource unit prices of all the computing powers, $np = (np_1, np_2, \cdots, np_k)$ is set of network resource unit prices. The ith grid application receives the resources proportional to its payment relative to the sum of the resource provider's revenue. Let e_i^l, i^j, y_i^k be the fraction of resource units allocated to grid application i by energy l, computing power j and network resource k.

The time taken by the ith grid application to complete nth job is:

$$t_i^n = \frac{cq_i^n cp_j}{Cc_j Pc_i^j} + \frac{bq_i^n np_k}{Cn_k Pn_i^k} + \frac{eq_i^n ep_l}{Ce_l Pe_i^l}$$

The energy dissipation used by the ith grid user to complete nth job is:

$$e_i^n = er.t_i^n = er \cdot \left(\frac{cq_i^n cp_j}{Cc_j Pc_i^j} + \frac{bq_i^n np_k}{Cn_k Pn_i^k} + \frac{eq_i^n ep_l}{Ce_l Pe_i^l} \right)$$

Grid user optimization can be reformulated as

$$MaxU_{\mathbf{GA}} = \left(B_i - \sum_{l=1}^{L} Pe_i^l - \sum_{j=1}^{J} Pc_i^j - \sum_{k=1}^{K} Pn_i^k \right) + \left(T_i - \sum_{n=1}^{N} \left(\frac{cq_i^n cp_j}{Cc_j Pc_i^j} + \frac{bq_i^n np_k}{Cn_k Pn_i^k} + \frac{eq_i^n ep_l}{Ce_l Pe_i^l} \right) \right)$$
$$+ \left(e_i^l - \sum_{n=1}^{N} er \left(\frac{cq_i^n cp_j}{Cc_j Pc_i^j} + \frac{bq_i^n np_k}{Cn_k Pn_i^k} + \frac{eq_i^n ep_l}{Ce_l Pe_i^l} \right) \right)$$

The Lagrangian for the grid application's utility is $L_1(Pe_i^l, Pc_i^j, Pn_i^k)$.

$$L_1(Pe_i^l, Pc_i^j, Pn_i^k) = \left(B_i - \sum_{l=1}^{L} Pe_i^l - \sum_{j=1}^{J} Pc_i^j - \sum_{k=1}^{K} Pn_i^k \right)$$
$$+ \left(T_i - \sum_{n=1}^{N} \left(\frac{cq_i^n cp_j}{Cc_j Pc_i^j} + \frac{bq_i^n np_k}{Cn_k Pn_i^k} + \frac{eq_i^n ep_l}{Ce_l Pe_i^l} \right) \right)$$
$$+ \left(e_i^l - \sum_{n=1}^{N} er \left(\frac{cq_i^n cp_j}{Cc_j Pc_i^j} + \frac{bq_i^n np_k}{Cn_k Pn_i^k} + \frac{eq_i^n ep_l}{Ce_l Pe_i^l} \right) \right)$$
$$+ \nu_i \left(B_i - \sum_{l=1}^{L} Pe_i^l - \sum_{j=1}^{J} Pc_i^j - \sum_{k=1}^{K} Pn_i^k \right)$$
$$+ \sigma_i \left(T_i - \sum_{n=1}^{N} \left(\frac{cq_i^n cp_j}{Cc_j Pc_i^j} + \frac{bq_i^n np_k}{Cn_k Pn_i^k} + \frac{eq_i^n ep_l}{Ce_l Pe_i^l} \right) \right)$$
$$+ \varepsilon_i \left(e_i^l - \sum_{n=1}^{N} er \left(\frac{cq_i^n cp_j}{Cc_j Pc_i^j} + \frac{bq_i^n np_k}{Cn_k Pn_i^k} + \frac{eq_i^n ep_l}{Ce_l Pe_i^l} \right) \right)$$

Where $\varepsilon_i, \sigma_i, \nu_i$ is the Lagrangian constant. From Karush-Kuhn-Tucker Theorem we know that the optimal solution is given $\partial L_1(Pe_i^l, Pc_i^j, Pn_i^k) \big/ \partial Pe_i^l = 0$ for $\varepsilon_i, \sigma_i, \nu_i > 0$.

$$\partial L_1(Pe_i^l, Pc_i^j, Pn_i^k) \big/ \partial Pe_i^l = -1 - \nu_i + \frac{eq_i^n ep_l}{Ce_l (Pe_i^l)^2} + er \frac{eq_i^n ep_l}{Ce_l (Pe_i^l)^2}$$
$$+ \sigma_i \frac{eq_i^n ep_l}{Ce_l (Pe_i^l)^2} + \varepsilon_i . er \frac{eq_i^n ep_l}{Ce_l (Pe_i^l)^2}$$

Let $\partial L_1(Pe_i^l, Pc_i^j, Pn_i^k)\big/\partial Pe_i^l = 0$ to obtain

$$Pe_i^l = \left(\frac{(1 + er + \sigma_i + \varepsilon_i \cdot er)eq_i^n \, ep_l}{(1 + \nu_i)Ce_l} \right)^{1/2}$$

Using this result in the constraint equation, we can determine $\theta = (1 + er + \sigma_i + \varepsilon_i \cdot er)\big/1 + \nu_i$ as

$$(\theta)^{-1/2} = \frac{T_i}{\sum\limits_{m=1}^{N} \left(\frac{ep_m eq_i^n}{Ce_m} \right)^{1/2}}$$

We obtain $Pe_i^{l^*}$

$$Pe_i^{l^*} = \left(\frac{eq_i^n \, ep_l}{Ce_l} \right)^{1/2} \frac{\sum\limits_{m=1}^{N} \left(\frac{eq_i^n \, ep_m}{Ce_m} \right)^{1/2}}{T_i}$$

It means that grid application wants to pay $Pe_i^{l^*}$ to energy resource l for needed energy consumed to execute grid jobs under completion time constraint.

$$\partial L_1(Pe_i^l, Pc_i^j, Pn_i^k)\big/\partial Pc_i^j = -1 + \frac{cq_i^n cp_j}{Cc_j(Pc_i^j)^2} + er_i^n \frac{cq_i^n cp_j}{Cc_j(Pc_i^j)^2} - \nu_i$$
$$+ \sigma_i \frac{cq_i^n cp_j}{Cc_j(Pc_i^j)^2} + \varepsilon_i \cdot er \frac{cq_i^n cp_j}{Cc_j(Pc_i^j)^2}$$

Let $\partial L_1(Pe_i^l, Pc_i^j, Pn_i^k)\big/\partial Pc_i^j = 0$ to obtain

$$Pc_i^j = \left(\frac{(1 + er + \sigma_i + \varepsilon_i.er)cq_i^n \, cp_j}{(1 + \nu_i)Cc_j} \right)^{1/2}$$

Using this result in the constraint equation, we can determine $\xi = (1 + er + \sigma_i + \varepsilon_i \cdot er)\big/1 + \nu_i$ as

$$(\xi)^{-1/2} = \frac{T_i}{\sum\limits_{m=1}^{N} \left(\frac{cp_m cq_i^n}{Cc_m} \right)^{1/2}}$$

We obtain $Pc_i^{j^*}$

$$Pc_i^{j^*} = \left(\frac{cq_i^n \, cp_j}{Cc_j} \right)^{1/2} \frac{\sum\limits_{m=1}^{N} \left(\frac{cq_i^n cp_m}{Cc_m} \right)^{1/2}}{T_i}$$

It means that grid application wants to pay $Pc_i^{j^*}$ to computing power j for needed resource to execute grid jobs under completion time constraint.

$$\partial L_1(Pe_i^l, Pc_i^j, Pn_i^k)\big/\partial Pn_i^k = -1 + \frac{bq_i^n np_k}{Cn_k(Pn_i^k)^2} + er_i^n \frac{bq_i^n np_k}{Cn_k(Pn_i^k)^2} - \nu_i$$

$$+\sigma_i \frac{bq_i^n np_k}{Cn_k(Pn_i^k)^2} + \varepsilon_i \frac{bq_i^n np_k}{Cn_k(Pn_i^k)^2}$$

Let $\partial L_1(Pe_i^l, Pc_i^j, Pn_i^k)\Big/\partial Pn_i^k = 0$ to obtain

$$Pn_i^k = \left(\frac{(1 + er + \sigma_i + er.\varepsilon_i)bq_i^n\, np_k}{(1 + \nu_i)Cn_k}\right)^{1/2}$$

Using this result in the constraint equation, we can determine $\tau = (1 + er + \sigma_i + er.\varepsilon_i)\big/1 + \nu_i$ as

$$(\tau)^{-1/2} = \frac{T_i}{\displaystyle\sum_{m=1}^{N}\left(\frac{np_m bq_i^n}{Cn_m}\right)^{1/2}}$$

We obtain Pn_i^{k*}

$$Pn_i^{k*} = \left(\frac{bq_i^n\, np_k}{Cn_k}\right)^{1/2} \frac{\displaystyle\sum_{m=1}^{N}\left(\frac{bq_i^n\, np_m}{Cn_m}\right)^{1/2}}{T_i}$$

It means that grid application wants to pay Pn_i^{k*} to network resource k for needed resource to execute grid jobs under completion time constraint.

In Eq. (5), different resource providers compute optimal resource allocation for maximizing the revenue of their own under constrains of resource capacity Ce_l, Cc_j, Cn_k, the objective of resource providers is to maximize

$$Pe_i^l \log e_i^l + Pc_i^j \log x_i^j + Pn_i^k \log y_i^k$$

under the constraints of their resource capacity.

We take first derivative and second derivative with respect to x_i:

$$U'_{\mathrm{GR}}(e_i^l) = Pe_i^l\big/e_i^j \qquad U''_{\mathrm{GR}}(e_i^l) = -Pe_i^l\big/e_i^{l2}$$

$U''_{\mathrm{GR}}(e_i^l) < 0$ is negative due to $0 < e_i^l$. The extreme point is the unique value maximizing the revenue of energy provider. The Lagrangian for (3.8) is $L_2(e_i^l, x_i^j, y_i^k)$.

$$\mathrm{L}_2(e_i^l, x_i^j, y_i^k) = \sum (Pe_i^l \log e_i^l + Pc_i^j \log x_i^j + Pn_i^k \log y_i^k)$$

$$+\lambda_i\left(Ce_l - \sum_i e_i^l\right) + \beta_i\left(Cc_j - \sum_i x_i^j\right) + \varphi_i\left(Cn_k - \sum_i y_i^k\right)$$

$$= \sum (Pe_i^l \log e_i^l + Pc_i^j \log x_i^j + Pn_i^k \log y_i^k - \lambda_i e_i^l - \beta_i x_i^j - \varphi_i y_i^k)$$

$$+\lambda_i Ce_l + \beta_i Cc_j + \varphi_i Cn_k$$

Where λ_i, β_i and φ_i, is the Lagrangian constant. From Karush-Kuhn-Tucker Theorem we know that the optimal solution is given $\partial L_2(e_i^l, x_i^j, y_i^k) / \partial e_i^l = 0$ for $\lambda_i, \beta_i, \varphi_i > 0$.

Let $\partial L_2(e_i^l, x_i^j, y_i^j) / \partial e_i^l = 0$ to obtain $\quad e_i^l = Pe_i^l / \lambda_i$

Using this result in the constraint equation $Ce_l \geqslant \sum e_i^l$, we can determine λ_i as

$$\lambda_i = \frac{\sum\limits_{d=1}^{n} Pe_i^d}{Ce_l}$$

We substitute λ into e_i^l to obtain

$$e_i^{l*} = \frac{Pe_i^l Ce_l}{\sum\limits_{d=1}^{n} Pe_i^d}$$

e_i^{l*} is the unique energy allocation for maximizing the revenue of energy provider l.

Using the similar method, we can solve computing power allocation optimization problem.

Let $\partial L_2(e_i^l, x_i^j, y_i^k) / \partial x_i^j = 0$ to obtain $\quad x_i^j = Pc_i^j / \beta_i$

Using this result in the constraint equation $Cc_j \geqslant \sum x_i^j$, we can determine β_i as

$$\beta_i = \frac{\sum\limits_{d=1}^{n} Pc_i^d}{Cc_j}$$

We substitute β into x_i^j to obtain

$$x_i^{j*} = \frac{Pc_i^j Cc_j}{\sum\limits_{d=1}^{n} Pc_i^d}$$

x_i^{j*} is the unique optimal computing power allocation for maximizing the revenue of computing power provider j.

Using the similar method, we can solve network resource allocation optimization problem.

Let $\partial L_2(e_i^l, x_i^j, y_i^k) / \partial y_i^k = 0$ to obtain $\quad y_i^k = Pn_i^k / \varphi_i$

Using this result in the constraint equation $Cn_k \geqslant \sum y_i^k$, we can determine φ_i as

$$\varphi_i = \frac{\sum\limits_{d=1}^{n} Pn_i^d}{Cn_k}$$

We substitute φ into y_i^k to obtain

$$y_i^{k*} = \frac{Pn_i^k Cn_k}{\sum\limits_{d=1}^{n} Pn_i^d}$$

$y_i^{k^*}$ is the unique optimal network resource allocation for maximizing the revenue of network resource provider k.

4. Energy efficient resource management algorithm in mobile grid

Energy efficient resource management algorithm decomposes energy consumption optimization problem into a sequence of sub-problems via an iterative algorithm. In each iteration, in the routine of grid user optimization, the grid application computes the unique optimal payment to resource provider under the energy budget, expense budget and the deadline constraint to maximize the grid application's satisfaction. The grid application individually solves its fees to pay for energy resources, computing power and network resource to complete its all jobs, adjusts its grid resource demand and notifies the grid resource provider about this change. In the routine of grid resource provider optimization, different resource providers compute optimal resource allocation for maximizing the revenue of their own. Grid resource provider updates its price according to optimal payments from grid application, and then sends the new prices to the grid applications and allocates the resource for grid application, and the cycle repeats. The algorithm that achieves energy efficient resource management in mobile grid is described as follows.

Algorithm 1 Energy Efficient Resource Management Algorithm (EERM)

Sub-algorithm 1

Step 1: Receives the new price ep_l from the energy provider l;

Step 2: $Pe_i^{l^*} = Max\{U_{app}(Pe_i^l, Pc_i^j, Pn_i^k)\}$;

Step 3: If $B_i \geqslant \sum\limits_j Pc_i^j + \sum\limits_k Pn_i^k + \sum\limits_l Pe_i^l$

Then Return $Pe_i^{l^*}$ to energy resource l;

Step 4: Else Return Null;

Step 5: Receives the new price cp_j from the computing power j;

Step 6: $Pc_i^{j^*} = Max\{U_{app}(Pe_i^l, Pc_i^j, Pn_i^k)\}$;

Step 7: If $B_i \geqslant \sum\limits_j Pc_i^j + \sum\limits_k Pn_i^k + \sum\limits_l Pe_i^l$

Then Return $Pc_i^{j^*}$ to computing power j;

Step 8: Else Return Null;

Step 9: Receives the new price np_k from the network resource provider k;

Step 10: $Pn_i^{k^*} = Max\{U_{app}(Pe_i^l, Pc_i^j, Pn_i^k)\}$;

Step 11: If $B_i \geqslant \sum\limits_j Pc_i^j + \sum\limits_k Pn_i^k + \sum\limits_l Pe_i^l$

Then Return $Pn_i^{k^*}$ to network resource k;

Step 12: Else Return Null;

Sub-algorithm 2

Step 1: Receives optimal payments $Pe_i^{l^*}, Pc_i^{j^*}, Pn_i^{k^*}$ from grid application i;

Step 2: If $Ce_l \geqslant \sum_{i=1}^{I} e_i^l$

Then

$$e_i^{l\,(n+1)*} = Max\{U_{resource}(e_i^l, x_i^j, y_i^k) = \sum_{i=1}^{I} (Pe_i^l \log e_i^l + Pc_i^j \log x_i^j + Pn_i^k \log y_i^k)\};$$

$$ep_l^{(n+1)} = \max\{\varepsilon, ep_l^{(n)} + \eta(e^l\, ep_l^{(n)} - Ce_l)\};$$

Return energy resource price $ep_l^{(n+1)}$ to all grid applications;

Step 3: Else Return Null;

Step 4: If $Cc_i \geqslant \sum_{i=1}^{I} x_i^j$

Then

$$x_i^{j\,(n+1)*} = Max\{U_{resource}(e_i^l, x_i^j, y_i^k) = \sum_{i=1}^{I} (Pe_i^l \log e_i^l + Pc_i^j \log x_i^j + Pn_i^k \log y_i^k)\};$$

// Calculates its optimal computing power $x_i^{j\,(n+1)*}$

$cp_j^{(n+1)} = \max\{\varepsilon, cp_j^{(n)} + \eta(x^j\, cp_j^{(n)} - Cc_j)\};$ // Computes a new price

// $x^j = \sum_i x_i^j, \eta > 0$ is a small step size parameter, n is iteration number

Return computing power price $cp_j^{(n+1)}$ to all grid applications;

Step 5: Else Return Null;

Step 6: If $Cn_k \geqslant \sum_{i=1}^{I} y_i^k$

Then

$$y_i^{k\,(n+1)*} = Max\{U_{resource}(e_i^l, x_i^j, y_i^k) = \sum_{i=1}^{I} (Pe_i^l \log e_i^l + Pc_i^j \log x_i^j + Pn_i^k \log y_i^k)\};$$

// Calculates its optimal network resource demand $y_i^{k\,(n+1)*}$

$np_k^{(n+1)} = \max\{\varepsilon, np_k^{(n)} + \eta(y^k\, np_k^{(n)} - Cn_k)\};$ // Computes a new price

// $y^k = \sum_i y_i^k, \eta > 0$ is a small step size parameter, n is iteration number

Return network resource price $np_k^{(n+1)}$ to all grid applications;

Step 7: Else Return Null;

5. Simulations

In this section, we present the performance evaluation of energy efficient resource management algorithm using the JAVASIM [15] simulator. Network generator BRITE [16] generates the computer network topology. We simulate a mobile grid environment with a 2 dimension area of 500m*500m to study mobile device's behavior. Each mobile device in the simulated environment has a maximal radio range of 100m, and moves following a random-walking mobility model. The average speed of each mobile device is 5 meters per second. The average distance between neighboring devices is 25 meters. Mobile devices dynamically enter and leave the mobile grid. There are a number of parameters associated with each device such as energy budget, expense budget, and a two-dimension position value. Each mobile device's battery capacity is initialized with a random value in the range of [700, 800], and reduced automatically by a random value in the range of [0, 5] in each iteration. There are a total of 150 resources and 600 applications are taken for experimental evaluation of the system. A LAN consists of 90 nodes all of which contribute resources to the grid. The LAN acts as the main Grid infrastructure into which we want to integrate mobile devices. Device schedulers residing in WLANs, acting as the interface point between the mobile devices. All Wi-Fi interfaces operate at a rate of 11Mb/s. All Ethernet interfaces operate at a rate of 10Gb/s. The selective grid applications for simulation are computation-intensive applications such as image processing applications and mpeg players. We assume that each grid application can use any of grid resources including computation, communication and energy resources. Processor capacity varies from 220 to 580 MIPS. The wireless network bandwidth is from 10Kbps to 1Mbps. The main memory is set by 128M, 256M, 512M, and 2G. The disk capacity is set by 80G, 30G, 20G. Energy consumption is represented as a percentage of the total energy required to meet all job deadlines. Assume that the maximum power, P_{max}, corresponds to running all jobs with the maximum processing frequency. The maximum frequency is assumed to be $f_{max} = 1$ and the maximum frequency-dependent power is $P_{max} = 1$. When the energy budget for each interval is limited, we can only consume a fraction of P_{max} when processing requests during a given interval. Jobs arrive at each site s_i, $i = 1, 2, \ldots, n$ according to a Poisson process with rate α. The capacities of the energy resources were chosen uniformly in the interval [50,500]. The energy cost can be expressed in grid dollar that can be defined as unit energy processing cost. The initial price of energy is set from 10 to 500 grid dollars. Users submit their jobs with varying deadlines. The deadlines of grid application are chosen from 100ms to 400 ms. The budgets of grid applications are set from 100 to 1500 grid dollars. Each experiment is repeated 6 times and 95% confidence intervals are obtained. The simulation results shown in the figures represent mean values.

The experiments are conducted to compare energy efficient resource management algorithm (EERM) with low-energy earliest deadline-first (LEDF) scheduling algorithm proposed by V. Swaminathan and K. Chakrabarty [7] studied scheduling workloads containing periodic tasks in real-time systems. The proposed approach minimizes the total energy consumed by the task set and guarantees that the deadline for every periodic task is met. They present a mixed-integer linear programming model for the NP-complete scheduling problem. They proposed a low-energy earliest deadline-first (LEDF) scheduling algorithm. The operation of the low-energy earliest deadline first (LEDF) is as follows: LEDF maintains a list of all released tasks, called the *ready list*. When tasks are released, the task with the nearest deadline is chosen to be executed. A check is performed to see if the task deadline can be met by executing it at the lower voltage (speed). If the deadline can be met, LEDF assigns the lower voltage to the task and the task begins execution. During the task's execution, other tasks may enter the system. These tasks are placed automatically on the *ready list*. LEDF again selects the task with the nearest deadline to be executed. As long as there are tasks waiting to be executed, LEDF does not keep the processor idle. This process is repeated until all the tasks have been scheduled. In the simulation, we compare EERM with LEDF by varying energy budget and price to study how they affect the performance of two algorithms. To investigate mobile grid settings and proposed energy efficient resource management algorithm, we evaluate them with respect to two criteria: application efficiency and resource efficiency which includes execution success ratio, energy consumption ratio, allocation efficiency and resource utilization. Energy consumption ratio is defined as the percentage of consumed energy among total

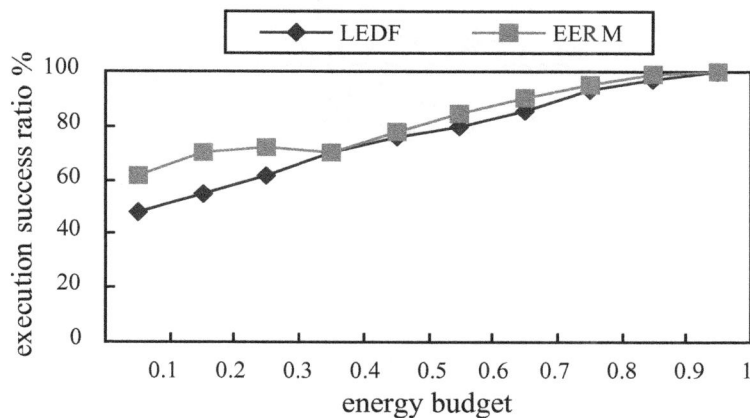

Fig. 1. Execution success ratio under various energy budget.

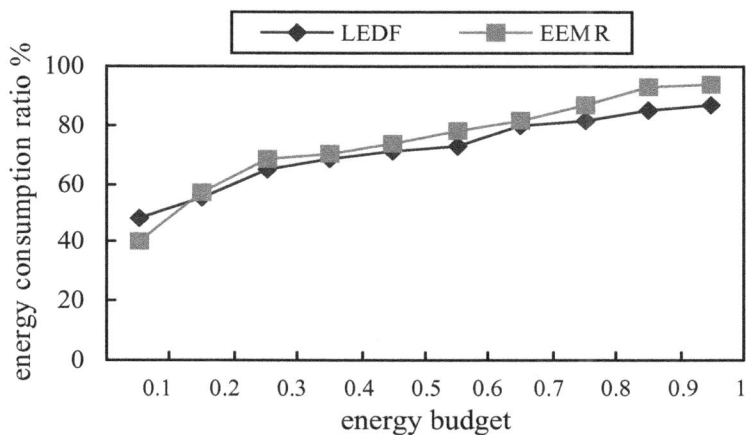

Fig. 2. Energy consumption ratio under various energy budget.

available energy resources. Execution success ratio is the percentage of tasks executed successfully before their deadline. Resource utilization is the ratio of the consumed resources to the total resources available as a percentage, commonly refers to the percent of time a resource is busy. Allocation efficiency is a measure of the efficiency of the allocation process, which is computed using the number of all requests and number of accepted requests.

The effects of energy budget on execution success ratio, energy consumption ratio, allocation efficiency and resource utilization were illustrated in Figs 1–4 respectively. Figure 1 is to show the effect of energy budget on execution success ratio. When increasing energy budget values, the execution success ratio becomes higher. A larger energy budget enables grid user to have enough energy to meet the deadlines and complete the task before its deadline. When energy budget increases ($E = 0.8$), execution success ratio is as much as 30% more than that with $E = 0.3$. Figure 2 shows the energy consumption ratio under different energy budgets. When the energy budget is high, the impact of different energy budget constraint on the energy consumption ratio is obvious; the energy consumption ratio is also high, because grid user tends to choose more energy-consuming resource to complete tasks within deadline. When $E =$ 0.8, the energy consumption ratio of EERM is 35% less than $E = 0.3$. Under same energy budget ($E =$

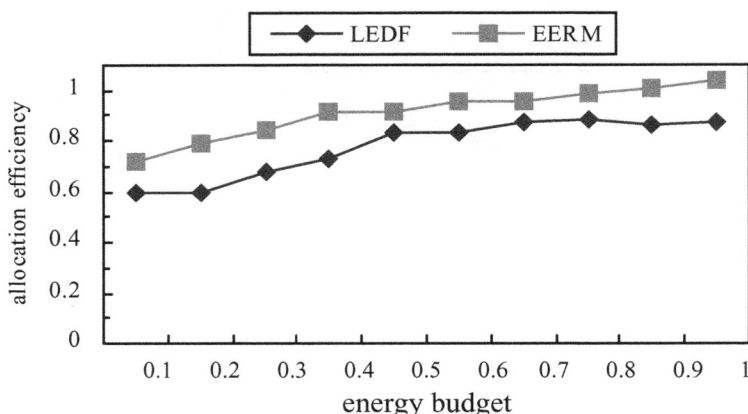

Fig. 3. Allocation efficiency under various energy budget.

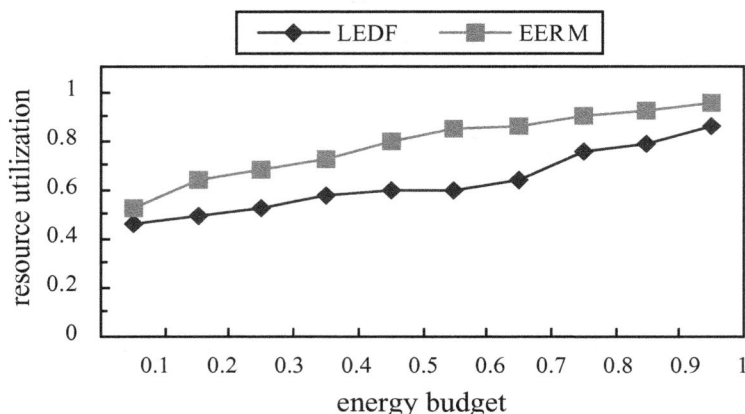

Fig. 4. Resource utilization under various energy budget.

0.3), the energy consumption ratio of LEDF is 6% less than EERM. Figure 3 shows the improvement of allocation efficiency as energy budget E increases. When the energy budget E is low, the system is very energy-constrained and it is crucial to utilize any excess energy due to achieve the performance objective on time. As energy budget E increases, the system becomes less energy-constrained; more jobs can be executed, the allocation efficiency is increased. When energy budget reaches maximum ($E = 100\%$), because the system has enough energy to meet all the deadlines and the allocation efficiency reaches its maximum value. Figure 4 shows the resource utilization under different energy budgets. As the energy budget is higher, the resource utilization becomes higher. When $E = 0.9$, the resource utilization is as much as 39% more than utilization by $E = 0.2$. Because when the energy budget decreases quickly, the users will be prevented from obtaining energy consuming resources. So, some energy consuming resources will be underutilized.

The impacts of the price on resource utilization, energy consumption ratio, execution success ratio, allocation efficiency were illustrated in Figs 5–8 respectively. The resource price (p) is set from 10 to 500 grid dollars. From the results in Fig. 5, as price is higher, the resource utilization becomes lower. When $p = 500$, the resource utilization of EERM is as much as 28% less than utilization by $p = 100$. Because

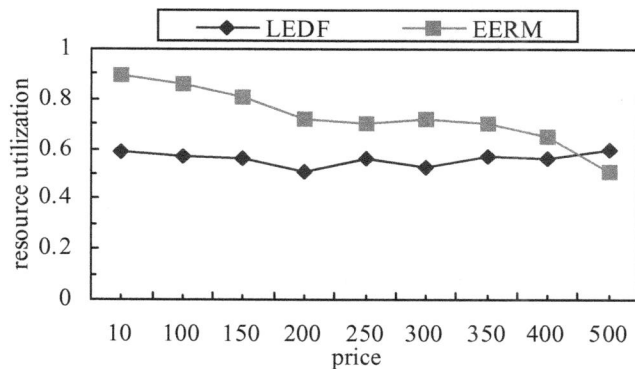

Fig. 5. Resource utilization vs. price.

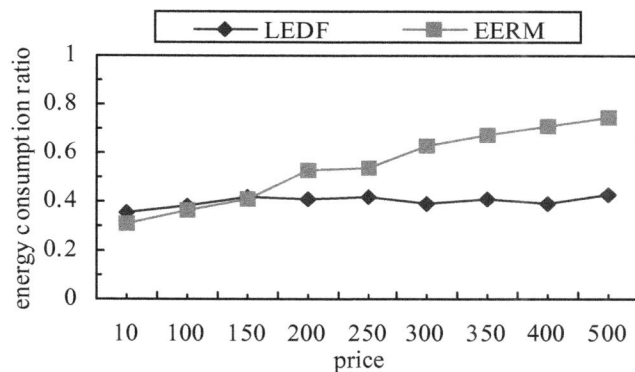

Fig. 6. Energy consumption ratio vs. price.

when the price increases quickly, the users with low expense budget will be prevented from obtaining resources. The smaller is p, the lower is the energy consumption ratio as shown in Fig. 6. Because when price becomes high, users will afford more payment to obtain energy-consuming resource, some tasks can't be completed before their deadlines. Price increasing quickly leads to some users with low budget can't be satisfied to fulfill their achievements. When $p = 500$, energy consumption ratio of EERM is as much as 34% more than that by $p = 100$. Considering the execution success ratio, the results of Fig. 7 show that when increasing price values, the execution success ratio become lower. Because when price becomes high, grid users will afford more payment to obtain the grid resource, some users with low budget will not complete tasks before their deadlines. When price increases ($p = 500$), execution success ratio of EERM is as much as 39% less than that with $p = 10$. Considering the allocation efficiency, the results of Fig. 8 show that when increasing p, the allocation efficiency become lower. Increasing prices of resource provider will prevent users from being admitted by the system, fewer users will exploit the resources. When $p = 500$, the allocation efficiency reduce to nearly 42% compared with $p = 10$.

6. Conclusions

This paper considers energy efficient resource management in mobile grid. The objective of energy efficient resource management in mobile grid is to maximize the utility of the mobile grid which is

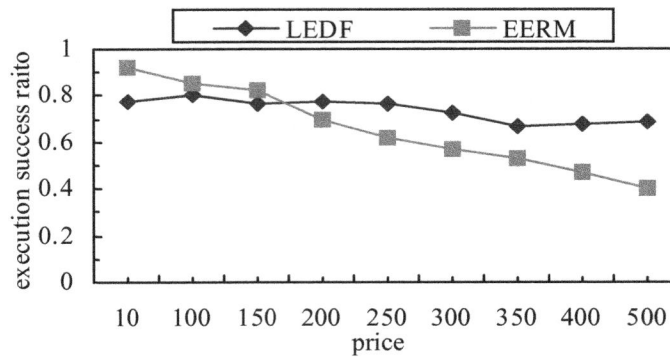

Fig. 7. Execution success ratio vs. price.

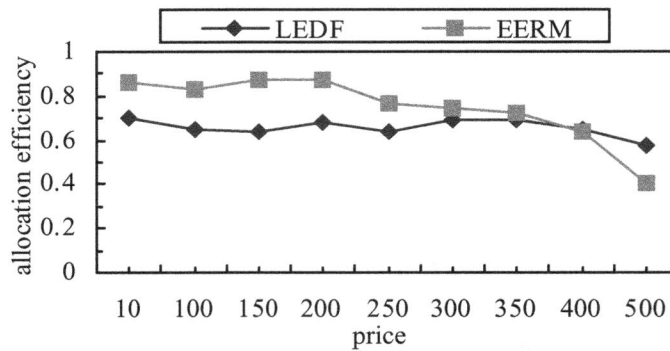

Fig. 8. Allocation efficiency vs. price.

denoted as the sum of grid application utility. The utility function models benefits of application and system. By using nonlinear optimization theory, energy efficient resource management in mobile grid can be formulated as multi objective optimization problem. In order to derive a distributed algorithm to solve global optimization problem in mobile grid, we decompose the problem into the sub problems. The proposed energy efficient resource management algorithm decomposes the optimization problem via iterative method. To test the performance of the proposed algorithm, the simulations are conducted to compare energy efficient resource management algorithm with other energy aware scheduling algorithm.

Acknowledgements

The work was supported by the National Natural Science Foundation of China (NSF) under grants (No. 60773211, No. 60970064), Program for New Century Excellent Talents in University, China (NCET-08-0806), the National Science Foundation of HuBei Province under Grant No. 2008CDB335, the Fundamental Research Funds for the Central Universities (2010-II-003), and Fok Ying-Tong Education Foundation, China (Grant No. 121067).

References

[1]   S.P. Ahuja and J.R. Myers, *A Survey on Wireless Grid Computing* **37**(1) (2006 7), 3–21.

[2] T. Xie, X. Qin and M. Nijim, Solving Energy-Latency Dilemma: Task Allocation for Parallel Applications in Heterogeneous Embedded Systems, *Proceedings of the 2006 International Conference on Parallel Processing* (ICPP'06), IEEE Press, 2006.

[3] Y. Huang, S. Mohapatra and N. Venkatasubramanian, An energy-efficient middleware for supporting multimedia services in mobile grid environments, *IEEE International Conference on Information Technology*, 2005.

[4] Z.L. Zong and X. Qin, Energy-Efficient Scheduling for Parallel Applications Running on Heterogeneous Clusters, *International Conference on Parallel Processing* (ICPP 2007), IEEE Press, 2007.

[5] T.A. AlEnawy and H. Aydin, Energy-Constrained Scheduling for Weakly-Hard Real-Time Systems, *Proceedings of the 26th IEEE International Real-Time Systems Symposium* (RTSS'05), IEEE Press, 2005.

[6] K.H. Kim, R. Buyya and J. Kim, Power Aware Scheduling of Bag-of-Tasks Applications with Deadline Constraints on DVS-enabled Clusters, Proceedings of the Seventh IEEE International Symposium on Cluster Computing and the Grid, *IEEE Computer Society Washington*, DC, USA, 2007, 541–548.

[7] V. Swaminathan and K. Chakrabarty, Real-time task scheduling for energy-aware embedded systems, *Journal of the Franklin Institute* **338** (September 2001), 729–750.

[8] E. Park and H. Shin, *Multimedia Service Composition for Context-Aware Mobile Computing*, T.-J. Cham et al., (eds.), MMM 2007, LNCS 4352, Part II, 2007, pp. 115–124.

[9] C. Li and L. Li, Joint QoS Optimization For Layered Computational Grid, *Information Sciences*, Vol 177/15, pp. 3038-3059, Elsevier, August 2007.

[10] C. Li and L. Li, Agent Framework to Support Computational Grid, *Journal of Systems and Software*, Elsevier, Vol 70/1–2, pp. 177–187, February, 2004.

[11] C. Li and L. Li, Multi Economic agent interaction for optimizing the aggregate utility of Grid Users In Computational Grid, Applied Intelligence, Springer-Verlag Heidelberg, Vol25/2, pp. 147–158, October, 2006.

[12] C. Li and L. Li, A Distributed Utility-based Two Level Market Solution For Optimal Resource Scheduling In Computational Grid, Parallel Computing, Elsevier, USA, Vol 31/3–4, pp. 332–351, March–April, 2005.

[13] C. Li and L. Li, Utility based QoS Optimisation Strategy For Multi-Criteria Scheduling on the Grid, Journal of Parallel and Distributed Computing, Vol67/2, pp. 142–153, Elsevier, USA, February, 2007.

[14] F. Kelly, A. Maulloo and D. Tan, Rate control for communication networks: shadow prices, proportional fairness and stability, *J of Operational Res Soc* **49**(3) (1998), 237–252.

[15] JAVASIM, http://javasim.ncl.ac.uk.

[16] BRITE, http://www.cs.bu.edu/brite.

[17] N. Chand, R.C. Joshi and M. Misra, Cooperative caching in mobile ad hoc networks based on data utility, *Mobile Information Systems* **3**(1) (2007), 19–37.

[18] M. Aleksy, T. Butter and M. Schader, Architecture for the development of context-sensitive mobile applications, *Mobile Information Systems* **4**(2) (2008), 105–117.

[19] V. Reinhard and J. Tomasik, A centralised control mechanism for network resource allocation in grid applications, *International Journal of Web and Grid Services* **4**(4) (2008), 461–475.

[20] P. Hellinckx, F. Arickx, J. Broeckhove and G. Stuer, The CoBRA grid: a highly configurable lightweight grid, *International Journal of Web and Grid Services* **3**(3) (2007), 267–286.

Li Chunlin received PhD degree in Computer Software and Theory from Huazhong University of Science and Technology in 2003. She now is a professor of Computer Science in Wuhan University of Technology. Her research interests include computational grid, distributed computing and mobile agent. She has published over 20 papers in international journals.

Li Layuan received BE degree in Communication Engineering from Harbin Institute of Military Engineering, China in 1970 and ME degree in Communication and Electrical Systems from Huazhong University of Science and Technology, China in 1982. He academically visited Massachusetts Institute of Technology, USA in 1985 and 1999, respectively. Since 1982, he has been with the Wuhan University of Technology, China, where he is currently a Professor and PhD tutor of Computer Science, and Editor in Chief of the Journal of WUT. He is Director of International Society of High-Technol. and Paper Reviewer of IEEE INFOCOM, ICCC and ISRSDC. He was the head of the Technical Group of Shaanxi Lonan P.O. Box 72, Ministry of Electrical Industry, China from 1970 to 1978. His research interests include high speed computer networks, protocol engineering and image processing. Professor Li has published over 150 technical papers and is the author of six books. He also was awarded the National Special Prize by the Chinese Government in 1993.

k-nearest neighbor search based on node density in MANETs

Yuka Komai*, Yuya Sasaki, Takahiro Hara and Shojiro Nishio
Department of Multimedia Engineering, Graduate School of Information Science and Technology, Osaka University, Osaka, Japan

Abstract. In a kNN query processing method, it is important to appropriately estimate the range that includes kNNs. While the range could be estimated based on the node density in the entire network, it is not always appropriate because the density of nodes in the network is not uniform. In this paper, we propose two kNN query processing methods in MANETs where the density of nodes is ununiform; the One-Hop (OH) method and the Query Log (QL) method. In the OH method, the nearest node from the point specified by the query acquires its neighbors' location and then determines the size of a circle region (the *estimated kNN circle*) which includes kNNs with high probability. In the QL method, a node which relays a reply of a kNN query stores the information on the query result for future queries.

Keywords: MANETs, kNN query, LBS

1. Introduction

A location-based service (LBS) [15] is a typical application in *mobile ad hoc networks* (*MANETs*) [1, 2,7,10,13,16,19,25]. In an LBS, it is common that a node issues queries to search information on a specific location held by a mobile node in real time. In such a case, it is effective to process the queries as k nearest neighbor (kNN) queries, which search the information on the k nearest neighbors (kNNs) from a specified location (*query point*) [4,6,14,17,21,26–28].

In our previous work, we proposed a kNN query processing method, the Explosion (EXP) method, which can reduce traffic and also maintain high accuracy of the query result in MANETs [12]. In the EXP method, the query-issuing node first transmits a kNN query using geo-routing to the nearest node from the query point (the *global coordinator*). Then, the global coordinator floods the kNN query to nodes within a specific circle region (the *estimated kNN circle*) whose center is the query point, which looks like a query message that explodes at the global coordinator. Each node that received the query replies with the information on itself to the global coordinator, and the global coordinator sends back kNNs to the query-issuing node.

It is very important to appropriately determine the size of the estimated kNN circle because it directly affects performance. If the estimated kNN circle is set too small, the information of all kNNs may not be acquired because there may be less than k nodes within the estimated kNN circle. On the other hand, if the estimated kNN circle is set too large, unnecessary traffic increases because nodes that are not

*Corresponding author: Yuka Komai, Department of Multimedia Engineering, Graduate School of Information Science and Technology, Osaka University, Yamadaoka 1-5, Suita-shi, Osaka, Japan. E-mail: komai.yuka@ist.osaka-u.ac.jp.

included in the result of the kNN query reply with their information. In the EXP method, the estimated kNN circle is determined based on the density of nodes in the entire MANET. However, in a real environment, it is not always easy to know the total number of nodes in the entire MANET and the area size beforehand. Moreover, since the density of nodes is generally not uniform in a MANET, the estimated kNN circle, which is set based on the density in the entire area, is not always appropriate.

In this paper, we propose two extended EXP methods; the One-Hop (OH) method and the Query Log (QL) method, for reducing traffic and also maintaining high accuracy of the query result in MANETs where the density of nodes is not uniform. In the OH method, the global coordinator acquires its neighbors' information (only one-hop nodes' information) by exchanging messages to know the density of nodes near the query point. If the number of neighbors exceeds k, the global coordinator can reply with kNNs to the query-issuing node. If not, the global coordinator sets the radius of the estimated kNN circle based on the density of nodes within its communication range, and acquires the information on nodes within the estimated kNN circle. In the QL method, a node that relays a reply for a kNN query stores the information on the query result to use it for determining the estimated kNN circle for future queries. During query forwarding, the query-issuing and query-relaying nodes attach some of the stored information to the query, which is used to estimate the density of nodes near the query point. The global coordinator then estimates the radius of the estimated kNN circle using some of the attached information, and acquires the information on nodes within the estimated kNN circle. These methods can set the size of the estimated kNN circle more appropriately using the information acquired during the query execution even if each node cannot know the information on the area size and the total number of nodes beforehand, and the density of nodes in the entire network is not uniform.

We also explain some experimental results to verify that our proposed methods can reduce traffic compared with the EXP method and also achieve high accuracy of the query result.

The contributions of this paper are as follows:

- Since the network bandwidth and batteries of mobile nodes are limited in MANETs, it is very important for kNN query processing to reduce unnecessary query messages and replies (i.e., traffic) as much as possible. We propose two effective kNN query processing methods (the One-Hop (OH) and Query Log (QL) method) for reducing traffic and also maintaining high accuracy of the query result.
- The performance of these methods is affected by several factors such as k and network topology. Thus, by adaptively choosing one of the two methods, we can adapt to various system situations.
- Through extensive simulations, we show that our proposed methods work very well in terms of both traffic reduction and high accuracy of the query result.

The remainder of this paper is organized as follows: In Section 2, we introduce related work. In Section 3, we explain our previous work. In Section 4, we present our proposed kNN query processing methods. In Section 5, we show the results of the simulation experiments. Finally, we summarize this paper in Section 6.

2. Related work

In [5,23], the authors proposed infrastructure-free kNN query processing methods. In these methods, a query is first transmitted to the nearest node from the query point [11], adding the information for setting the search range that contains kNNs with high probability. Then, the nearest node from the query point estimates the size of the search range based on the information attached to the query. In [23], a relaying node sends the query with a list including its location and the number of newly encountered neighbors. The nearest node from the query point determines the size of the search range based on the information on the list. On the other hand, in [5], a relaying node updates the query message including the total area covered by the communication ranges of all relaying nodes and the total number of nodes within the area. Then, the nearest node from the query point determines the search range based on the calculated density of nodes within the area. After that, the search range is partitioned into some sectors. With respect to each sector, a node collects partial results that contain information on nodes within its communication range and propagates the query to the next node along a well devised itinerary structure.

However, these methods basically assume a location-aware sensor network. More specifically, in these methods, each sensor node must precisely know its neighbors (e.g., by frequently exchanging beacon messages), which causes too much overhead in highly dynamic MANETs. On the other hand, in our proposed methods, each mobile node does not have to know the network topology or its neighbors beforehand, which is more suitable for MANETs.

In [24], the authors proposed methods for processing kNN queries in location-aware sensor networks; the GRT, KBT and IKNN algorithms. In the KBT algorithm, a tree infrastructure composed of sensor nodes is constructed and a kNN query propagates along it. The nearest node from the query point determines the search range using for example an approach based on the number of hops during geo-routing. However, in these approaches, because the radius of the search range is set large enough in order not to miss kNNs, unnecessary replies are sent back from nodes that are in the circle but not kNNs. Moreover, a static sensor network is basically assumed with these methods; therefore, they cannot be directly applied to highly dynamic MANETs.

In [9], the authors proposed a kNN query processing method in a 3D sensor network. This method adopts a data collector that efficiently tours kNNs. Because it is assumed that nodes are uniformly distributed in the network and each node knows the area size and number of nodes in the entire network with this method, the search range is set based on the average density of nodes in the network similar to our EXP method. However, as mentioned, in a real environment, it is not always easy to know the total number of nodes in the entire MANET and the area size beforehand. Moreover, the density of nodes is generally not uniform in a MANET.

3. Previous work: EXP method

In this section, we describe the EXP method that we previously proposed [12].

3.1. Geo-routing for forwarding query to query point

In the EXP method, the query-issuing node first forwards a kNN query using our geo-routing method (an extension of the protocol proposed in [8]) to the global coordinator. Our geo-routing method adopts a three-way handshake protocol to send a query to the neighboring node closest to the query point among its neighboring nodes. By repeating this procedure, the query is forwarded to the global coordinator.

More specifically, in our geo-routing method, the query-issuing node first broadcasts a neighbor searching message. Then, when a node receives the neighbor searching message, if it is closer to the query point than the source node, it sets the waiting time for sending a reply. Because nodes closer to the query point transmit a reply message after a shorter waiting time, the nearest node from the query point among the neighbors firstly transmits a reply message to the source node. Then the source node that received reply messages from its neighbors sends a (forwards the) kNN query message only to the node that firstly sent the reply. The node that received the forwarded kNN query broadcasts a neighbor searching message in the same procedure. Finally, if the node that sent a neighbor searching message does not receive any reply messages when the query point is included in its communication range, it recognizes itself as the global coordinator and starts acquiring kNNs.

Therefore, the query-issuing node forwards a kNN query to the global coordinator with little traffic because this geo-routing method neither uses beacon messages nor constructs multi-paths.

3.2. Forwarding kNN query and replying result

In the EXP method, we assume each node knows the total number of nodes and the size of the entire area in which nodes exist. First, the query-issuing node determines the size of the estimated kNN circle based on the density of nodes in the entire area. After receiving a query transmitted using the geo-routing method described in Section 3.1, the global coordinator floods a local query message to nodes within the estimated kNN circle. Then, each node that received the local query message stores the identifier of the source node as its $EXP\ parent$ and sets the waiting time, WT, for sending a reply. Here, WT gets decreases as the distance between the node and the query point increases. When WT has passed, the node transmits a reply message attached with its information to its EXP parent. Finally, after collecting replies from nodes in the estimated kNN circle, the global coordinator replies with the kNN result to the query-issuing node.

Figure 1 shows an example of executing the EXP method where M_1 is the global coordinator. When M_5 receives a local query message from M_1, it stores M_1's ID as its EXP parent and broadcasts a local query message to its neighboring nodes because it is within the estimated kNN circle. In the same way, upon receiving the query message, M_6 stores M_5's ID as its EXP parent and broadcasts a local query message to its neighboring nodes. On the other hand, upon receiving the query message, M_7 discards the message because it is not within the estimated kNN circle. When WT has passed at M_6, M_6 transmits a reply message attached with M_6's information to M_5. M_5 that received the reply message from M_6 transmits a reply message attached with M_5's and M_6's information when WT has passed. Such procedures are performed in the entire MANET. Finally, M_1 acquires the information on all nodes within the estimated kNN circle. When WT has passed at M_1, M_1 transmits the kNN result to the query-issuing node.

The EXP method can reduce the traffic for collecting the kNN result because a tree structure is dynamically constructed during transmissions of local query messages, and the information of nodes in only a specific region (the estimated kNN circle) is efficiently collected along the tree. In this method, determining the radius of the estimated kNN circle, R, is based on the density of nodes in the entire MANET. However, in a real environment, it is not always easy to know the total number of nodes in the entire MANET and the area size beforehand. Moreover, the density of nodes is generally not uniform in a MANET.

4. KNN query processing methods

In this section, first we describe the design policy of our proposed methods and the assumed environment. Then, we give an overview of our proposed methods. Finally, we describe in detail of how to process a kNN query with our methods.

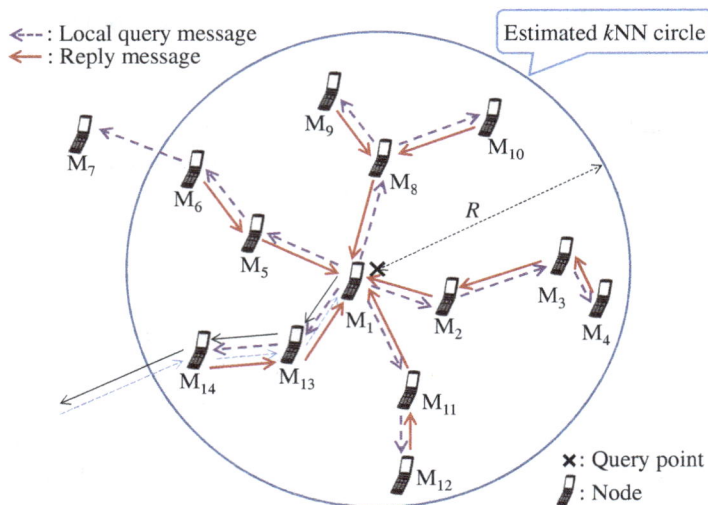

Fig. 1. EXP method.

4.1. Design policy

In MANETs, it is very important to reduce as much traffic as possible due to limitations of network bandwidth and battery of mobile nodes. If each node periodically broadcasts a beacon message even when no node searches kNNs in the network, this causes unnecessary traffic. A method without using beacon messages is suitable for MANETs. However, without exchanging beacon messages, each node cannot know its neighboring nodes' information beforehand. Therefore, we design our methods as beacon-less methods that perform on-demand search.

Moreover, in MANETs, the k nearest nodes from the query point change during the search because a node moves freely. Therefore, the query-issuing node should acquire the query result in a short time. For this aim, we design our methods to execute a query by just one round of message transmissions.

Since mobile nodes consume limited communication bandwidth for data transmission, packet loss and packet retransmission may occur when the network is congested, i.e., some information cannot be transmitted. Thus, the amount of information transmitted by each mobile node should be reduced as much as possible, so the estimated kNN circle should be appropriately set. When each node knows the total number of nodes in the entire network and the area size, and the density of nodes is uniform, the density of nodes in the entire network can be the optimal estimated kNN circle. However, these assumptions are not always true in a real environment. Therefore, we propose the methods without these assumptions.

4.2. Assumptions

The system environment is assumed to be a MANET in which all mobile nodes have the same radio communication facility, in which the communication range is a circle with a fixed size. We assume that the MANET is sufficiently dense so that network partitioning does not occur and geo-routing can be performed between any pair of nodes. In the MANET, mobile nodes retrieve the information on mobile nodes using kNN queries. The query-issuing node transmits a query message associated with the query point and acquires the information on the k nearest nodes from the query point among all nodes in the entire network.

We assign a unique *node identifier* to each mobile node in the system. The set of all mobile nodes in the system is denoted as $M = \{M_1, M_2, \cdots, M_n\}$, where n is the total number of mobile nodes and M_i $(1 \leqslant i \leqslant n)$ is a node identifier. Each mobile node moves freely. Every mobile node knows its current location by using positioning devices such as GPS.

4.3. Overview of our methods

To appropriately set the estimated kNN circle, the node should efficiently know the density of nodes near the query point because there should be kNNs near the query point. When the density of nodes is not uniform in the entire network, it is more effective to know the density of nodes near the query point than the average density of nodes in the entire network. However, it is costly to widely acquire the information on locations of many nodes to calculate the density of nodes. Therefore, the global coordinator first acquires only the number of nodes within its communication range (one-hop) by broadcasting the query to its neighbors. By doing so, the global coordinator can calculate the density of nodes near the query point.

According to the above policy, after a query is transmitted to the global coordinator using geo-routing with the OH method, the global coordinator acquires its neighbors' information (one-hop nodes' information) by exchanging messages to know the density of nodes near the query point. If the number of neighbors exceeds k, the global coordinator can reply with kNNs to the query-issuing node. If not, the global coordinator sets the radius of the estimated kNN circle based on the density of nodes within its communication range and acquires the information on nodes within the estimated kNN circle.

The problem in which a node can estimate the density of nodes only from its (one-hop) neighboring nodes is evident with the OH method. Thus, when k is large or the density of nodes is sparse, i.e., the range where kNNs exist is large, the accuracy of the estimated kNN circle, which is estimated by the information obtained from its neighboring nodes, is expected to decrease. If a node stores the information on the query result when receiving or relaying the query reply, it can use the information to know the density of nodes near the query point in a wider range than with the OH method. Since there are various kNN queries issued in the network, a node can widely determine the density of nodes in the entire network by storing the information on the query result. When a new query is issued, such information on the density of nodes can be used for determining the estimated kNN circle. That information can also be collected from multiple nodes that relay the query message during geo-routing, which is helpful to enhance the quality of the information on the node density. After receiving the query, the global coordinator can determine the radius of the estimated kNN circle based on the density information attached to the query.

According to the above policy, a node that relays a reply for a kNN query in the QL method stores the information on the query result as the *query log*, which includes the density of nodes around the query point, to use it for future queries. During query forwarding for a new query, the query-issuing and query-relaying nodes attach some of the stored information to the query message, which is used to estimate the density of nodes near the query point. Then, the global coordinator estimates the radius of the estimated kNN circle based on some of the attached information. The information on the node density attached to the query message is expected to be more accurate when its query point is closer to that of the current query, and when its query-issuing time is closer to that of the current query. Thus, the QL method takes this fact into account when determining the estimated kNN circle.

4.4. Forwarding kNN query and replying with result: One-Hop (OH) method

The behaviors of the query-issuing node, M_s, and mobile nodes that receive the query message are as follows. After step 6 except for step 10, each node mostly behaves in the same way as in the EXP method.

1. M_s specifies the requested number of kNNs, k, and the query point. Then, similar to the EXP method, M_s transmits a kNN query message to the global coordinator using the geo-routing method described in Section 3.1. In the query message, the query-issuing node's ID and location are respectively set as M_s and its location, the requested number of kNN is set as k, and the query point is set as the location specified by the query.

2. Through the procedures described in Section 3.1, the global coordinator, M_p, is selected. Then it broadcasts a *one-hop query message* to its neighboring mobile nodes. In the message, the global coordinator's ID and location are respectively set as M_p and its location.

3. Each mobile node, M_q, that received the one-hop query message replies with a *one-hop reply message* to M_p if the distance between the query point and M_q is shorter than the radius of the communication range. In the message, the source node's ID and location are respectively set as M_q and its location.

4. M_p that received the one-hop reply messages from its neighboring nodes stores the *tentative kNN result* by adding the information on nodes that replied with the one-hop reply messages.
   If the number of nodes included in the tentative kNN result exceeds k, the information on nodes that are not kNNs from the query point is removed from the tentative result, and the procedure continues to step 11.

5. M_p determines the radius of the estimated kNN circle, R, which contains kNNs with high probability, based on the tentative kNN result by the following equation:

$$R = \alpha \cdot l \cdot \sqrt{\frac{k}{n'}}. \tag{1}$$

   For determining R for the first time, l is the radius of the communication range, n' is the number of nodes included in the tentative kNN result (including M_p), and α is a margin for safely setting the estimated kNN circle. After the second time, (returned from step 10), l' is the distance from the farthest node to the query point in the tentative kNN result and n' and α are same. As Eq. (1) shows, R is set based on the density of nodes acquired by the global coordinator.

6. M_p broadcasts a local query message to its neighboring mobile nodes. In the message, the requested number of kNNs is set as k, the radius of the estimated kNN circle is set as R, the query point is set as that in the received query message, and the global coordinator's ID and location are respectively set as M_p and its location.

7. Each mobile node, M_q, that received the local query message the first time stores the identifier of the source node as its EXP *parent*. If M_q is within the estimated kNN circle, it sets the waiting time, *WT*, for sending a reply by the following equation:

$$WT = \beta \cdot \left(\frac{R}{r}\right) \cdot \left(1 - \frac{a}{R+b}\right). \tag{2}$$

   a is the distance between M_p and M_q, b is the distance between the query point and M_p, β is a parameter decided by a system designer to prevent collisions of messages, and r is the communication range. As Eq. (2) shows, *WT* decreases as the distance between M_p and M_q increases.
   At the same time (without *WT*), M_q broadcasts a local query message to its neighboring mobile nodes.

If M_q has already received a local query message before or it is not within the estimated kNN circle, it discards the message and does nothing.

8. The node that has set the minimum *WT* starts to transmit a reply message (after *WT*) attached with the information on itself including its location to its EXP parent. This attached information is also called the tentative kNN result.

9. Each node that received the reply message (from its EXP child) updates the tentative kNN result attached in the reply message by adding the information on itself. If the number of nodes whose information is included in the tentative kNN result exceeds k, the information on the node which is the farthest from the query point is removed from the tentative kNN result.

   When *WT* has passed, it transmits a reply message attached with the updated tentative kNN result to its EXP parent if it is not the global coordinator. Then, the procedure returns to step 8. Otherwise, if it is the global coordinator, the procedure continues to step 10.

10. When *WT* has passed, the global coordinator, M_p, behaves as follows. If the number of nodes included in the tentative kNN result exceeds k, or the times of retransmitting the local query exceeds T, the procedure goes to step 11. Otherwise, the procedure returns to step 5, i.e., the global coordinator re-estimates R and re-does the same process.

11. M_p replies with the tentative kNN result as the final result to the query-issuing node using the geo-routing method described in Section 3.1, where the query point is set as the location of the query-issuing node. If M_p or a relaying node incidentally knows a node on the query path from the query-issuing node to the global coordinator, it forwards the kNN result to the node and the kNN result is sent back to the query-issuing node along the query path. If some nodes along the query path do not connect with their parents due to link disconnection, they again transmit the kNN result using the geo-routing method.

By using this method, the radius of the estimated kNN circle can be appropriately estimated because it is calculated based on the density of nodes near the query point. Therefore, unnecessary transmissions of queries and replies can be suppressed.

4.5. Forwarding kNN query and replying with result: Query Log (QL) method

The behaviors of the query-issuing node, M_s, and mobile nodes that receive the query message are as follows.

1. M_s specifies the requested number of kNNs, k, and the query point. If M_s has some query logs, it attaches the query logs (L) that satisfy the following condition:

$$\forall i \ \{L.t \leqslant L(i).t \ \cup \ L.d \leqslant L(i).d\}. \tag{3}$$

   Here, $L(i)$ is the i-th query log, $L(i).t$ is the time interval between the query-issuing time of the current query and that for $L(i)$, and $L(i).d$ is the distance between the query point of the current query and that for $L(i)$. This condition shows that only query logs of queries issued recently and near the query point of the current query can be used to determine the estimated kNN circle. Because all query logs under this condition are attached to the query, they can be used to select query logs for estimating the radius of the estimated kNN circle. Although the size of the query message is slightly large since some query logs are attached, this condition can prevent many query logs from being attached.

   M_s transmits a kNN query message to the global coordinator using the geo-routing method described in Section 3.1. In the query message, the query-issuing node's ID and location are respectively set as M_s and its location, the requested number of kNN is set as k, the query point is set as the location specified by the query, and the query log list is set as the list of query logs each of which contains the information on the query point, the query-issuing time, the requested number of kNN, k_past, and the distance from the query point to the k_past-th nearest node of the corresponding query.

2. During geo-routing, if M_t that received the query message has some query logs, it adds some of the query logs to the query log list in the query message, which are chosen by using the method described in step 1, and broadcasts the message.

3. Through the process described in Section 3.1, the global coordinator, M_p, is selected. If the query log list in the received query message is empty and M_p does not store any query logs that satisfy condition (3), it performs steps 2 to 9 in Section 4.4, and the procedure continues to step 4. Otherwise, M_p determines the radius of the estimated kNN circle by using query logs in the query log list and that are stored on M_p. First, the estimated kNN circle, R, for each of the query logs that satisfy condition (3) is calculated by the following equation:

$$R' = l' \cdot \sqrt{\frac{k}{k\_past}}. \tag{4}$$

$$R = \sqrt{R'^2 + \frac{d}{\gamma} + \frac{t}{\theta}}. \tag{5}$$

We call the query corresponding to the query log *previous query*. k_past is the requested number of kNNs in the previous query, l' is the distance from the query point to the k_past-th nearest node in the previous query, d is the distance between the query point of the current query and that of the previous query, t is the time interval between the query-issuing time of the current query and that of the previous query, and γ and θ are weighting parameters to adjust the impact of d and t. R' in Eq. (4) is set based on the density of nodes in the query log. When the query point of the previous query is farther than that of the current query, and when the query-issuing time of the previous query is older, as shown in Eq. (5), R increases. This is effective in preventing the accuracy of the query result from decreasing by an error estimation of the estimated kNN circle.

Then, the query log that has the smallest R among all the query logs is selected, and its R becomes the radius of the estimated kNN circle. Then, steps 6 to 9 in Section 4.4 are performed, and the procedure continues to step 4.

4. When *WT* has passed, the nearest node from the query point, M_p (the global coordinator) behaves as follows. If the number of nodes included in the tentative kNN result exceeds k, or the times of retransmitting the local query exceeds T, the procedure continues to step 5. Otherwise, it performs steps 5 to 9 in Section 4.4 again, i.e., the global coordinator re-estimates R and re-does the same process. Here, l' in step 5 is set to R determined in step 3.

5. M_p replies with the tentative kNN result as the final result to the query-issuing node using the geo-routing method (same as step 11 in Section 4.4).

During geo-routing, M_s, M_p, and the relaying nodes store the information on the query result as a query log, which contains the query point, the query-issuing time, the requested number of kNNs, k, and the distance from the query point to the k-th nearest node.

Since the QL method determines the radius of the estimated kNN circle using the stored information on previous queries, it does not require extra message exchanges. Moreover, it can estimate the node density in a wider area than the OH method, which estimate the node density based on the number of one-hop neighbors. To reduce errors in estimation, the QL method gives higher priority to queries in the query log list that are newer and specify the query point closer to that of the current query.

5. Simulation experiments

In this section, we explain the results of simulation experiments regarding the performance evaluation of our proposed methods. For the simulation experiments, we used the network simulator QualNet5.2 [22].

5.1. Simulation model

The number of mobile nodes in the entire system is 400 (except in Section 5.7 setting on the number of nodes as 800 nodes). These mobile nodes exist in an area of 800×800 m² and their initial positions are randomly selected. These nodes move according to the random walk model [3] where nodes select a random direction and random speed from 0.5 to 1.0 m/sec every minute.[1] We also conducted simulations with other mobility models: the random waypoint and random waypoint models with a home area. In the latter model, the entire area is partitioned into four square regions of equal size, and each node selects its next destination either from the region in which it resides (90% probability) or from another region (10%). The results show that our proposed methods achieve roughly the same performance in all the three mobility models, and the differences in performance between our methods and comparative methods are almost same in the three mobility models. Thus, we only show here the results with the random walk model.

Each mobile node transmits messages using an IEEE 802.11b device whose data transmission rate is 11 Mbps. The transmission power of each mobile node is determined so that the radio communication range becomes about 100 m. Packet losses and delays occur due to radio interference. We assume that each node knows its current location. The query point specified by a kNN query is randomly selected within the entire area, and α in Eq. (1), β in Eq. (2), γ in Eq. (5), and θ in Eq. (5) are respectively set to 1, 1, 0.01 and 10 based on our preliminary experiments. The requested number of kNNs, k, is randomly selected from 1 to 100 by the query-issuing node.

We compare the performance of our proposed methods with that of two different methods. The first method is the EXP method [12]. We assume that each node can know the total number of nodes, n, and the area size. Thus, in the EXP method, R (radius of the estimated kNN circle) is determined by the following equation:

$$R = \sqrt{\frac{k \cdot area}{\pi \cdot n}}. \tag{6}$$

$area$ is the area size ($area = 800 \times 800$) and n is the total number of nodes. We adopt three different values of n for the EXP method, $n = 400, 200,$ and 800. It should be noted that the real number of nodes in the simulations is 400 as described above. Here, we assume that each node in the EXP method misunderstands the number of nodes when $n = 200$ and 800. These two cases respectively represent situations in which nodes overestimate and underestimate the estimated kNN circle because of the difference in node density between the entire area and the target region of kNN queries. From this, we can verify the impact of misestimation of node density in the EXP method. In the graphs of the experimental results, we show the results when $n = 400, 200,$ and 800 as "EXP (just)", "EXP (over)", and "EXP (under)", respectively. The other method for comparison is called the "optimal method" (denoted as "Optimal" in the graphs). In this method, we assume that the global coordinator completely knows

---

[1] We change the simulation setting on initial positions and movement of nodes in Section 5.5.

Table 1
Message types and sizes

Type	Size [B]
Neighbor searching (geo-routing)	48
Reply (geo-routing)	8
Query (OH and EXP methods)	56
Query (QL method)	$64+32p$
One-hop query	32
One-hop reply	32
Local query	64
Local reply	$16+16q$
Reply (OH and EXP methods)	$32+16q$
Reply (QL method)	$60+16q$
Ack to a received reply	16

the position of the k-th nearest node from the query point; thus, it can set the optimal search range (i.e., estimated kNN circle) to acquire kNNs with the smallest traffic. After setting the search range, it performs in the same way as the EXP method. Of course, this is an ideal method, which cannot be implemented in reality. We show the performance of this method as the upper-bound of performance, which is just a guideline.

After one minute passes since the simulation started, the query-issuing node is randomly chosen among all nodes and it issues a kNN query. We repeat this process 1,000 times (i.e., 1,000 queries) every 20 seconds and evaluate the following three criteria.

– Traffic

   We examine the total volume of query messages and replies exchanged for processing a query. Table 1 lists the size of each message used in our methods and the comparative methods, where p denotes the number of query logs attached to a query and q denotes the number of nodes whose information is included in the reply. We define "traffic" as the average total volumes for all queries issued.

– Response time

   We examine the time from transmitting a query message by the query-issuing node until receiving the kNN result. We define "response time" as the average times for all queries issued.

– Accuracy of query result

   We examine the ratio of the number of kNNs whose information is included in the kNN result acquired by the query-issuing node to the requested number of kNNs, k. We define "accuracy of query result" as the mean average precision (MAP) value which measures the performance of the result with a ranking [18]. MAP is an average of the average precision (AP) for each query. AP and MAP are determined by the following equations.

$$AP_i = \frac{1}{k} \sum_{j=1}^{k} \frac{g}{j} \cdot e \tag{7}$$

$$MAP = \frac{1}{querynum} \sum_{i=1}^{querynum} AP_i \tag{8}$$

AP_i is the AP on the i-th issued query, g is the number of nodes that are included in the query result among the top-j nearest nodes, $querynum$ is the total number of issued queries (i.e., 1,000 in this simulation), and e is determined by the following equation:

$$e = \begin{cases} 1 \ (j\text{-th nearest node is included in } k\text{NN result}). \\ 0 \ (\text{otherwise}). \end{cases} \tag{9}$$

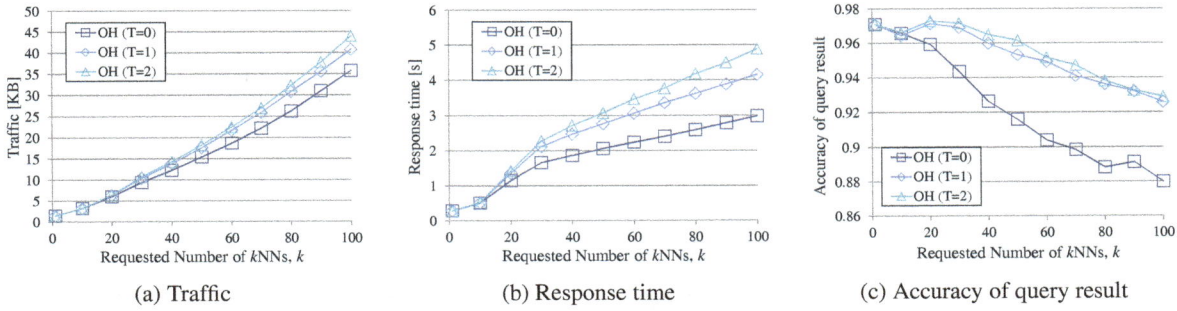

(a) Traffic   (b) Response time   (c) Accuracy of query result

Fig. 2. Impact of number of re-estimations in OH method.

Thus, MAP becomes higher as the query-issuing node obtains the information on nodes closer to the query point.

5.2. Impact of number of re-estimations of R

In our methods, the number of re-estimations of R occurring in processing a kNN query affects the performance. Therefore, we first explain the results of simulations where the maximum number of re-estimations of R (denoted as T) is set to 0 (no re-estimation), 1 (up to one re-estimation) and 2 (up to two re-estimations).

5.2.1. OH method

First, we examine the performance of the OH method when varying the number of re-estimations of R. Figure 2 shows the simulation results. In the graphs, the horizontal axis indicates the requested number of kNNs, k, and the vertical axes indicate the traffic in Fig. 2(a), the response time in Fig. 2(b), and the accuracy of query result in Fig. 2(c).

From Fig. 2(a), when k is smaller than 10, the traffic does not increase rapidly as k increases. This is because the global coordinator can acquire the information on more than k nodes by one-hop replies since it has at least 10 neighboring nodes. This fact can be confirmed from the results where there are no differences in traffic when $T = 0$, 1, and 2. When k is larger than 20, the traffic for $T = 1$ and $T = 2$ is much larger than that for $T = 0$. This is because retransmissions of queries and replies occur due to misestimation of R, which increase traffic. This suggests a disadvantage with the OH method, which estimates the density of nodes only from the information on one-hop neighbors. When k is large, the search range increases (larger than the communication range); thus, errors in estimation also increase. The traffic for $T = 1$ and $T = 2$ is almost the same, which shows that kNNs can be acquired by re-estimating R only once in most cases.

From Fig. 2(b), when k is smaller than 10, the response time is very short in all cases. This is because the global coordinator can acquire the information on more than k nodes by one-hop replies as mentioned. Moreover, the response times for $T = 1$ and $T = 2$ are longer than that for $T = 0$ because it takes time to acquire the information on remaining kNNs after performing re-estimations of R.

From Fig. 2(c), the OH method where $T = 0$ can maintain high accuracy of the query result when k is small. However, the accuracy decreases as k increases. This is because, as mentioned above, errors in estimation of the estimated kNN circle increase as k increases, and the search range increases. In the OH method, where $T = 1$ and $T = 2$, the accuracy of query result remains high regardless of k (better than that for $T = 0$). This shows the effectiveness of re-estimating R. The accuracy of query result is

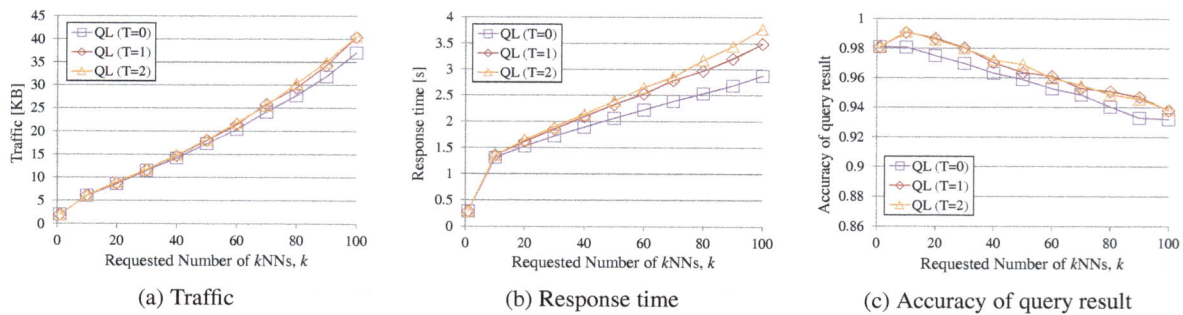

Fig. 3. Impact of number of re-estimations in QL method.

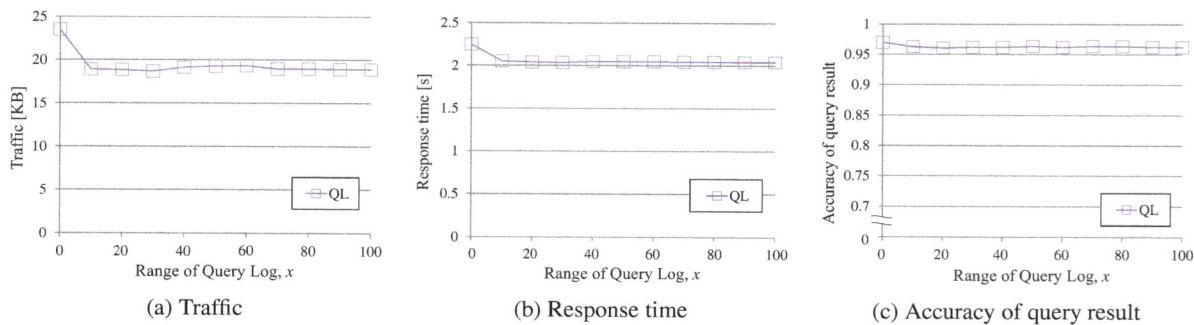

Fig. 4. Impact of k_past in query logs.

almost the same for $T = 1$ and $T = 2$. Since the traffic and the response time increase as T increase, we can conclude that the number of re-estimations of R should be at most once.

5.2.2. QL method

We also examine the performance of the QL method when varying the number of re-estimations of R. Figure 3 shows the simulation results. In the graphs, the horizontal axis indicates the requested number of kNNs, k, and the vertical axes indicate the traffic in Fig. 3(a), the response time in Fig. 3(b), and the accuracy of query result in Fig. 3(c).

From Fig. 3(a), the traffic is almost the same when $T = 0$, 1, and 2. From Fig. 3(b), the response times for $T = 1$ and 2 are longer than that for $T = 0$. This is due to the same reason as that in Section 5.2. However, the differences in response time are smaller than the result in Fig. 2(b).

From Fig. 3(c), the QL method, in which $T = 0$, can maintain high accuracy of query result even when k is large (unlike the OH method). This shows the advantage of the QL method in using the information on node density in a wider area than the OH method, obtained from query logs. The accuracy of query result decreases as k becomes larger. This is because packet losses often occur due to message collisions when the search range increases. In the QL method where $T = 1$ and $T = 2$, the accuracy of query result remains high, which is similar to the case of the OH method. This shows the effectiveness of re-estimating R. The accuracy of query result is almost same for $T = 1$ and $T = 2$. Since the traffic and the response time increase as T increases, we can conclude that the number of re-estimations of R should be at most once in the QL method, similar to the OH method.

Table 2
Number of queries using query logs

x	0	10	20	30	40	50	60	70	80	90	100
♯ of queries using query logs	402	933	953	961	961	961	962	963	963	964	964

5.3. Impact of k_past in selected query logs

In the QL method, the global coordinator estimates the search range based on query logs attached to a query, which are selected among query logs stored on nodes through which the query is transmitted. This makes the global coordinator know the density of nodes in a wider area with a low traffic. Here, even if two queries which respectively specify the same query point are issued at the same time, the densities of nodes stored in their query logs are basically not same when k (k_past in query logs) specified by the two queries is different. This is because ranges where the k_past nearest neighbors exist are different.

Therefore, in the QL method, k_past in query logs selected for estimation of the kNN circle affects the performance, and we examine the impacts of k_past in selected query logs. For this aim, we conducted a simulation where nodes transmitting a query select only query logs among those with k_past within "the current $k \pm x$". The range of query logs, x, is varied from 0 to 100 in this simulation; when x is 0, query logs are selected among those with the same k_past as the current query, and when x is 100, query logs are selected among all query logs.

Figure 4 and Table 2 show the simulation results. In the graphs, the horizontal axis indicates the range of query logs, x, and the vertical axes indicate the traffic in Fig. 4(a), the response time in Fig. 4(b), and the accuracy of query result in Fig. 4(c). In Table 2, each number in the lower column indicate the number of queries which are processed by using query logs for kNN circle estimation out of 1,000 queries. In this subsection, the accuracy of query result is calculated only for queries processed by using query logs for kNN circle estimation.

From Figs 4(a) and 4(b), the traffic and response time are large when x is 0. This is because the global coordinator often overestimates the search range since it estimates it based on the density of nodes in a few query logs whose k_past is the same as that of the current query. More specifically, since it does not often happen that the query point and query-issuing time of the attached query logs are close to that of the current, the search range tends to be estimated largely for safety especially for the case of $x = 0$, where only a few query logs are available. Table 2 shows that when x is 0, only less than half of all queries (402) are processed by using query logs. This is because it often happens that nodes transmitting a query do not store any query logs whose k_past is the same as that of the current query. Except for the case of $x = 0$, the traffic and response time are nearly constant. This is because the global coordinator can estimate the search range using enough number of query logs in most cases.

From Fig. 4(c), the accuracy of query result is slightly higher when x is 0. This is because the search range tends to be estimated largely for safety. However, improvement of accuracy of query result is less than 1% while the traffic and response time increase. Therefore, we can conclude that the performance of the QL method is not sensitive to x (except for the case of $x = 0$). Based on this, in all simulations, nodes use all query logs (i.e., $x = 100$) in the QL method.

5.4. Impact of requested number of nodes, k

Next, we compare our proposed methods with the optimal and the EXP methods. Figure 5 shows the simulation results. In the graphs, the horizontal axis indicates the requested number of kNNs, k, and the vertical axes indicate the traffic in Fig. 5(a), the response time in Fig. 5(b), and the accuracy of query result in Fig. 5(c). In our proposed methods; the OH method and the QL method, T is set to 0.

From Fig. 5(a), as k increases, the traffic increases in all methods. This is because the searching area for processing a kNN query and the data volume of the reply increase. The traffic in the EXP method (just) is almost same as that in the optimal method. This is because the estimated kNN circle is appropriately set when the density of nodes near the query point is the same as that in the entire area. However, in the EXP method (over), the traffic becomes higher due to the overestimation of the kNN circle, i.e., the search range is too large. On the other hand, in the EXP method (under), the traffic is very small due to the underestimation of the estimated kNN circle, which can be seen from the result in Fig. 5(c) where the information on only about half of kNNs is acquired. This result suggests that the EXP method cannot appropriately estimate the search range (the estimated kNN circle) when the density of nodes is different from that near the query point. In the OH method, the traffic is slightly larger than that in the optimal method. This is because extra message exchanges are necessary to acquire the neighboring nodes' information. The traffic in the QL method is also slightly larger than that in the optimal method. This is because the estimated kNN circle is sometimes set to much larger than the optimal range for safety (e.g., when new query logs for queries issued near the query point cannot be found).

From Fig. 5(b), the response time in all methods increases as k increases. This is because in all methods, the waiting time, WT, increases as the estimated kNN circle increases. In particular, since the EXP method (over) overestimates the search range, it sets a longer WT than other methods. Meanwhile, the EXP method (under) gives the shortest response time since the search range is the smallest, i.e., WT is the smallest. In the OH method, the response time is very short when k is smaller than 20. This is because the global coordinator can acquire the information on about 20 nodes by one-hop replies. When k is large, the response time of the OH method is longer than that in the optimal method. This is because the OH method requires at least two rounds of message exchanges; (i) to acquire the neighboring nodes' information and (ii) to acquire the information on kNNs. The response time in the QL method is slightly longer than that in the optimal method. This is because the estimated kNN circle is sometimes set to much larger, as mentioned above.

From Fig. 5(c), the accuracy of query result is very high (nearly 1) in the QL, EXP (over), and optimal methods. This suggests that the estimated kNN circle can be appropriately set in the QL method. In the optimal method (though the estimated kNN circle is optimally set), the accuracy of query result slightly decrease as k increases, which also occurs in our proposed methods and EXP method (over). This is because many replies are sent back to the query-issuing node and collisions of messages often occur. In the OH method, the accuracy of query result slightly decreases, but it can remain high with low traffic by re-estimating R, as described in Section 5.2. On the other hand, the EXP method (just) gives lower accuracy of query result. This is because R is sometimes set smaller than the optimal one when the density of nodes near the query point is incidentally lower than that in the entire area. The accuracy of query result in the EXP method (under) is much lower than other methods due to the underestimation of R.

5.5. Impact of requested number of nodes, k, in skewed network

Finally, we change the simulation setting on distribution and movement of nodes and compare our proposed methods with the optimal and EXP methods. We aim to examine the impact of skewed node density on our methods. Specifically, the simulation area is partitioned into four square sub-areas with the same size and 50, 150, 50, and 150 nodes are deployed in top-left, top-right, bottom-left, and bottom-right sub-areas, respectively. Nodes move according to the random walk model [3] in their own sub-areas, where each node selects a random direction and random speed from 0.5 to 1.0 m/sec every minute.

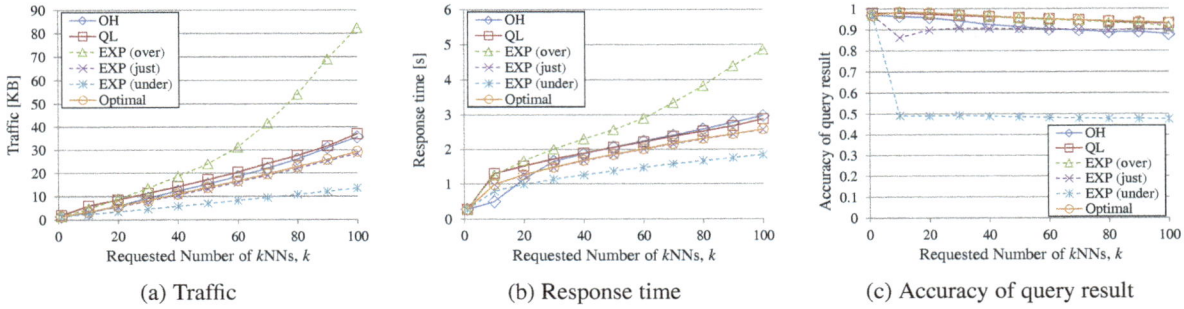

(a) Traffic (b) Response time (c) Accuracy of query result

Fig. 5. Impact of requested number of kNNs, k.

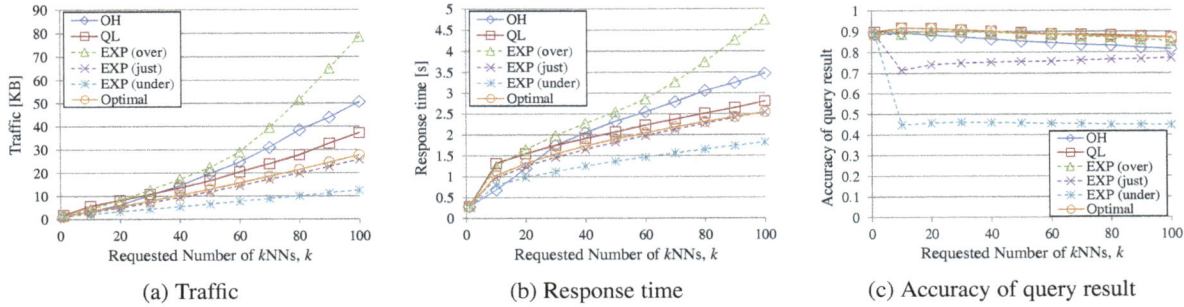

(a) Traffic (b) Response time (c) Accuracy of query result

Fig. 6. Impact of requested number of kNNs, k, in skewed network.

This represents a situation where sub-areas with 150 nodes (dense sub-areas, $150/(400 \times 400)$ $1/m^2$) are much more dense than that with 50 nodes (sparse sub-areas, $50/(400 \times 400)$ $1/m^2$). In our proposed methods, T is set to 0, and γ and θ in Eq. (5) are respectively set to 0.1 and 1 based on our preliminary experiments. Figure 6 shows the simulation results. In the graphs, the horizontal axis indicates the requested number of kNNs, k, and the vertical axes indicate the traffic in Fig. 6(a), the response time in Fig. 6(b), and the accuracy of query result in Fig. 6(c).

From Fig. 6(a), in the OH method, the traffic is much larger than that in Fig. 5(a). This is because the estimated kNN circle is set larger than necessary due to an error in estimation when the query point is set as a point in a sparse sub-area, which is near the border of a dense sub-area (i.e., the OH method does not take into account the density of nodes in a dense area). On the other hand, the traffic in the QL method is almost the same as that in Fig. 5(a). This shows that the QL method can set the estimated kNN circle appropriately even in a skewed network where the density of nodes is not uniform.

From Fig. 6(b), in the OH method, the response time is longer than that in Fig. 5(b) when k is large. This is because the waiting time, WT, increases as the estimated kNN circle increases.

From Fig. 6(c), the accuracy of query result is lower than that in Fig. 5(c) in all methods because geo-routing sometimes does not work well in a sparse area. More specifically, a node that relays a message sometimes cannot find any nodes closer to the query point than itself. In the OH method, the accuracy of query result remains high, while it produces much traffic, as described above. On the other hand, in the QL method, the accuracy of query result remains high, while the traffic is not so large. From these facts, we can confirm that using query logs is effective in estimating the estimated kNN circle in a network where the density of nodes is not uniform.

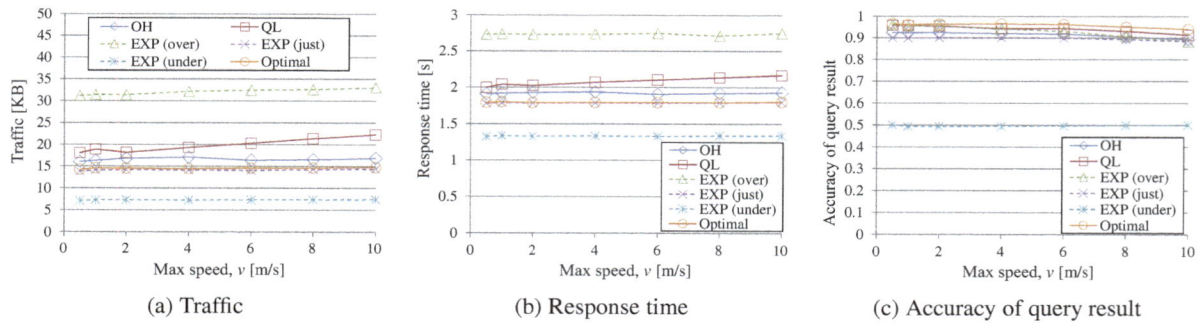

(a) Traffic   (b) Response time   (c) Accuracy of query result

Fig. 7. Impact of node speed.

5.6. Impact of node speed, v

We vary the maximum speed of nodes to examine the impact of node speed. Figure 7 shows the simulation results. In the graphs, the horizontal axis indicates the maximum speed of nodes, v, and the vertical axes indicate the traffic in Fig. 7(a), the response time in Fig. 7(b), and the accuracy of query result in Fig. 7(c). In our proposed methods; the OH method and the QL method, T is set to 0.

From Figs 7(a) and 7(b), in the QL method, the traffic and response time slightly increase as v increases. This is because the search range tends to be overestimated more since the density of nodes more dynamically changes as the node speed increases. In the other methods, the traffic and response time are nearly constant, regardless of nodes' speed. This shows the size of the search range is almost same even if nodes move faster.

From Fig. 7(c), the accuracy of query result slightly decreases due to more packet losses. However, even if the max speed is 10 m/s, our methods still keep high accuracy of query result.

5.7. How to choose a method

As shown above, our proposed methods, the OH and QL methods, show different performance in different situations, e.g., k and the density of nodes, and thus, which is the best method changes depending on a situation. More specifically, when k is small and the density of nodes is high, the OH method outperforms the QL method. This is because the estimation based on the density of the global coordinator's neighboring nodes works well since the area where k nearest nodes exist is relatively small. In addition, since more than k nodes are often neighbors of the global coordinator, the global coordinator can acquire the information on k nearest nodes without setting the search range in the OH method. On the other hand, when k is large and the density of nodes is low, the QL method outperforms because it is useful to estimate the search range based on the density of nodes in a wider area. Therefore, the global coordinator can select either the OH method or the QL method based on the size of k or the density of nodes in order to effectively process a query.

To examine how to partially achieve this, we conducted an experiment where we introduce a new system parameter *switching-k* as the border line of selecting the two methods. If the specified k is less than *switching-k*, the OH method is chosen, i.e., the global coordinator estimates the search range based on the density of the global coordinator's neighbors. Otherwise, the QL method is chosen, i.e., the global coordinator estimates the search range based on query logs.

Only in this experiment, we set the number of nodes as 800 because it shows clearer characteristics than other setting. Figure 8 shows the simulation results. In the graphs, the horizontal axis indicates the

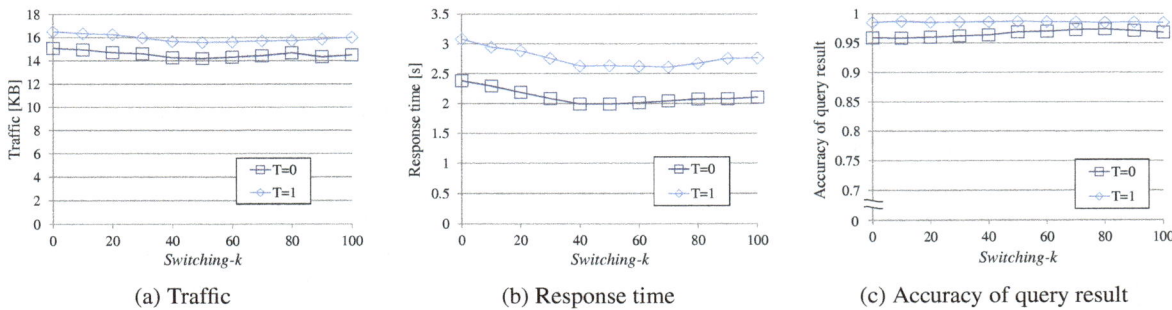

(a) Traffic (b) Response time (c) Accuracy of query result

Fig. 8. Impact of method selection.

switching-k, and the vertical axes indicate the traffic in Fig. 8(a), the response time in Fig. 8(b), and the accuracy of query result in Fig. 8(c). In our proposed methods, T is set to 0 and 1.

From Figs 8(a) and 8(b), the traffic and response time are small when *switching-k* is around 50. This shows that it is somewhat effective to switch between the OH and QL methods based on the value of k. The traffic and response time slightly increase as T increases as described in Section 5.2. Here, the traffic and response time show almost the same tendencies in both cases of T ($T = 0$ and 1), i.e., the optimal value of *switching-k* is same regardless of T.

From Fig. 8(c), in the case of $T = 0$, the accuracy of query result is the highest when the *switching-k* is 80. When the *switching-k* is small, the accuracy of query result decreases because the search range is sometimes underestimated based on inaccurate past information on density of nodes in the QL method. This shows that the OH method should be selected when k is small, i.e., the estimation based on the density of the global coordinator's neighbors works well. When the *switching-k* is significantly large, the accuracy of query result also decreases. This is because when k is large, the density of nodes near the global coordinator is no more reliable for estimation, i.e, the QL method outperforms the OH method. The accuracy of query result where $T = 1$ is higher than that where $T = 0$, and is nearly constant regardless of *switching-k*.

In summary, to efficiently reduce the traffic and response time in this simulation setting (e.g., the number of nodes is 800), it is effective *switching-k* is set to 50 and the global coordinator re-estimates the search range once if needed.

6. Conclusions

In this paper, we proposed two kNN query methods; the One-Hop (OH) method and the Query Log (QL) method, for reducing traffic and also maintaining high accuracy of the query result in MANETs, assuming that the density of nodes is not always uniform. In the OH method, the global coordinator acquires its neighbors' information (only one-hop nodes' information) by exchanging messages to know the density of nodes near the query point. If the number of neighbors exceeds k, the global coordinator can reply with the information on kNNs to the query-issuing node. If not, the global coordinator sets the radius of the estimated kNN circle based on the density of nodes within its communication range and acquires the information on nodes within the estimated kNN circle. In the QL method, a node which relays a reply for a kNN query stores the information on the query result to use it for determining the estimated kNN circle for future queries. During query forwarding, the query-issuing and query-relaying nodes attach some of the stored information to the query, which is used to estimate the density of nodes near the query point. Then, the global coordinator estimates the radius of the estimated kNN circle using some of the attached information and acquires the information on nodes within the estimated kNN circle. These methods can set the size of the estimated kNN circle more appropriately using the information that is acquired during the query execution even if each node cannot know the information on the area size and total number of nodes beforehand, and the density of nodes in the entire network is not uniform.

The experimental results show that our proposed methods produce similar traffic for processing kNN queries as the optimal method and also achieve high accuracy of the query result. The EXP method, which calculates the estimated kNN circle based on the average density of nodes, sometimes cannot appropriately set the estimated kNN circle since the density of nodes near the query point is not always the same as that in the entire area. In the OH method, when k is small, the estimated kNN circle can be appropriately set; however, when k is large, i.e., the search range is large, the accuracy of the estimated kNN circle decreases. In the QL method, the accuracy of the query result sometimes decreases because the QL method estimates R based on the density of nodes using past query logs, which contain some errors.

We also assumed that kNNs are the k nearest nodes from the query point (i.e., nodes are the targets of search). We plan to extend our proposed methods to search general objects associated with locations (e.g., location-based data).

Acknowledgment

This research is partially supported by the Grant-in-Aid for Scientific Research (S) (21220002), (B) (24300037), and (A) (26240013) of MEXT, Japan.

References

[1]   D.J. Baker, J. Wieselthier and A. Ephremides, A Distributed Algorithm for Scheduling the Activation of Links in a Self-Organizing, Mobile, Radio Network, *Proc. IEEE ICC*, 1982, pp. 2F6.1–2F6.5.

[2]   J. Broch, D.A. Maltz, D.B. Johnson, Y.C. Hu and J. Jetcheva, A Performance Comparison of Multi-Hop Wireless Ad Hoc Network Routing Protocols, *Proc. Int. Conf. on MobiCom*, 1998, pp. 159–164.

[3]   T. Camp, J. Boleng and V. Davies, A Survey of Mobility Models for Ad Hoc Network Research, *Wireless Communications and Mobile Computing* **2**(5) (2002), 483–502.

[4]   M. Demirbas and H. Ferhatosmanoglu, Peer-to-Peer Spatial Queries in Sensor Networks, *Proc. Int. Conf. on Peer-to-Peer Computing* (2003), 32–39.

[5]   T.Y. Fu, W.C. Peng and W.C. Lee, Parallelizing Itinerary-Based KNN Query Processing in Wireless Sensor Networks, *IEEE Transactions on Knowledge and Data Engineering* **22**(5) (2010), 711–729.

[6]   Y. Gao, B. Zheng, G. Chen and Q. Li, Algorithms for Constrained K-Nearest Neighbor Queries over Moving Object Trajectories, *GeoInformatica* **14**(2) (2010), 241–276.

[7]   T. Hara and S.K. Madria, Consistency Management Strategies for Data Replication in Mobile Ad Hoc Networks, *IEEE Transactions on Mobile Computing* **8**(7) (2009), 950–967.

[8]   M. Heissenbüttel, T. Braun, T. Bernoulli and M. Walchli, BLR: Beacon-Less Routing Algorithm for Mobile Ad-Hoc Networks, *Computer Communications* **27**(11) (2004), 1076–1086.

[9]   P.P. Jayaraman, A. Zaslavsky and J. Delsing, Cost-Efficient Data Collection Approach Using K-Nearest Neighbors in a 3D Sensor Network, *Proc. Int. Conf. on Mobile Data Management*, 2010, pp. 183–188.

[10] D.B. Johnson, Routing in Ad Hoc Networks of Mobile Hosts, *Proc. IEEE Workshop on Mobile Computing Systems and Applications*, 1994, pp. 158–163.

[11] B. Karp and H.T. Kung, GPSR: Greedy Perimeter Stateless Routing for Wireless Networks, *Proc. Int. Conf. on MobiCom'00*, 2000, pp. 243–254.

[12] Y. Komai, Y. Sasaki, T. Hara and S. Nishio, A kNN Query Processing Method in Mobile Ad Hoc Networks, *Proc. Int. Conf. on Mobile Data Management*, 2011, pp. 287–288.

[13] F. Kuhn, R. Wattenhofer, Y. Zhang and A. Zollinger, Geometric Ad-Hoc Routing: Of Theory and Pracitice, *Proc. Symposium on PODC*, 2003, pp. 63–72.

[14] W.-C. Lee and B. Zheng, DSI: A Fully Distributed Spatial Index for Location-Based Wireless Broadcast Services, *Proc. Int. Conf. on Distributed Computing Systems*, 2005, pp. 349–358.

[15] D.L. Lee, M. Zhu and H. Hu, When Location-Based Services Meet Databases, *Mobile Information Systems* **1**(2) (2005), 81–90.

[16] J. Li, J. Jannotti, D.S.J.D. Couto, D.R. Karger and R. Morris, A Scalable Location Service for Geographic Ad Hoc Routing, *Proc. Int. Conf. on MobiCom*, 2000, pp. 120–130.

[17] B. Liu, W.-C. Lee and D.L. Lee, Distributed Caching of Multi-Dimensional Data in Mobile Environments, *Proc. Int. Conf. on Mobile Data Management*, 2005, pp. 229–233.

[18] C.D. Manning, P. Raghavan and H. Schütze, Introduction to Information Retrieval, *Cambridge University Press*, 2008.

[19] C.E. Perkins and E.M. Royer, Ad Hoc On Demand Distance Vector Routing, *Proc. IEEE Workshop on Mobile Computing Systems and Applications*, 1999, pp. 90–100.

[20] N. Roussopoulos, S. Kelley and F. Vincent, Nearest Neighbor Queries, *Proc. ACM SIGMOD*, 1995, pp. 71–79.

[21] M. Safar, K Nearest Neighbor Search in Navigation Systems, *Mobile Information Systems* **1**(3) (2005), 207–224.

[22] Scalable Network Technologies: "Qualnet," http://www.scalable-networks.com/.

[23] S.H. Wu, K.T. Chuang, C.M. Chen and M.S. Chen, Toward the Optimal Itinerary-Based KNN Query Processing in Mobile Sensor Networks, *IEEE Transactions on Knowledge and Data Engineering* **20**(12) (2008), 1655–1668.

[24] Y. Xu, T.Y. Fu, W.C. Lee and J. Winter, Processing K Nearest Neighbor Queries in Location-Aware Sensor Networks, *Signal Processing* **87**(12) (2007), 2861–2881.

[25] Y. Xu, J. Heidemann and D. Estrin, Geography-Informed Energy Conservation for Ad Hoc Routing, *Proc. Int. Conf. on MobiCom*, 2001, pp. 70–84.

[26] B. Xu, F. Vafaee and O. Wolfson, In-Network Query Processing in Mobile P2P Databases, *Proc. Int. Symposium on Advances in Geographic Information SyStems*, 2009, pp. 207–216.

[27] Y. Yao, X. Tang and E.-P. Lim, Localized Monitoring of *k*NN Queries in Wireless Sensor Networks, *VLDB Journal* **18**(1) (2009), 99–117.

[28] C. Yu, B.C. Ooi, K.-L. Tan and H.V. Jagadish, Indexing the Distance: An Efficient Method to KNN processing, *Proc. Int. Conf. on VLDB*, 2001, pp. 421–430.

Yuka Komai received the B.E. degree in the Multimedia Engineering and the M.E. degree in the Information Science and Technology from Osaka University, Osaka, Japan, in 2011 and 2013, respectively. Currently, she is a Ph.D. candidate in the Information Science and Technology from Osaka University, Osaka, Japan. Her research interests include distributed databases, mobile networks, and mobile computing systems.

Yuya Sasaki received the B.E. degree in the Multimedia Engineering and the M.E. degree in the Information Science and Technology from Osaka University, Osaka, Japan, in 2009 and 2011, respectively. Currently, he is a Ph.D. candidate in the Information Science and Technology from Osaka University, Osaka, Japan. His research interests include data search and replication mechanisms in mobile computing environments.

Takahiro Hara received the B.E, M.E, and Dr.E. degrees in Information Systems Engineering from Osaka University, Osaka, Japan, in 1995, 1997, and 2000, respectively. Currently, he is an Associate Professor of the Department of Multimedia Engineering, Osaka University. He has published more than 300 international Journal and conference papers in the areas of databases, mobile computing, peer-to-peer systems, WWW, and wireless networking. He served and is serving as a Program Chair of IEEE International Conferences on Mobile Data Management (MDM'06 and 10) and Advanced Information Networking and Applications (AINA'09 and 14), and IEEE International Symposium on Reliable Distributed Systems (SRDS'12). He guest edited IEEE Journal on Selected Areas in Communications, Sp. Issues on Peer-to-Peer Communications and Applications. His research interests include distributed databases, peer-to-peer systems, mobile networks, and mobile computing systems. He is a senior member of IEEE and ACM and a member of three other learned societies.

Shojiro Nishio received his B.E., M.E., and Ph.D. degrees from Kyoto University in Japan, in 1975, 1977, and 1980, respectively. He has been a full professor at Osaka University since August 1992, and is currently a distinguished professor of Osaka University. He served as a Vice President and Trustee of Osaka University from August 2007 to August 2011. He also acted as the Program Director in the Area of Information and Networking, Ministry of Education, Culture, Sports, Science and Technology (MEXT), Japan from April 2001 to March 2008. His research interests include database systems and multimedia systems for advanced networks such as broadband networks and mobile computing environment. Dr. Nishio has co-authored or co-edited more than 55 books, and authored or co-authored more than 600 refereed journal or conference papers. He served as the Program Committee Co-Chairs for several international conferences including DOOD 1989, VLDB 1995, and IEEE ICDE

2005. He has served as an editor of international journals including IEEE Trans. on Knowledge and Data Engineering, VLDB Journal, ACM Trans. on Internet Technology, and Data & Knowledge Engineering. Dr. Nishio has received numerous awards during his research career, including the Medal with Purple Ribbon from the Government of Japan in 2011, which is awarded to people who have made outstanding and important contributions in academic fields, arts and sports. He is also a fellow of IEEE, IEICE and IPSJ, and is a member of four learned societies, including ACM.

7

A multi-hop advertising discovery and delivering protocol for multi administrative domain MANET

Federico Mari[a], Igor Melatti[a], Enrico Tronci[a] and Alberto Finzi[b,*]

[a]DI, University of Roma "La Sapienza", Roma, Italy
[b]DSF, University of Napoli "Federico II", Complesso Universitario di Monte Sant'Angelo, Napoli, Italy

Abstract. A *Mobile Ad-hoc NETwork* (MANET) is *Multi Administrative Domain* (MAD) if each network node belongs to an independent authority, that is each node owns its resources and there is no central authority owning all network nodes. One of the main obstructions in designing *Service Advertising, Discovery and Delivery* (SADD) protocol for MAD MANETs is the fact that, in an attempt to increase their own visibility, network nodes tend to flood the network with their advertisements. In this paper, we present a SADD protocol for MAD MANET, based on Bloom filters, that effectively prevents advertising floods due to such misbehaving nodes.

Our results with the ns-2 simulator show that our SADD protocol is effective in counteracting advertising floods, it keeps low the collision rate as well as the energy consumption while ensuring that each peer receives all messages broadcasted by other peers.

Keywords: Wireless network, Mobile Ad-hoc NETwork (MANET), Service Advertising, Discovery and Delivery (SADD)

1. Introduction

A *Mobile Ad-hoc NETwork* (MANET) is *Single Administrative Domain* (SAD) if all its nodes belong to a single authority (administrative domain). For example, a *Wireless Network* (WN) consisting of mobile sensors moving in a given area and gathering data (e.g. temperature) is a SAD-MANET since all sensors fall under the same administrative domain. Following [2], a MANET is *Multi Administrative Domain* (MAD) if each network node belongs to an *independent* authority. In other words, in a MAD-MANET each node owns its resources and there is no central authority owning all network nodes. For example, a network consisting of PDAs, laptops, and other WiFi capable devices each belonging to a different user is a MAD-MANET since each node has a different owner (its user). Note that, both network mentioned above are *Peer-to-Peer* (P2P) networks. However, the first one (mobile sensors) is a SAD-MANET whereas the second one (PDAs, laptops, etc.) is a MAD-MANET. Although the devices forming a SAD-MANET and a MAD-MANET may physically be the same, the dynamics of the two networks may be quite different. In fact in a MAD-MANET each node owns its resources (software as well as hardware). For example, a node in a MAD-MANET may modify the software running on its hardware or even modify the hardware if this is at its advantage (*selfish* behavior).

---

*Corresponding author: Alberto Finzi, DSF, University of Napoli "Federico II", Complesso Universitario di Monte Sant'Angelo, Via Cinthia – 80126 Napoli, Italy. E-mail: finzi@na.infn.it.

1.1. Motivations

WiFi (IEEE 802.11, e.g. see [31]) MANETs consisting of mobile devices such as laptops, cell-phones, PDAs are more and more widespread [56]. This opens up opportunities for many interesting applications. Typical examples are: mobile commerce (*m-commerce*), entertainment, content sharing, emergency management. Here are a few typical scenarios.

1. *Sell:* While you are having a nice walk your handheld device periodically *advertises* a service you may offer, e.g. piano lessons. Upon *discovering* that you offer piano lessons someone (possibly many network *hops* away from you) may ask you for more details that your handheld device will promptly *deliver* to your potential pupil.
2. *Buy:* During the very same walk your handheld device discovers that someone is advertising for math lessons. Your device knows you are interested in math lessons (since you told it) so it will ask for more details that will be delivered by the peer device of your potential teacher.
3. *P2P:* Of course, in much the same way content can be advertised and exchanged between peers thus supporting entertainment as well as emergency management applications.

Protocols supporting the above activities are often called *Service Advertising, Discovery and Delivery* (SADD) protocols. SADD protocols have been extensively studied. For example see [15,22,25,26,28,33, 35–37,51]. However, to the best of our knowledge, all SADD protocols proposed in the literature target SAD-MANETs.

Unfortunately, SADD protocols designed for SAD-MANETs may not work for MAD-MANETs. For example, a WiFi MANET consisting of handheld devices each belonging to a different user is indeed a MAD-MANET. If network nodes behave selfishly they may deviate from the specified protocol if this is at their advantage. Thus, a selfish PDA (user) may refuse to forward packets (to save energy) or may decrease its backoff time (to increase its bandwidth) or may increase its advertisement frequency (to increase its own visibility). Hence, a SADD protocol for MAD-MANETs must deploy suitable countermeasures to protect the network from node misbehaviors that may eventually kill any networking activity.

1.2. Node behavior

In order to design protocols for MAD-MANETs, reasonable hypotheses on node behaviours are needed. Here are some well known classes of node behaviours.

Malicious nodes are willing to spend their resources just to damage the network. For example, malicious nodes may perpetrate a *Denial of Service* (DoS) attack by flooding the network (or part of it) with their messages. Malicious nodes may do so even if this will use up all of their energy without actually doing them any real service.

Selfish nodes act in their best interest. For example, rather than spending its energy flooding the network with messages without getting any reward, a selfish node may refuse to forward packets (to save energy) or may decrease its backoff time after a collision (to increase its bandwidth).

Altruistic or *obedient* nodes (e.g. see [2]) just follow the given protocol. One may think that altruistic nodes do not exist. However it is just a matter of fact that most nodes (agents) in a MAD-MANET are indeed altruistic. That is one can often safely assume that most (although not all) network nodes are altruistic.

Assuming that all nodes are malicious is a too pessimistic hypothesis. No protocol for MAD-MANETs can be designed under such an hypothesis. As for MAD-MANETs the typical approach is to assume that most nodes are selfish or altruistic (e.g. see [11]). In some cases malicious node can be tolerated following the approach in [2,38]. As for SADD protocols, the main obstruction to overcome is *flooding*. In fact, all nodes in the network will be eager to broadcast their advertisements (otherwise they would not participate in the protocol to begin with). This may result in flooding which, in turn, leads to a *Denial of Service* (DoS) attack. In fact, if *too many* nodes flood the network with their advertisements eventually no one will be able to send anything (DoS attack). Note that flooding is an *attractive* misbehavior for malicious as well as selfish nodes. In fact a selfish node may be interested in increasing its advertising frequency to increase its visibility. On the other hand, a malicious node may increase its advertising frequency since it is an easy way to flood the network and carry out a DoS attack. Of course, flooding is not the only possible misbehavior for nodes in a MAD-MANET. We note, however, that flooding is something that any node participating in the protocol will desire to do and, last but not least, can easily do by simply changing a protocol parameter (namely, the advertisement frequency). This is not the case with other attacks. For example, packet dropping may be desirable for a node, but usually requires some nontrivial work on the protocol implementation.

Resting on the above considerations, in this paper we assume that all network nodes follow the given protocol and may deviate from it only by increasing their advertisement frequency. This models the fact that nodes participating in the protocol are eager to broadcast their advertisements. Accordingly, our goal is to devise suitable countermeasures to guarantee that node attempts to increase their advertisement frequency do not result in an advertisement flood destroying any networking activity.

1.3. Our contribution

We present *MAD-SADD*, a SADD protocol for MAD-MANETs. From a functional point of view our protocol is similar to the SADD protocols for SAD-MANETs proposed in the literature (see Section 1.1). Our main contribution here is in the mechanism that allows our protocol to counteract advertisement *flooding*. In our setting, each node has an *advertisement*, a *profile* and a *full service description*. The advertisement and the profile define the services (e.g. content, resources, consultancy, etc.). The full service description (just *full description* in the following) gives full information about the advertised service. Here are examples of full descriptions. If the advertising node is offering a movie then the full description will be the advertised movie file. If the advertising node is offering, say, piano lessons, then the full description will be a file with the maestro address.

Each node periodically *broadcasts* its advertisement to the network nodes. All nodes cooperate in spreading the advertisement by *forwarding* it to their neighbors (*advertising* phase). Upon receiving an advertisement, a node uses its own profile to evaluate its interest in the received advertisement. If the received advertisement is considered interesting, the interested node starts a *unicast* transmission asking the advertising node for the *full description* (*discovery* phase). Finally, the advertising node delivers such full description using again a unicast transmission (*delivery* phase).

To limit collisions, nodes should avoid forwarding *recently* forwarded packets. In a SAD-MANET this is typically done by endowing packets with a *sequence number* field (e.g. see [39,53]). However, in a MAD-MANET sequence numbers do not work since a misbehaving node may broadcast many times the same packet (advertisement) using higher and higher sequence numbers. In this way, packets coming from that node will appear to other nodes as new packets and, accordingly, will be always forwarded. This increases bandwidth usage and visibility of the *misbehaving* node, but decreases bandwidth and visibility of all other nodes.

To counteract such misbehavior, nodes may store in RAM the forwarded messages. In this way, each node will be able to detect if an incoming message (from a certain node) is new or recently seen (and processed). However, usually handheld devices do not have enough RAM to effectively store all recently seen messages.

We propose a trade-off between not storing messages (which results in an unacceptably high number of collisions) and storing them all (which results in an unacceptably high RAM usage).

More specifically, we propose to store advertisement signatures by using a *Bloom filter* [17]. A Bloom filter is a data structure that can effectively store message signatures. For example, using only 1.2 Kbytes of RAM (easily available on any handheld device) we can store signatures for more than 1000 messages with a false positive probability (i.e. the probability of considering as old an advertise that is actually new) of 9.4×10^{-3}. The Bloom filter is periodically cleared, thus only *recent* advertisements are kept. This allows nodes to propose old advertisements to newcomers.

Our experimental results with the ns-2 simulator confirm the effectiveness of our protocol in counter-acting advertisement flooding DoS attacks. Namely, our protocol keeps low the collision rate as well as the energy consumption (*safety*) while ensuring that each peer receives all messages broadcasted from the other peers reachable in one or more hops (*liveness*).

1.4. Comparison with related works

SADD protocols have been widely studied. See for example [14,15,22,28,33–37,45,51]. However all mentioned papers address the problem of SADD protocols for SAD-MANETs, that is, they do not account for selfish or malicious *misbehaviors* of nodes. On the other hand, by exploiting the *obedient* nature of SAD-MANET nodes, the previously mentioned protocols propose routing (e.g. see [4] for a survey) schemas much more sophisticated than ours.

To the best of our knowledge no previously published paper addresses the problem of designing a SADD protocol for MAD-MANETs. There are however protocols for MAD networks (i.e. networks consisting of selfish nodes). An example is the BAR Gossip [38] protocol which allows an altruistic (i.e. following the protocol) broadcaster to stream data to a pool of possibly selfish or even malicious clients. Other approaches rely on specific domain policies and architectures [19].

Flooding of advertisements can be considered as a particular case (an easy one to carry out) of *Denial of Service* (DoS) attack. For this reason we will compare our protocol with those striving to counteract DoS attacks in MANETs. *Denial of Service* (DoS) attacks for ad hoc networks have been studied in [1, 32,48], but in these works advertisement flooding is not addressed.

The SEAD protocol in [29] makes use of elements from a one-way hash chain to provide authentication for both the sequence number and the metric in each entry. The SRP protocol in [47] is based on multiple routes and relies exclusively on the mutual authentication of the end nodes (source and destination). In [45], a probabilistic routing approach is proposed to drive the on-demand discovery process and to reduce the control overhead. Note that the above secure protocols do not solve our problem since they are not able to prevent flooding of legitimate messages (advertising) from legitimate nodes participating in the protocol. The Ariadne protocol [30] enables to secure the routing discovery phase and ensures that all forwarded packets follow the secure route. Here, request-flooding attacks are considered and the proposed solution consists of a rate limit for each node the route requests it is asked to relay. Even if mitigating the effect of flooding at the network layer, this solution does not solve our problem. Indeed, following the approach in [30] the more advertisement messages a node sends, the more it will get broadcasted at the expense of the nodes sending less advertisements. As a result, in our framework the Ariadne approach not only is not sufficient to discourage advertising flood, but encourages it since nodes that are not flooding may never see their advertisement broadcasted.

MANETs have been highly vulnerable to attacks due to the dynamic nature of their network infrastructure. A discussion on security attacks and techniques applicable to MANETs is presented in [5]. A risk aware response mechanism to systematically cope with routing attacks is proposed in [60]. In [46] the authors present an authentication, authorization and security assessment strategy in which, once a device enters a MANET, it is immediately taken out if considered dangerous by the infrastructure. None of the aforementioned approaches can be applied to our context since the DoS attack we are considering here (advertisement flood) does not fall in the class of attacks studied in [5,46,60].

In [58] the authors introduce and analyze *Ad Hoc mobile networks Flooding Attacks* (AHFA). The proposed countermeasures use neighbor suppression to counteract route request flooding attacks and path cutoff to counteract data flooding attacks. Note that this problem is different from the one considered in our paper. In fact, AHFA works at the network layer whereas in our case flooding stems from message advertisement flooding at the application layer and it is perfectly compatible with a legal behavior at the network level.

Since in a MANET nodes are usually constrained by limited computation resources, selfish nodes may refuse to cooperate. Consensus in sparse MANETs is discussed in [3]. In [40], the authors address the noncooperation problem following game-theoretic approach combining reputation and price-based systems. Cooperation in the context of Vehicular Ad-Hoc Networks (VANET) – specializing from MANETs – is studied in [42]. None of these approaches apply to our context where network flooding is the main misbehavior.

The use of bloom filters in networks is discussed in [8]. The deployment of bloom filters in MANET was proposed in [23,52] but only to store requested services or to guide service requests respectively, not as a method to counteract DoS attacks. Furthermore, differently form our approach, in [23] the authors analyze static nodes organized in a grid, while the service discovery protocol in [52] relies on a backbone of directories constituting a virtual network. Instead, in [57], the bloom filter is used as a mechanism for distributed call admission control in MANET.

2. Our scenario

The main assumptions underlying our MAD-SADD design are the following.

Misbehavior. The only possible misbehavior for a node participating in the protocol is network flooding with its own advertisements. As discussed in Section 1.2 this is the most relevant problem to consider in our context.

Strong Identity. Each node has a strong identity. This identity is a unique identifier (ID) for the node. This ID could be a node MAC address or, if available, an IP obtained as in [16,24].

Slow moving. We assume that nodes move *slowly enough* with respect to messages travel time in the network. In particular, nodes move slowly enough so that they can be considered standing still during the *unicast* communications (discovery and delivery). This guarantees that once an advertisement is found to be interesting and more data are requested the path between the source and destination nodes found during the broadcast phase remains valid. This hypothesis is quite reasonable in our setting. In fact we are considering MANETs consisting of handheld devices in a (downtown) metropolitan area such as a square, a mall, a shopping street, a restaurant, etc. In short, relatively crowded scenarios where people move slowly or just stand steel.

Fixed Bandwidth. The bandwidth that each node dedicates to our SADD protocol is fixed. This limits the node resources used by the SADD protocol. For example, even in a densely populated area a node will not spend all of its networking resources forwarding advertisements. This is a typical approach in P2P systems.

3. Some more misbehaviors

In this paper we focus on advertisements *flooding* in MANET, of course there are many other possible misbehaviors in our scenario. Here are some of them.

To begin with, MAD-MANETs protocols have been studied at the infrastructure level in order to counteract misbehavior of selfish nodes at the MAC and network layers. As for the *MAC layer*, selfish nodes may try to increase their bandwidth by decreasing their backoff time (or variations thereof). Paper studying such issues and proposing countermeasures are, for example [6,12,13,41,49,59]. As for the *Network layer*, a selfish node may decide not to forward one or more packets, thus saving energy. There are basically two approaches to encourage nodes to participate in the network operations: micropayments schemes and reputation mechanisms. In a *micropayment* schema honest nodes forwarding packets are remunerated with some suitable form of currency. Micropayments have been studied in [9, 10,55,61,62]. In a *reputation* system schema nodes that refuse to forward packets are punished by denying them service. Reputation systems are studied, for example, in [20,21]. It is worth noting that all works on cooperative (MAD) networks rest on game theory to model node *selfishness*. For a more complete discussion on these topics see [11]. Note that in our setting each node will have many applications running at the same time. For example, beside our SADD protocol, a node may run a VoIP protocol or a file transfer protocol. For this reason the protection mechanisms used at the network layer level cannot directly be used at the application layer level. In particular we cannot avoid advertisement flooding working at the network layer. Finally, *malicious nodes* may modify or even forge *transit traffic* [54]. For example, *malicious* nodes may modify messages when they are forwarding them. More in general, *malicious* nodes are willing to spend their own resources just to damage other nodes.

4. Bloom filter background

A *Bloom filter* (e.g. see [17]) of *size m* and *signature k* consists of a bit-vector B of size m and k suitably (e.g. see [17]) chosen hash functions mapping strings into B entries.

Two operations are possible on a Bloom filter: `query()` and `insert()`. Operation `insert(u)` stores a k-bit signature of message u in B. Operation `query(u)` returns `true` if the k-bit signature of message u is in B, `false` otherwise. If `query(u)` returns true we conclude that message u has been stored in B. Of course *false positives* are possible. That is, `query(u)` may return true even when message u has never been stored in B. However, by choosing suitable values for m and k the probability of getting a false positive can be made very small [17].

More specifically, for a filter of size m and signature k after γ insertions the probability of getting a false positive is approximately $p = (1 - e^{-\frac{k\gamma}{m}})^k$. For example, if we plan for γ insertions, by taking $m = 10\,\gamma$ with $k = 5$ (our best choice from [17]) we get $p = 9.4 \times 10^{-3}$. Thus if we want to store say 10^3 messages with the above value of p we may use a bit-vector of size $m = 10\,\gamma = 10^4$, that is about 1250 bytes of RAM. As a result, even a small handheld device can easily recognize 1000 *recently forwarded* messages.

5. Protocol description

An overview for our protocol is in Section 1.3. A detailed presentation follows.

5.1. Communication environment

As for the *link layer*, wireless communication links between network nodes are implemented using the WiFi (IEEE 802.11b) protocol. As for the *transport layer*, nodes (processes) communicate using the UDP protocol. As for our protocol, MAD-SADD, it is an *application layer* protocol. Note that since MAD-SADD takes care of routing we do not use the routing protocols provided at the *network layer* level. As for the *network layer* we assume (see assumptions 2 and 3 of Section 2) that countermeasures are implemented to counteract network layer selfish misbehaviors.

5.2. Message header

MAD-SADD messages are organized into packets. A packet is organized as a record consisting of administrative fields (*header*) and a data field. The administrative fields are the following: 1. *time*: packet time stamp; 2. *source_address*: ID of the node (see assumption 4 of Section 2); 3. *destination_address*: ID of final destination node; 4. *packet_from_address*: ID of node from which the packet has been received; 5. *next_hop_address*: ID of node to which the packet will be forwarded; 6. *seq_number*: sequence number; 7. *type*: packet type (i.e. *short-description*, *query-unicast*, *data-info*, *data-info-ack*, *data*, *ack*, *no-route*).

5.3. Advertising phase: Broadcast

In our scenario, each node is eager to transmit its own advertisement (see assumption 1 of Section 2). In order to meet this requirement, each node periodically sends its own advertisement in broadcast to all the reachable neighbors (see Fig. 1(a)). This happens with a fixed *advertisement frequency* f, i.e. a node broadcasts its own advertisement every $1/f$ seconds.

The advertisement must fit into one packet. For this reason, we also call it *short description* (of the service). Moreover *short-description* is also the type of a packet containing a short description (i.e. an advertisement). A node receiving a short description will request (in unicast) the *full description* of the advertisement only if it is interested in the advertised service.

As an example, in Fig. 1(a) node a (*advertising node*) broadcasts its advertising, reaching the two nodes within its transmission range.

5.4. Advertising phase: Forwarding

When receiving a *short-description* packet, a node is required to forward it in broadcast to all the reachable neighbors (see Fig. 1(b)). Because of assumptions 1 and 3 of Section 2 we can assume that when requested a node will actually forward a packet, without modifying it. However, if all nodes forward in broadcast all the advertising they receive, there would be many collisions, resulting in poor overall performances. On the other hand, as discussed in Section 1.3, in this phase we cannot use sequence numbers as it is usually done in MANETs. We address this problem by forwarding only *new* (i.e. not *recently* received) *short-description* packets. In this way, we reduce the number of forward operations for each node, thus saving energy and decreasing the number of collisions. In order to decide if a newly arrived *short-description* packet m was already received in the past or is new, we proceed as follows. Each node maintains a cache BF with the *signatures* of the received short descriptions. The data structure used to implement BF is a Bloom filter (see Section 4). Thus, when node i (*interested node*) receives from node j a *short-description* packet m, node i checks whether (the signature of) m is already in BF (m is old) or not (m is new). If m is in BF, then i checks if m was *recently* forwarded. More specifically, let f be the advertising frequency. If the difference between the time stamp of m and the time stamp of the last message from j is less than $1/f$, then m is discarded, otherwise it is forwarded. This means that j cannot send to i message m more than once within $1/f$ seconds. Of course j can send to i other messages within the same period of time. The above approach allows us to avoid advertisement *flooding*. Note that, since the Bloom filter stores advertisement signatures, we may have false positives, i.e. we may decide that a short description is old when it is indeed new. However, this is very unlikely to happen (see Section 4). As an example, in Fig. 1(b) the advertising is propagated through the network.

Remark: One may be tempted to simplify the above Bloom filter based schema by just saying that after having forwarded a packet a node has an *inhibition* time of $1/f$ seconds where it does not forward any packet. Unfortunately this approach does not work. In fact, if incoming packets are queued then advertisement flooding results in a buffer overflow attack. If incoming packets are not buffered then advertisements from a flooding node will have a greater chance of being forwarded thus making flooding quite interesting (almost needed if a node wants its advertisement to be actually broadcasted). We also note that the Ariadne protocol [30] takes into account flooding based DoS attacks (e.g. route request flooding) and proposes mechanisms to counteract such attacks. However the mechanisms proposed in [30] aim at counteracting DoS attacks stemming from malicious nodes forging route requests. In our case however flooding takes place in the broadcast phase (no routing needed) and stems from a perfectly legal node behavior: sending advertisements. Note also that we may use the mechanisms proposed in Ariadne [30] to secure the discovery (Section 5.5) and delivery (Sects. 5.6, 5.7) phases of our protocol.

5.5. Discovery phase

When node i receives a *new* advertisement message m (i.e. m is not in BF), it evaluates its interest for advertisement m using its profile. For example, this can be done by using a data mining algorithm (e.g. *cosine similarity* [27]) when messages are free format or exploiting the message format (e.g. as in [15, 22,28,33,35,36,51]). Of course any of the above approaches can be used within MAD-SADD. To carry out our simulations (Section 6) here we use a *free format* for profiles and advertisements, and *cosine similarity* to evaluate advertisements.

If node i deems m to be interesting, it discards any other incoming *short-description* packet and behaves as follows.

First of all, node i stores in a k-dimensional vector, say undo-vector, the entry values of BF for the k indexes computed by BF hash functions on argument m. Then i inserts the signature of m into BF. Finally, i starts a *unicast* communication with the advertiser of m, call it a. Namely, using a unicast transmission i asks a for the full description M of the advertisement m.

The Bloom filter BF is cleared when the γ-th message is added, where γ is the maximum number of insertions BF has been designed for. In this way, BF will contain only *recently* seen advertisements, thus allowing to forward old advertisements to newly arrived nodes. Routing for the discovery phase will be discussed in Section 5.9. Here, we simply remark that, once the intermediate nodes between i and a are selected, all such nodes will cooperate in forwarding the *query-unicast* packet from i to a (see assumption 3 of Section 2). As an example, in Fig. 1(c) node i sends a *query-unicast* packet to a.

5.6. Delivery phase: Communication setup

When an advertising node a receives a *query-unicast* packet q from an interested node i, q is immediately served, if a is not busy, otherwise q is stored in a queue Q_queries.

In order to serve a *query-unicast* packet coming from a node i, the first operation to carry out is the setup of the communication between i and a. This is done in the following way. Node a sends to i a *data-info* packet, containing the total length of M and the number of *data* packets which are needed to completely transmit M. Then, node i sends to a a *data-info-ack* message. All the intermediate nodes in the path between i and a simply forward these packets (by assumption 3 of Section 2, we can assume that the path is still active). Note that during this phase other queries unicast may be received by a. These queries are all enqueued in `Q_queries`.

When a receives the *data-info-ack* packet, a can start sending M. To this end, M is divided into segments (*data* packets). Each of these *data* packets is then stored in a queue `Q_segments`. Routing for the delivery phase will be discussed in Section 5.9.

5.7. Delivery phase: Sending the full description

Once queue `Q_segments` contains all the needed *data* packets, each one of them is sent to the requiring node i. Namely, for each *data* packet p in `Q_segments`, a sends p to i and waits for the *ack* for p from i. Once the *ack* packet is received, the next packet from `Q_segments` is considered.

When `Q_segments` becomes empty, `Q_queries` is checked again. If `Q_queries` is not empty, i.e. if there are pending unicast queries for a, then node a extracts a new unicast query from `Q_queries` and serves it as explained above. In this way, we force nodes to complete the transmission of full descriptions, before serving new unicast queries.

As for the *data-info* and *data-info-ack* packets, all the intermediate nodes in the path between i and a simply forward the *data* and *ack* packets.

As an example, in Fig. 1(d) node a sends to i a *data-info* packet, which is acknowledged by i with a *data-info-ack* packet. Finally, a sends the *data* packets to i. Each *data* packet is properly acknowledged by i.

At this point, the unicast communication is finished and node i can consider new incoming *short-description* packets.

From Section 2 follows that, as far as we are concerned, no selfish misbehavior takes place during a unicast communication. Thus, as usual, we can use sequence numbers to avoid forwarding many times the same *data* packets.

5.8. Handling link failures

Since nodes are mobile, links may fail. For example, a node may leave the network or run out of energy. Note that link failures may cause problems only during the unicast transmissions, that is during the discovery and delivery phases.

If a node in the transmission chain is not able to deliver a packet (either *data* or *ack*) to the next hop, it notifies this link failure to the previous hop with a *no-route* packet, which will be sent back in the chain to the source. As usual, a receiving node uses a *timeout* to detect transmission failures.

Furthermore, upon a failure, node i (i.e. the one that has received the interesting advertisement m) restores the previous status of the Bloom filter `BF` by using the values stored in the `undo-vector` during the discovery phase (see Section 5.5).

One may wonder why we need to restore the previous values of `BF` in case of communication failure. Suppose that we do not use the `undo-vector` of Section 5.5 and we simply store new advertisements in `BF`. Then a node may never get the *full description* of an advertisement it is interested in. Here is how. Node a broadcasts its advertisement m. Node i receives m, stores it in `BF` and, finding m interesting,

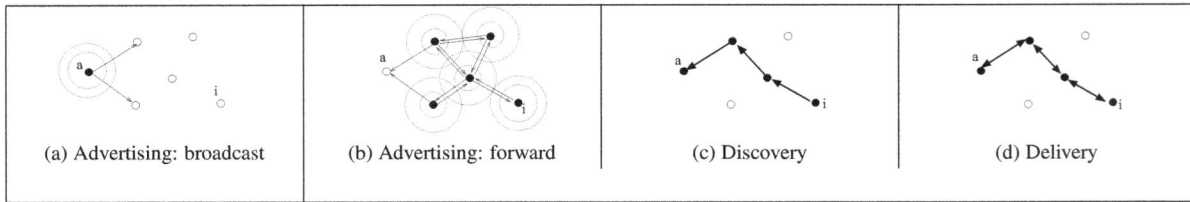

Fig. 1. An example of MAD-SADD execution.

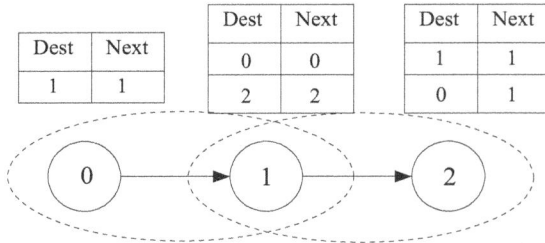

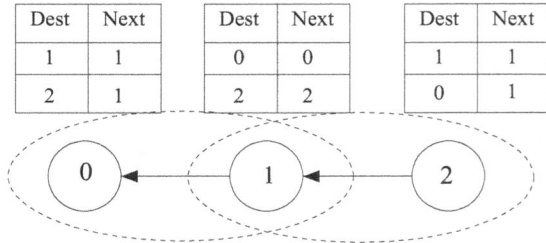

Fig. 2. Routing: advertising phase.   Fig. 3. Routing: discovery phase.

sends a *query-unicast* to a asking for a full description M of m. The unicast communication fails so i never gets M from a. After a while (depending on the advertisement frequency) node a will broadcast advertisement m again. Now i has m in BF so it will not check its interest for m and will just forward m. Thus, as long as m stays in BF, node i will never consider m any more, so missing the full description of an advertisement i is actually interested in. For this reason, upon a failure, node i must *erase* m from its BF. This is achieved by using the `undo-vector` to restore the previous values of BF entries modified by storing m in BF.

5.9. Routing protocol

While routing is not needed during broadcast communication, when unicast communication takes place we have to address the problem of reaching a specified destination.

Our main contribution concerns the mechanism to thwart node selfish behavior, thus avoiding "advertising flooding". Thus, as for routing, we may use any of the many routing protocols available (e.g. *AODV* [7] and *DSR* [18]). However, by exploiting the advertising and discovery (unicast) phases of MAD-SADD we use here a sort of "lightweight" AODV during the delivery (unicast) phase. In fact, we can piggyback routing information in MAD-SADD packets thus avoiding the overhead of additional routing messages.

Namely, our protocol extracts routing information during both the broadcast and the unicast phase. This is done by extracting the information contained in the headers of *short-description* and *query-unicast* packets in order to set up message routing. An example will clarify the matter.

Consider the situation shown in Fig. 2. Node 0 broadcasts its *short-description* packet, which can be received by node 1 only. Upon receiving the packet, node 1 updates its routing table adding a link to node 0. Then, node 1 broadcasts the same message. Since node 2 is reachable from node 1, node 2 can now receive the *short-description* packet and update its routing table as well. Note that Fig. 2 shows the routing tables after 2 has forwarded the advertisement too.

The routing tables of Fig. 2 are partial and do not contain enough information to establish a bidirectional unicast tunnel between 0 and 2. In fact, whenever node 2 queries node 0 for detailed data, the latter is not able to reroute the requested information towards node 2, since the routing table in 0 does not contain 2 as a possible destination.

This problem can be solved by considering also the information exchanged during the unicast phase. Namely, each node i receiving a *query-unicast* packet p extracts from p the source address s and the previous hop address h. Then, i adds to its routing table an entry with s as destination and h as next hop. For example, in Fig. 3 node 0 adds the last entry (destination $= 2$, next hop $= 1$) by extracting this information from the *query-unicast* packet forwarded by 1.

6. Simulation results

In order to evaluate MAD-SADD performances we implemented it within the ns-2 simulator [43,44]. Of course our goal is not to evaluate performances of the routing protocols. We just use a known one (AODV) which performance has been already widely studied [50].

Our goal here is to evaluate effectiveness of our approach in counteracting DoS attacks consisting of advertisement floods. This entails checking that indeed flooding does not take place (*safety*) and that legitimate messages get indeed transmitted (*liveness*).

We run the following sets of simulations.

1. *Bloom filter performance:* This set of simulations (Section 6.3) aims at evaluating Bloom filter effectiveness in our context.
2. *Advertisement frequency:* This set of simulations (Section 6.5) aims at estimating the best value of MAD-SADD *advertisement frequency*, that is how often an advertisement should be broadcasted. This is the most important MAD-SADD parameter.

6.1. IEEE 802.11b on ns-2 simulator

Our goal is to use MAD-SADD with WiFi (IEEE 802.11b) devices. Accordingly, we analyzed a few WiFi chipset and used their parameters to define ns-2 physical and MAC layer configurations. In order to allow exchange of data between nodes, we extended ns-2 classes with suitable methods allowing data exchange between the application layer and the network layer. In fact, ns-2 primitives *send()* and *receive()* only take as argument the number of bytes to be exchanged but do not directly support data exchange.

6.2. Simulation environment

Initially, we load each node with an advertisement and a profile that will be kept unchanged during all our simulation campaign. Advertisements and profiles have been chosen so as to have a reasonable overlap between them in order to trigger the delivery phase.

Nodes are positioned in a square region. Each node initial position is chosen at random with a uniform distribution. In the Bloom filter performance simulations (Section 6.3) nodes do not move, while in advertisement frequency simulations (Section 6.5) nodes may move with constant speed in a *waypoint* fashion. We simulate one hour of MAD-SADD execution (*simulation horizon*) in each simulation run. Each node initial energy (i.e. at the beginning of each simulation run) is 3000 J.

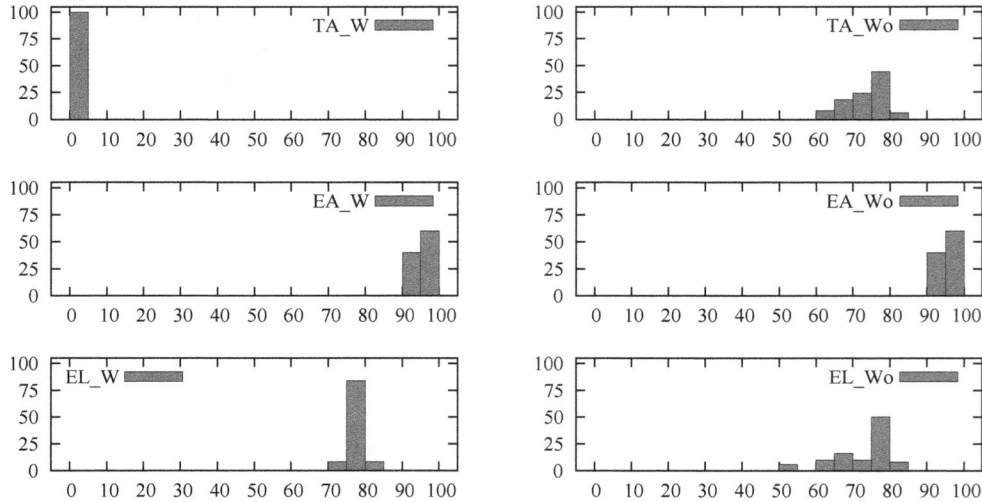

Fig. 4. Bloom filter performances on advertising and energy. The y axis represents percentages of (the total number of) nodes. The x axis represents: *total advestising percentage, effective advertising percentage, energy left percentage* respectively in the first, second, and third row.

The main settings for ns-2 simulation parameters are the following. The *Device Tx/Rx power* has been chosen so as to have a transmission range of 60 meters. The *Antenna type* has been set to OmniAntenna (omnidirectional). The *Radio propagation model* has been set to `TwoRayGround`. The *Packet length* has been set to 1 Kbyte. The *Full description length* has been set to 3 Kbytes, thus to send an advertisement full description we need 3 packets.

For all graphics in this Section the y axis represents percentages of (the total number of) nodes.

6.3. Bloom filter simulations

We compare MAD-SADD protocol performances with those of the MAD-SADD protocol *without* the Bloom filter (MAD-SADD-NO-BF). In other words, we obtain MAD-SADD-NO-BF from MAD-SADD by skipping MAD-SADD tests checking for *recently* received advertisement.

Here are the simulation parameters we used in our experiments in this section.

1. *Density:* We employ 80 nodes, which are located in a $300x300m^2$-wide area. Each node position is chosen uniformly at random within the given area.
2. *No mobility:* Nodes do not move from their initial position.

This is a reasonable setting, since node mobility only affects MAD-SADD ability to reach a node, while here we are only interested in studying Bloom filter performances within MAD-SADD.

Our results are in Figs 4 and 5. Figure 4 shows 6 graphics, divided in three rows and two columns. Graphics in the first (leftmost) column show simulation results obtained with MAD-SADD (i.e. *with* Bloom filter) while graphics on the second column (rightmost) show simulation results obtained with MAD-SADD-NO-BF (i.e. MAD-SADD *without* Bloom filter).

As for Fig. 4, the y axis represents percentages of nodes, while the x axis meaning is the following.

1. *First row:* (Labels *TA_W* and *TA_Wo*) The x axis represents the *total advertising percentage*, i.e. the percentage of received *short-description* packets with respect to the *total* number of sent *short-description* packets. Note that a node broadcasts its advertisement many times (depending on the

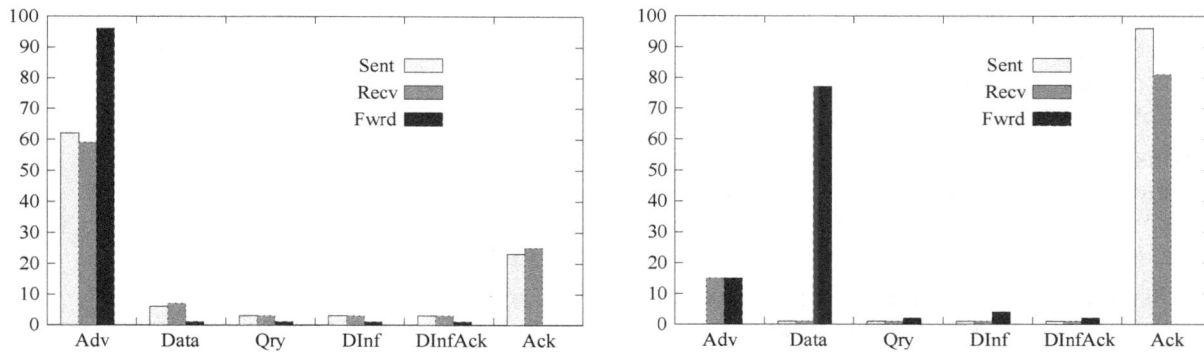

Fig. 5. MAD-SADD performances on packets with (left) and without (right) Bloom filter. The y axis shows percentages of (the total number of) nodes.

advertisement frequency). On the other hand, the same advertisement message can reach the same node many times because of multiple paths. Typically the number of sent *short-description* packets is much greater than the number of advertisement messages.

2. *Second row:* (Labels *EA_W* and *EA_Wo*) The x axis represents the *effective advertising percentage*, i.e. the percentage of received advertisement messages with respect to the sent advertisement messages. Note that here we are counting advertisement messages and not packets, thus each message is counted once (the first time it is sent/received).

3. *Third row:* (Labels *EL_W* and *EL_Wo*) The x axis represents the *energy left percentage*, i.e. the percentage of energy remaining in the devices at the end of the simulation run.

As an example, from the first row of Fig. 4 we can see that without Bloom filter (right column) about 80% of the total advertising is received by about 44% of the nodes.

Figure 5 shows the statistics on the type of exchanged packets *with* Bloom filter (left side) and *without* Bloom filter (right side). Namely, we detail how *sent*, *received* or *forwarded* packets are distributed (in percentage on the total number of packets) among the possible types of packets, i.e. *short-description* (column *Adv*), *data* (column *Data*), *query-unicast* (column *Qry*), *data-info* (column *DInfo*), *data-info-ack* (column *DInfAck*), and *ack* (column *Ack*).

6.4. Reading bloom filter simulations

The simulation results in Section 6.3 show the following.

1. *Total advertising:* From the first row of Fig. 4 we see that when the Bloom filter is used each node receives just a small percentage of the total number of sent *short-description* packets (which includes already seen, i.e. undesired, advertisement messages). In fact, graphics *TA_W* shows that 100% of the nodes receives at most 5% of the generated advertising. On the other hand, if the Bloom filter is turned off, most of the nodes receive most of the (undesired) advertising. For example graphics *TA_Wo* shows that 74% of the nodes receives at least 71% of the (undesired) advertising, and that 50% of the nodes receives at least 76% of the (undesired) advertising.

2. *Effective advertising:* From the second row of Fig. 4 we see that the Bloom filter is accurate enough so that no *desired* advertisement message is lost. In fact, graphics *EA_W* and *EA_Wo* are equal. This means that there is no difference between turning on and off the Bloom filter when considering only *new* advertisement messages. Namely, in both cases all the nodes receive at least 91% of the desired advertising.

3. *Energy left:* From the third row of Fig. 4 we see that Bloom filter usage allows MAD-SADD to save energy. In fact, graphics *EL_W* shows that by enabling the Bloom filter 100% of the nodes have at least 71% of their initial energy at the end of the simulation. On the other hand, if the Bloom filter is turned off, only 68% of the nodes achieve the same result.

One may wonder why using Bloom filter leads to energy saving, since Bloom filter usage has a significant computational cost on each node. The answer is that such higher computational cost incurred by each node is more than compensated by a reduction in the number of packets exchanged between nodes. This translates into a remarkable energy saving. More specifically from Fig. 5, we can see the following.

1. *Acks:* When the Bloom filter is not used (right side), nearly all of the sent packets and about 81% of the received packets are acks. On the other hand, only 23% of sent packets and 25% of received packets are acks when the Bloom filter is used (left side).

2. *Advertisements:* When the Bloom filter is used (left side), most of sent and received packets have type *short-description* (i.e. are advertisement messages), while only 1% of the sent packets (and 15% of the received ones) have type *short-description* when the Bloom filter is turned off.

3. *Forward:* When the Bloom filter is used, nearly all the forwarded packets have type *short-description*, while 77% of the forwarded packets have type *data* when the Bloom filter is turned off.

Shortly, Fig. 5 shows that using a Bloom filter most of the packets are advertisements (*short-description*) while the percentage of "administrative" packets (e.g. *ack*) and data packets is small. On the other hand, turning off the Bloom filter most of the packets are administrative (namely *ack*) and data packets.

Note in fact that each time a node receives an "interesting" advertisement it starts the discovery and delivery phases, that in turn trigger date exchange. Thus when Bloom filter is turned off the discovery and delivery phases are triggered many times for the very same "interesting" advertisement. This increases the (undesired) data traffic.

6.5. Advertisement frequency simulations

We present simulation results aiming at finding an effective value for MAD-SADD *advertisement frequency* in our scenario. Of course if we change the node density or the node deployment area the results will be different but nevertheless they can be obtained following the procedure outlined here.

First of all, note that, in order to save as much energy as possible, one may want to set the advertisement frequency as low as possible. However, since nodes are moving, in order to reach as many nodes as possible with its advertisement message, a node should set the advertisement frequency to a large enough value. Thus an advertisement frequency that is a reasonable trade-off between energy saving and node coverage has to be found.

To this end we proceed as follows. Our *candidate* frequencies are (in Hz): $f_1 = 1/30$, $f_2 = 1/60$, $f_3 = 1/120$, $f_4 = 1/180$.

For each f_i ($i = 1, 2, 3, 4$) we run three sets of simulations, each set representing a given scenario in a squared area: *low density scenario*, i.e. 35 nodes in a $300m^2$-wide area; *medium density scenario*, i.e. 60 nodes in a $300m^2$-wide area; and *high density scenario*, i.e. 60 nodes in a $100m^2$-wide area.

In all these scenarios, the initial position of each node is picked uniformly at random. However, since here we are interested in MAD-SADD nodes coverage, we have to consider moving nodes. To this end, at the beginning of each simulation run, each node i picks at random a time $t_i \in [0h, 1h]$, a (*waypoint*) position p_i (within the given area) and a speed $v_i \in [0.5m/s, 1.5m/s]$ (that is the typical speed for a

Table 1
Summary for the low density scenario

	Data sent	Data recvd	Collisions	Del perc
f_1	42	40	37,965	92.89
f_2	44	42	14,524	93.99
f_3	42	38	6,098	92.44
f_4	34	30	4,000	88.52

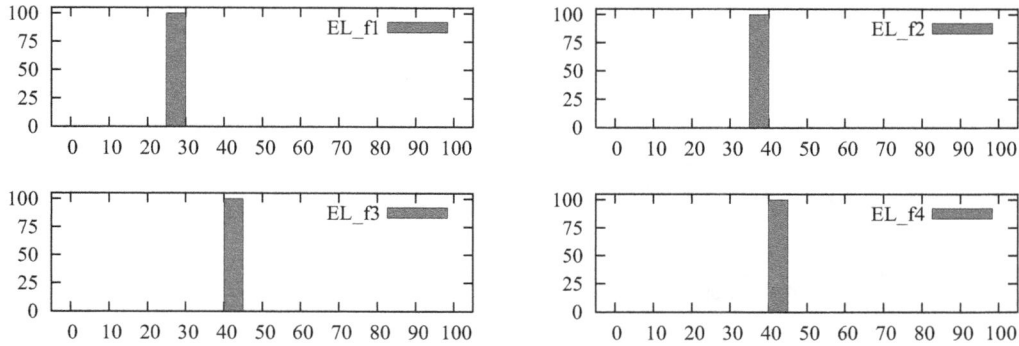

Fig. 6. MAD-SADD performances in the low density scenario. The y axis represents percentages of (the total number of) nodes; in the x axis we have *energy left percentage* (EL).

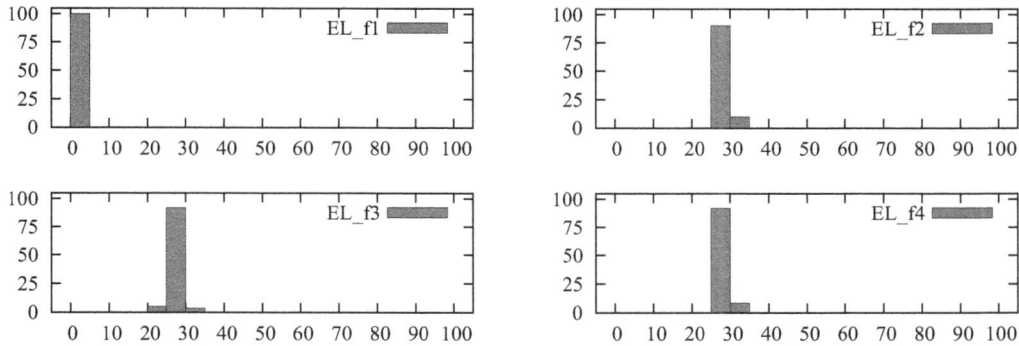

Fig. 7. MAD-SADD performances in the medium density scenario. The y axis represents percentages of (the total number of) nodes; in the x axis we have *energy left percentage* (EL).

walking person). Then, at time t_i from the beginning of the simulation, node i will begin moving towards p_i with speed v_i.

Our results are shown in Fig. 6 and Table 1 for the low density scenario; in Fig. 7 and Table 2 for the medium density scenario; in Fig. 8 and Table 3 for the high density scenario.

Figure 6 contains 4 graphics, which represent the *energy left percentage* for the four possible advertising frequencies.

Columns in Table 1 have the following meaning. Column *Data Sent* (resp. *Data Recvd*) shows the number of *data* packets sent (resp. received) during the simulations. Column *Collisions* shows the number of collisions detected. Finally, column *Del Perc* shows the fraction between the number of all received packets received and the number of all sent packets.

Table 2
Summary for the medium density scenario

	Data sent	Data recvd	Collisions	Del perc
f_1	180	168	1,599,865	80.47
f_2	806	767	624,019	70.83
f_3	757	706	591,936	58.96
f_4	747	688	599,373	59.09

Table 3
Summary for the high density scenario

	Data sent	Data recvd	Collisions	Del perc
f_1	2,041	1,790	15,982,439	80
f_2	2,233	2,120	6,515,836	86.12
f_3	2,294	2,171	2,718,257	87.03
f_4	2,396	2,249	1,596,387	87.23

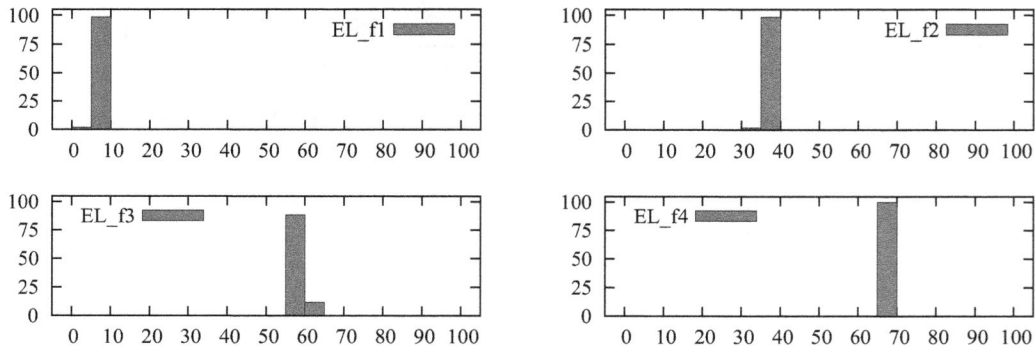

Fig. 8. MAD-SADD performances in the high density scenario. The y axis represents percentages of (the total number of) nodes; in the x axis we have *energy left percentage* (EL).

The meaning for the graphics of Figs 7 and 8 is the same as that for the graphics of Fig. 6. Analogously, the meaning for the columns of Tables 2 and 3 is the same as that for the columns of Table 1.

6.6. Reading advertisement frequency simulations

Using the simulation results of Section 6.5 we may suggest an effective value for the advertisement frequency.

First of all, note that we are interested in maximizing the packet delivery percentage and the energy left in each node, while minimizing the number of collisions. These are conflicting requirements, thus a trade-off has to be found.

With these targets, from Fig. 6 and Table 1 we can see that f_2 and f_3 are to be preferred for the *low density* scenario. In fact, f_1 leads to a high number of collisions, while all the nodes have no more than 30% of energy left at the end of the simulation runs. On the other hand, f_4 leads to a low packet delivery percentage. A similar reasoning may be done for the *medium density* scenario (Fig. 7 and Table 2), where f_2 turns out to be our best choice. Finally, in the *high density* scenario (Fig. 8 and Table 3), f_3 and f_4 turn out to be preferable to the other frequencies.

As for Table 2, one may be puzzled by the values in columns *Data Sent* and *Data Received* corresponding to row f_1. In fact, such values are much smaller than the other values in the same columns. The reason for this is that in the medium density scenario the nodes quickly run out of energy when using advertising frequency f_1. This can be seen from Fig. 7 which shows that for frequency f_1 (upper left graph) at the end of the simulation run each node has no more than 5% of its initial energy left whereas for frequencies f_2, f_3, f_4 nodes have much more energy left. For this reason the number of data packets exchanged in the f_1 scenario is much smaller than those of scenarios f_2, f_3, f_4.

Since f_3 is one of the best choices in 2 out of 3 scenarios (and leads to acceptable value for number of collisions and energy left in the remaining scenario), we finally choose f_3 as the best choice for MAD-SADD.

7. Conclusions

We presented a *Service Advertising, Discovery and Delivery* (SADD) protocol for *Multi Administrative Domain* MANETs. The main obstacle to overcome in designing SADD protocols for MAD MANETs is devising effective mechanisms to prevent selfish or malicious nodes from *flooding* the network with their advertisements. We have shown that by using *Bloom filters* it is possible to effectively counteract advertising floods. Our results with the ns-2 simulator show that our protocol keeps low the collision rate as well as the energy consumption (*safety*) while ensuring that each peer receives all messages broadcasted by other peers (*liveness*). Extending the proposed protocol to MAD-MANETs with malicious nodes able to modify or forge messages appears to be an interesting future work.

References

[1] I. Aad, J.P. Hubaux and E. Knightly, Impact of denial of service attacks on ad hoc networks, *IEEE Transactions on Networking* **16**(4) (2008), 791–802.

[2] A.S. Aiyer, L. Alvisi, A. Clement, M. Dahlin, J.P. Martin and C. Porth, Bar fault tolerance for cooperative services, *SOSP '05* (2005), 45–58.

[3] K. Alekeish and P. Ezhilchelvan, Consensus in sparse, mobile ad hoc networks, *IEEE Transactions on Parallel and Distributed Systems* **23**(3) (2012), 467–474.

[4] J. Al-Karaki and A. Kamal, Routing techniques in wireless sensor networks: A survey, *IEEE Wireless Communications* **11**(6) (2004), 6–28.

[5] M.S. Al-Mazrouei and S. Narayanaswami, Mobile adhoc networks: A simulation based security evaluation and intrusion prevention, *ICITST '11* (2011), 308–313.

[6] E. Altman, R. El-Azouzi and T. Jimenez, Slotted aloha as a stochastic game with partial information, *WiOpt'03: Modeling and Optimization in Mobile, Ad Hoc and Wireless Networks*, 2003.

[7] AODV. http://tools.ietf.org/html/rfc3561 (2007).

[8] A. Broder and M. Mitzenmacher, Network applications of bloom filters: A survey, *Internet Mathematics* (2002), 636–646.

[9] L. Buttyán and J.P. Hubaux, Enforcing service availability in mobile ad-hoc wans, *MobiHoc '00*, IEEE Press (2000), 87–96.

[10] L. Buttyán and J.P. Hubaux, Stimulating cooperation in self-organizing mobile ad hoc networks, *Mob Netw Appl* **8**(5) (2003), 579–592.

[11] L. Buttyán and J.P. Hubaux, Security and Cooperation in Wireless Networks: Thwarting Malicious and Selfish Behavior in the Age of Ubiquitous Computing, *Cambridge University Press*, 2007.

[12] M. Cagalj, Thwarting selfish and malicious behavior in wireless networks, *Ph.D. thesis*, Lausanne (2006).

[13] L. Cao and H. Zheng, Spectrum allocation in ad hoc networks via local bargaining, *IEEE SECON '05* (2005), 475–486.

[14] Z. Chen, H.T. Shen, Q. Xu and X. Zhou, Instant advertising in mobile peer-to-peer networks, *ICDE '09* (2009), 736–747.

[15] L. Cheng, Service advertisement and discovery in mobile ad hoc networks. *CSCW '02* (2002), 16–20.

[16] S. Cheshire, B. Aboba and E. Guttman, Dynamic configuration of IPv4 link-local addresses, *RFC 3927*, (2005) http://www.ietf.org/rfc/rfc3927.txt.

[17] P.C. Dillinger and P. Manolios, Bloom filters in probabilistic verification, *FMCAD, LNCS* **3312** (2004), 367–381.

[18] DSR. http://tools.ietf.org/html/rfc4728 (2007).

[19] A. Durresi, P. Zhang, M. Durresi and L. Barolli, Architecture for mobile heterogeneous multi domain networks, *Mob Inf Syst* **6**(1) (2010), 49–63.

[20] S. Eidenbenz, G. Resta and P. Santi, Commit: A sender-centric truthful and energy-efficient routing protocol for ad hoc networks with selfish nodes, *IPDPS '05: IEEE Intl. Parallel and Distributed Processing Symposium*, 2005.

[21] D. Figueiredo, M. Garetto and D. Towsley, Exploiting mobility in ad-hoc wireless networks with incentives, *Tech. Rep. 04-66*, University of Massachussetts, Computer Science, 2004.

[22] C. Frank and H. Karl, Consistency challenges of service discovery in mobile ad hoc networks, *MSWiM '04*, ACM Press, New York, NY, USA (2004), 105–114.

[23] P. Goering and G. Heijenk, Service discovery using Bloom filters, *Twelfth Annual Conference of the Advanced School for Computing and Imaging* (2006), 219–227.

[24] E. Guttman, C. Perkins and J. Kempf, Service templates and service: Schemes, RFC Editor, 1999.

[25] A.M. Hanashi, I. Awan and M. Woodward, Performance evaluation with different mobility models for dynamic probabilistic flooding in MANETs, *Mob Inf Syst* **5**(1) (2009), 65–80.

[26] J. Haillot and F. Guidec, A protocol for content-based communication in disconnected mobile ad hoc networks, *Mob Inf Syst* **6**(2) (2010), 123–154.

[27] D. Hand, H. Mannila and P. Smyth, Principles of Data Mining, *The MIT Press*, 2001.

[28] S. Helal, N. Desai, V. Verma and C. Lee, Konark – a service discovery and delivery protocol for ad-hoc networks, *WCNC 2003* **3** (2003), 2107–2113.

[29] Y.C. Hu, D.B. Johnson and A. Perrig, Sead: Secure efficient distance vector routing for mobile wireless ad hoc networks, *In Ad Hoc Networks Journal* **1**(1), (2003), 175–192.

[30] Y.C. Hu, A. Perrig and D.B. Johnson, Ariadne: A secure on-demand routing protocol for ad hoc networks, *Wireless Networks* **11**(1–2), (2005), 21–38.

[31] IEEE: IEEE standard 802.11. IEEE (1999).

[32] X. Jin, Y. Zhang, Y. Pan and Y. Zhou, Zsbt, A novel algorithm for tracing dos attackers in manets, *EURASIP Journal on Wireless Communications and Networking* **2006** (2006), 1–9.

[33] M. Khambatti, K.D. Ryu and P. Dasgupta, Push-pull gossiping for information sharing in peer-to-peer communities, *PDPTA-03*, CSREA Press (2003), 1393–1399.

[34] Y.S. Kim, Y.S. Shim and K.H. Lee1, A cluster-based web service discovery in MANET environments, *Mob Inf Syst* **7**(4) (2011), 299–315.

[35] N. Klimin, W. Enkelmann, H. Karl and A. Wolisz, A hybrid approach for location-based service discovery in vehicular ad hoc networks, *WIT '04: Workshop on Intelligent Transportation*, 2004.

[36] U.C. Kozat and L. Tassiulas, Network layer support for service discovery in mobile ad hoc networks, *INFOCOM-03* **3** (2003), 1965–1975.

[37] E. Kulla, M. Hiyama, M. Ikeda, L. Barolli, V. Kolici and R. Miho, MANET performance for source and destination moving scenarios considering OLSR and AODV protocols, *Mob. Inf. Syst.* **6**(4) (2010), 325–339.

[38] H.C. Li, A. Clement, E.L. Wong, J. Napper, I. Roy, L. Alvisi and M. Dahlin, Bar gossip, *OSDI '06: USENIX Operating Systems Design and Implementation*, USENIX Association (2006), 191–204.

[39] A. Liu, H. Yu and L. Li, An energy-efficiency and collision-free mac protocol for wireless sensor networks, *Proc. of Vehicular Technology Conference 2005* **2** IEEE (2005), 1317–1322.

[40] Z. Li and H. Shen, Game-Theoretic Analysis of Cooperation Incentive Strategies in Mobile Ad Hoc Networks, *IEEE Transactions on Mobile Computing* **11**(8) (2012), 1287–1303.

[41] A. MacKenzie and S.B. Wicker, Stability of multipacket slotted aloha with selfish users and perfect information, *INFOCOM-03* **3** (2003), 1583–1590.

[42] H. Mousannif, I. Khalil and S. Olariu, Cooperation as a service in VANET: Implementation and simulation results, *Mob Inf Syst* **8**(2) (2012), 153–172.

[43] K. Fall and K. Varadhan, ns notes and documentation: http://www.monarch.cs.rice.edu/ftp/monarch/wireless-sim/nsDoc.pdf (2007).

[44] ns-2. http://www.isi.edu/nsnam/ns/ (2007).

[45] F. Palmieri and A. Castiglione, Condensation-Based Routing in Mobile Ad-Hoc Networks, *Mob Inf Syst* **8**(3) (2012), 199–211.

[46] F. Palmieri, U. Fiore and A. Castiglione, Automatic security assessment for next generation wireless mobile networks, *Mob Inf Syst* **7**(3) (2011), 217–239.

[47] P. Papadimitratos and Z.J. Haas, Secure data transmission in mobile ad hoc networks, *WiSe '03: Proceedings of the 2nd ACM workshop on Wireless security*, ACM (2003), 41–50.

[48] S. Parvin, F.K. Hussain and S. Ali, A methodology to counter DoS attacks in mobile IP communication, *Mob Inf Syst* **8**(2) (2012), 127–152.

[49] C. Peng, H. Zheng and B. Zhao, Utilization and fairness in spectrum assignment for opportunistic spectrum access, *MONET* **11**(4) (2006), 555–576.

[50] C.E. Perkins and E.M. Royer, Ad hoc on-demand distance vector routing. *2nd IEEE Workshop on Mobile Computing Systems and Applications* (1999), 90–100.

[51] O. Ratsimor, D. Chakraborty, A. Joshi and T. Finin, Allia: alliance-based service discovery for ad-hoc environments, *WMC '02*, ACM Press (2002), 1–9.

[52] F. Sailhan and V. Issarny, Scalable service discovery for manet, *PERCOM '05: Proceedings of the Third IEEE International Conference on Pervasive Computing and Communications*, IEEE Computer Society (2005), 235–244.

[53] G. Thomas, Capacity of the wireless packet collision channel without feedback. *IEEE Transactions on Information Theory* **46**(3) (2000), 1141–1144.

[54] D. Venugopal and G. Hu, Efficient signature based malware detection on mobile devices, *Mob Inf Syst* **4**(1) (2008), 33–49.

[55] W. Wang, X.Y. Li, S. Eidenbenz and Y. Wang, Ours: optimal unicast routing systems in non-cooperative wireless networks, *MobiCom '06*, ACM Press (2006), 402–413.

[56] D. Woelk, B. Haskell, J.L. Carter, R. Brice, Rusin and A.A. Helal, Any Time, Anywhere Computing: Mobile Computing Concepts and Technology, *Kluwer Academic Publishers*, Norwell, MA, USA (1999).

[57] D. Yi, A Novel Call Admission Control Routing Mechanism Using Bloom Filter in MANET, *NSWCTC 2009*, 2009, 675–678.

[58] P. Yi, Z. Dai, Y. Zhong and S. Zhang, Resisting flooding attacks in ad hoc networks, *ITCC '05: Proceedings of the International Conference on Information Technology: Coding and Computing*, vol. 2, IEEE Computer Society, (2005), 657–662.

[59] J. Zander, Jamming games in slotted aloha packet radio networks, *Proc. of Military Communications Conference '90* (1990), 830–834.

[60] Z. Zhao, H. Hu, G.J. Ahn and R. Wu, Risk-Aware Mitigation for MANET Routing Attacks, *IEEE Transactions on Dependable and Secure Computing* **9**(2) (2012), 250–260.

[61] S. Zhong, J. Chen and Y.R. Yang, Sprite: A simple, cheat-proof, credit-based system for mobile ad-hoc networks, *INFOCOM-03* **3** (2003), 1987–1997.

[62] S. Zhong, L.E. Li, Y.G. Liu and Y.R. Yang, On designing incentive-compatible routing and forwarding protocols in wireless ad-hoc networks: An integrated approach using game theoretical and cryptographic techniques, *MobiCom '05*, ACM Press (2005), 117–131.

Enrico Tronci is currently an Associate Professor with the Computer Science (CS) Department of Sapienza University of Rome (Italy). Previously he was: a researcher with the CS Department of the University of L'aquila (Italy), a Post-Doct at LIP (Laboratoire pour l'Informatique du Parallelisme) at the ENS (Ecole Normal Superior) of Lyon (France). He received his Ph.D degree from Carnegie Mellon University, Pittsburgh, USA and his Master degree in Electrical Engineering from Sapienza University of Rome. His current research interests comprise: Model checking algorithms for automatic verification and synthesis of reactive systems. He has served as conference chair, program committee member, reviewer in many International Journals and Conferences. He has authored more than 50 scientific papers on International Journals and Conferences. He has been recently involved in more than a dozen research projects sponsored by the European Community, European Space Agency, as well as private companies.

Alberto Finzi is Assistant Professor at DSF, Università degli Studi di Napoli "Federico II" (Italy). He received his Ph.D degree in Computer Engineering from Sapienza Università di Roma (Italy). His research interests include: V & V methods for autonomous systems, multi-agent systems, autonomous and adaptive systems, planning and scheduling systems. He has been recently involved with several research projects sponsored by the EC (European Community), NASA (National Aeronautics and Space Administration), ESA (European Space Agency), ASI (Italian Space Agency), FWF (Austrian Science Fund), MIUR (Italian Ministry for University and Research), and private industries.

Igor Melatti is currently Researcher at at the Computer Science Department of the Sapienza University of Rome. He obtained his Ph.D. in Computer Science and Applications from the University of L'Aquila in 2001, after having graduated in the same institution in 2005. He held Post-Doc positions at the School of Computing of the University of Utah (2005) and at the Computer Science Department of the Sapienza University of Rome (2006–2010). His main research interests comprise: Formal methods, Automatic synthesis of reactive programs from formal specifications, Hybrid systems, Automatic verification algorithms, Model checking, Software Verification.

Federico Meli holds a Post-Doc position at the Computer Science Department of Sapienza University of Rome (Italy). In 2009, under the supervision of Prof. Enrico Tronci, he received his Ph.D degree from Sapienza University of Rome. His current research interests comprise: formal methods, automatic verification algorithms, model checking, hybrid systems verification, automatic synthesis of control software from formal specifications, automatic verification of Nash Equilibria in multi administrative distributed systems.

8

Event sharing in vehicular networks using geographic vectors and maps

Thierry Delot[a,*], Sergio Ilarri[b], Nicolas Cenerario[a] and Thomas Hien[a]
[a]*University Lille North of France, Valenciennes, France*
[b]*University of Zaragoza, Zaragoza, Spain*

Abstract. By exchanging events in a vehicular ad hoc network (VANET), drivers can receive information that allows them to find relevant places (e.g., parking spaces) or avoid dangerous/undesirable situations (e.g., an accident or a traffic jam). However, developing this kind of information services for drivers calls for new data management approaches, such as an appropriate dissemination protocol and some mechanism to decide when a driver should be alerted.

In this paper, we present a data management solution for event exchange in vehicular networks and compare two different approaches for relevance assessment. The first approach relies on the computation of geographic vectors to estimate the relevance of events, whereas the second approach exploits digital road maps. We also describe a prototype that has allowed us to test our proposals in a real environment. Moreover, we present an exhaustive simulation-based experimental evaluation that proves the usefulness of exploiting the information stored in digital road maps for data management and sharing in vehicular networks, which is an important novelty regarding existing works. The experiments also show that the first approach can also be used with a good accuracy, although smaller in some situations, in cars where maps are not available.

Keywords: Vehicular ad hoc networks, data management, event relevance estimation

1. Introduction

In the last few years, intensive research efforts are being developed in the area of transportation, mainly motivated by safety issues and technological improvements. Besides research focusing on enhancing vehicle applications such as navigation systems for vehicles (e.g., see [4,31,40]), exchanging dynamic data (i.e., data whose relevance can change very quickly) in a vehicular network using *Inter-Vehicle Communications (IVC)* is a hot topic nowadays [23,32,37]. Thus, thanks to the development of wireless networks and portable computers/devices, two vehicles nearby (within communication range of each other) can exchange interesting data, such as information about both static and mobile events (e.g., an emergency braking, an available parking space, a driver exhibiting risky behavior, etc.). Other communication schemes can also be considered, based on a fixed infrastructure or mobile telephony networks (e.g., 3G). Thus, even if it may be unrealistic to assume the availability of a generalized wide-area fixed infrastructure in the next years, mobile telephony networks already offer new perspectives for the development of applications to assist drivers. Anyway, such solutions, based on a centralization of the data and decision processes, still suffer from issues such as poor scalability or low reaction time available when dealing with some events like an emergency braking. Therefore, we focus in the following on

---

*Corresponding author: Thierry Delot, Univ. Lille North of France, UVHC/LAMIH FRE CNRS 3304, Le Mont Houy, 59313 Valenciennes, France. E-mail: Thierry.Delot@univ-valenciennes.fr.

vehicular ad hoc networks, although it is important to emphasize that this does not prevent the possibility to benefit also from a wired backbone (when available) as proposed in [30].

As an example, *VESPA* (*Vehicular Event Sharing with a mobile P2P Architecture*) is a system developed to share information about events in inter-vehicle ad hoc networks [13]. Data are received from other vehicles and stored locally in a *data cache*. Then, query evaluation techniques are used to sift through the stored information to determine what is relevant according to that time and location, and issue a warning or transmit information to the driver when necessary. Two main challenges in VESPA, as well as in other data sharing approaches for vehicular networks [25,28], are how to decide which data are relevant to the driver and how to transmit data to potentially interested vehicles.

In this paper, we focus on the concept of Encounter Probability (EP), whose goal is to determine whether an *event* (e.g., a traffic congestion) or a *resource* (e.g., an available parking space) is relevant to a vehicle, by using both spatial and temporal criteria. Thus, the EP is a measure of the likelihood that the vehicle will meet the event in the future. Such an EP can be used to filter, among the events received, those events that are relevant to the vehicle and so may be also relevant to the driver [12]. The EP can also be used by data dissemination protocols. In [6], we introduced a dissemination protocol able to handle the diffusion of a wide variety of events in the network. This protocol relies on the assumption that an event relevant to a vehicle may also be relevant to its neighbors. The EP is then used at each hop in the network to determine the relevance of the event. Summing up, the main contributions of this paper are the following:

- *We propose a general data management architecture for vehicular networks.* The architecture proposed allows exchanging data in vehicular networks and processing the data exchanged on the vehicles. One of the important components of the architecture is an *Encounter Probability Evaluator*, which computes the EP between a vehicle and the events received.
- *We propose a method to compute the EP based on the use of geographic vectors* (approach 1). This method simply considers the Euclidean space and thus does not rely on the availability of road maps or other information about the environment.
- *We propose a method to compute the EP based on the use of digital road maps* (approach 2). This method exploits the information available in digital road maps to try to compute the EP in a way that can be used more effectively. It is important to emphasize that the use of digital road maps for relevance assessment has not been considered so far in the literature.
- *We perform an experimental evaluation to test and compare both proposals.* The experimental results show the interest of the approaches and allow us to draw conclusions about the advantages and disadvantages of each alternative.

The structure of the rest of this paper is as follows. In Section 2, we describe the basics of the data management approach that we propose, which relies on some technique to compute the EP between a vehicle and an event. In Section 3, we summarize an approach to compute the EP based on the management of geographic vectors. In Section 4, we describe a new proposal that benefits from the use of digital road maps. In Section 5, we evaluate and compare experimentally both proposals. In Section 6, we present some related works. Finally, in Section 7 we summarize our conclusions and indicate some ideas for future research.

2. General data management approach

In this section, we describe the basic aspects of the general data management approach that we propose for exchanging data in vehicular networks and processing the data exchanged on the vehicles. Firstly, in Section 2.1 we present the different types of events that we consider. Secondly, in Section 2.2 we describe the way the information about the events is represented. Thirdly, in Section 2.3 we present the general architecture proposed for data management in vehicular networks. Finally, in Section 2.4 we explain how the events are disseminated and processed once received by the vehicles.

2.1. Types of events

Based on mobility features, we distinguish different types of events: direction-dependent vs. non-direction-dependent, and mobile vs. stationary. As opposed to other proposals, our proposed system not only supports *stationary events* (e.g., the presence of available parking spaces) but also *mobile events* (e.g., an emergency vehicle asking preceding vehicles to yield the right of way). When supporting such mobile events, the set of vehicles for which the event is relevant evolves according to both the movements of the vehicle generating the event (in the example, the emergency vehicle) and the other vehicles involved (in the example, the preceding vehicles). The direction of traffic is also of major importance in establishing the relevance of shared information, even for non-mobile events (e.g., consider a traffic jam affecting only the vehicles moving in one direction). So, some events are *direction-dependent events* and others are *non-direction-dependent events*.

Besides, an orthogonal classification of events, considering attraction and repulsion events, is also proposed. *Attraction events* are events that the driver would like to meet (e.g., parking spaces, petrol stations, etc.) according to her/his current interests/goals, even if this means that s/he has to change her/his current route. As an example, a driver approaching downtown for a business meeting would be interested in parking spaces nearby even if they are not just in front of her/his final destination. As another example, an unavailable taxi driver could release an event reporting other taxis about a person looking for a taxi, and this event could potentially be relevant to any taxi nearby (independently of its direction). On the contrary, *repulsion events* are events that should be avoided whenever possible because they imply driving difficulties (e.g., accidents, traffic jams, a slippery road, fire on the road, a vehicle driving in the wrong direction, etc.).

Finally, we could also distinguish between *events* and *resources*. An event and a resource differ in the sense that competition between vehicles may appear in the case of resources (e.g., parking spaces). In the rest of the paper, we will use the term event to refer to all types of events, since we do not want to focus on competition management here. The interested reader can consult [14,15] for more information about parking spaces allocation and competition management in VANETs.

2.2. Representation of events

The different types of events mentioned previously are represented uniformly in our approach. Specifically, the following attributes are used to represent the events and generate messages exchanged between vehicles to assist the drivers when necessary:

- A *Key* (composed of a unique identifier of the vehicle, such as its MAC address, plus a local event identifier) identifies the event.
- A *Version* number allows to distinguish between different updates of the same event (e.g., used to refresh the location of a mobile event or to remind that a long-lived event still exists).
- An *Importance* value helps to determine the urgency of presenting that information to the driver (e.g., an emergency braking has a higher importance than an available parking space).

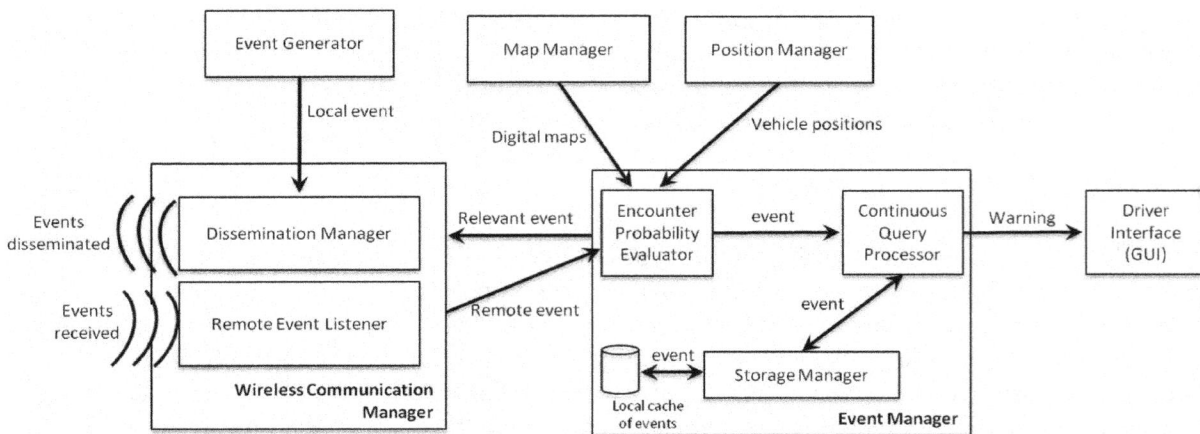

Fig. 1. General data management architecture.

- A *CurrentPosition* field indicates the time and place corresponding to the event. Using the GPS time on each vehicle allows to avoid synchronisation problems between the clocks of the different vehicles.
- A *Description* field contains further information for the driver.

Other additional attributes will be introduced throughout the paper depending on the relevance assessment approach considered. Specifically, in Section 3.2.1 we introduce the attributes *DirectionRefPosition* and *MobilityRefPosition* for the approach based on geographic vectors, and in Section 4.2 we introduce the attribute *initialTTL* for the approach based on digital road maps. All these attributes could be part of the representation of an event, which would enable vehicles to use any of the two relevance assessment approaches proposed.

It could be interesting to mention that we plan to enrich the structure of events by adding semantic information (by considering the RDF/XML exchange syntax for the Web Ontology Language OWL), which will help the vehicles to interpret unambiguously the information exchanged (e.g., the *Description* field) and could also facilitate the interoperability between different systems (e.g., developed by different car manufacturers). However, for the purposes of this paper the structure of events presented in this section is enough.

2.3. General architecture for data management in vehicular networks

The general data management architecture that we propose, which is deployed on every equipped vehicle, is presented in Fig. 1, where the following main elements can be distinguished:

- The *Wireless Communication Manager* is in charge of the reception and transmission of events. This module is composed by the *Dissemination Manager*, which allows the vehicle to broadcast events, and the *Remote Event Listener*, which is responsible for the reception of events transmitted by neighboring vehicles.
- The *Event Manager* handles the events received by the vehicle. It is composed of the *Continuous Query Processor*, which processes *active continuous queries* representing the driver's interests (e.g., a driver is informed about available parking spaces only if s/he specified her/his interest in that type of event) by using an *Encounter Probability Evaluator* (based on geographic vectors or digital road

maps, as described in Sections 3 and 4, respectively), and the *Storage Manager*, which is in charge of deciding about the storage and removal of events in a local cache.

- The *Driver Interface* is the graphical user interface used to interact with the driver (e.g., showing information about relevant events).
- The *Position Manager* interacts with the GPS receiver of the vehicle to retrieve information regarding the location of the vehicle.
- The *Map Manager* is in charge of managing digital road maps containing information about the roads in the surroundings of the vehicle. This module is only needed when the approach based on digital road maps is used to compute the *Encounter Probability* (see Section 4).
- Finally, the *Event Generator* releases events detected by the vehicle. The generation of many events could be initiated using the numerous sensors embedded in modern cars (for example, by coupling the airbag system with the creation of an event representing an accident) or via other static data sources (e.g., sensors on a road). This will prevent a driver from disseminating false information to her/his own benefit (ensuring the reliability of messages manually generated by drivers is out of the scope of this paper and considered in works such as [29]).

In the following, we explain briefly the way the different modules interact:

1. An event received by the Remote Event Listener is communicated to the Encounter Probability Evaluator. The Encounter Probability Evaluator computes the *Encounter Probability* between the vehicle and the event, which is a measure of the likelihood that the vehicle will meet the event in the future, based on information provided by the Position Manager. Additionally, the information provided by the Map Manager can also be used if the information stored in digital road maps is exploited as described in Section 4. The relevance of an event may change continuously due to the different dynamic factors affecting the computation of the Encounter Probability (such as the distance to the event), whatever the method used to compute it. Therefore, the Continuous Query Processor, using the Encounter Probability Evaluator, evaluates periodically the active continuous queries to verify which events must be reported to the driver through the Driver Interface. For this, each event for which the Encounter Probability is higher than a certain *relevance threshold* (see Section 2.4) must be checked against the set of active continuous queries. Additionally, some events representing dangers on the road (identified by a high value of the *Importance* field, described in Section 2.2) are reported to the driver immediately, even if there is no query asking for those data.

2. The Storage Manager is informed by the Query Processor about the probabilities computed by the Encounter Probability Evaluator. If the Encounter Probability of a previously stored event is smaller than a certain *storage threshold* (see Section 2.4), then the Storage Manager removes the event from the local cache. On the contrary, if the Encounter Probability of a new event is greater than the storage threshold, the event is stored.

3. For a new event received, in case its Encounter Probability is higher than a certain *diffusion threshold* (see Section 2.4), the Dissemination Manager is contacted by the Event Manager to broadcast the event and inform other vehicles.

It should be emphasized that our focus on this paper is on the software side, and more specifically on the data management issues involving the *Event Manager* component, instead of on networking aspects. However, in other previous works, we have also proposed a data dissemination approach aiming at minimizing the network overload [6].

2.4. Management of events

To share information in VANETs, different protocols have been proposed over the last years. All these protocols have to control both the number of messages exchanged and the delivery of the information to the interested vehicles. For example, in [6], we proposed a dissemination protocol whose goal is to carry different types of events in the vehicular network (e.g., available parking spaces, emergency vehicles, traffic congestions, etc.) to the potentially interested vehicles. To achieve this, it is necessary to adapt the dissemination of an event according to its type. Thus, for example, the information about an available parking space may be interesting for all vehicles around, whatever their direction. On the contrary, the information relative to an emergency braking or a traffic congestion should be delivered only to the vehicles driving towards that event. In [6], we used the concept of Encounter Probability (EP) to adapt the dissemination chain of an event according to its type. Thus, each vehicle receiving an event estimates its relevance by computing the EP and then relays it only if such event is considered relevant. In addition, we introduced in the proposed protocol features to limit the number of times a single message can be relayed, in order to avoid network flooding.

Thus, once an event is received by a vehicle, this vehicle reacts based on the comparison of its EP with three different thresholds, the *relevance threshold* (RT), the *storage threshold* $(ST$, with $ST \leqslant RT)$, and the *diffusion threshold* (DT):

- If $EP < ST$ then the (potential) relevance of the event for the vehicle is not enough. Therefore, the event is discarded.
- If $EP \geqslant ST$ then the event is considered relevant to the vehicle and it is stored in the vehicle's data cache. Additionally, a warning will be communicated to the driver if and only if: 1) $EP \geqslant RT$ (which means that *the event is relevant to the vehicle* at that moment), and 2) either the *Importance* of the event is high (e.g., it is an accident) or the driver has specified her/his interest in that type of event (i.e., *the event is relevant to the driver*).
- Besides, if $EP \geqslant DT$ then the vehicle relays the message it received about the event. The diffusion threshold is used by the dissemination protocol proposed in [6]. The basic idea of this protocol is that a vehicle should diffuse a message about an event if it estimates it relevant enough. In that case, the probability that the event is also relevant to its neighbors is indeed high. Thus, an event will keep being disseminated to neighboring vehicles while it is considered relevant for dissemination in a particular area. Besides, this dissemination strategy relying on the EP also ensures the adaptation of the *diffusion chain* according to the type of event. For example, an event representing a traffic congestion will only be diffused to the vehicles driving towards it; on the contrary, a notification about an available parking space will be relayed to all the vehicles in the vicinity of that resource, whatever their direction.

It should be noted that a distinction between events that are *relevant to the vehicle* and events that are *relevant to the driver* is made, emphasizing the importance of cooperation between the vehicles. Moreover, an event that is initially relevant only to the vehicle could become also relevant to the driver (e.g., for parking spaces the driver may decide at any time that s/he wants to park).

3. First approach: Using geographic vectors

In this section, we focus on the approach using geographic vectors [13]. Firstly, in Section 3.1 we intuitively explain the motivation of this approach. Then, in Section 3.2 we describe how the EP is computed by using geographic vectors.

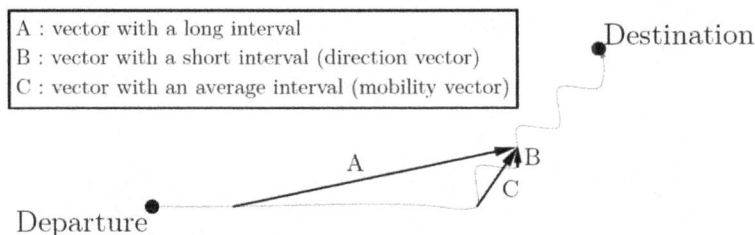

Fig. 2. Mobility and direction vectors.

3.1. Motivation for using geographic vectors

The main motivation for using geographic vectors to estimate the EP is that the direction of a vehicle (and/or a mobile event) can be used to predict its future positions, and so it is possible to compute a probability that estimates if the vehicle is going to meet an event.

Thus, when no digital road maps are available, the Encounter Probability between a vehicle and an event is obtained by considering the estimated future positions of the vehicle. For this purpose, two movement vectors are defined for a vehicle: the direction vector and the mobility vector (see Fig. 2). The *direction vector* allows to estimate future positions of the vehicle quite precisely on a short term, whereas the *mobility vector* captures an overall impression of the direction of the vehicle and allows to estimate future positions on the long term. Each vehicle can compute its direction vector and its mobility vector easily. Similarly, each vehicle can compute the *mobility and direction vectors of the events* it receives, since not only vehicles but also events (more specifically, *mobile events*) can move. By comparing the direction and mobility vectors of a vehicle and an event it is possible to compute the Encounter Probability, as we explain in the following. It should be noted that, as these vectors are defined in the Euclidean space, no digital road maps are needed with this approach to compute the Encounter Probability.

3.2. EP with geographic vectors

In this section, we explain our approach to compute the EP based on geographic vectors. First, we present the main aspects of the approach in Section 3.2.1. Then, in Section 3.2.2 we detail some ideas about the use of *penalty coefficients* to adjust the weights of the different spatio-temporal parameters involved in the computation of the EP.

3.2.1. Computation of the EP by using geographic vectors

As explained before, this approach for the computation of the EP is based on the *mobility and direction vectors* of vehicles and events. A vehicle can compute its mobility and direction vectors by sampling its location periodically. To allow computing the mobility and direction vectors of an event by a vehicle, we introduce two additional attributes in the event description presented in Section 2.2: the *DirectionRefPosition* and the *MobilityRefPosition*, which store two preceding reference positions (according to what is described in Section 3.1).

For each event, a *mobility vector (and direction vector) of the vehicle in relation to the event* is computed by the vehicle by changing the frame of reference. Figure 3 illustrates this change, explained in detail in [13]. The mobility vectors of one vehicle and one event are represented on the left side of the figure, and the resulting vector after the frame of reference has been changed is shown on the right

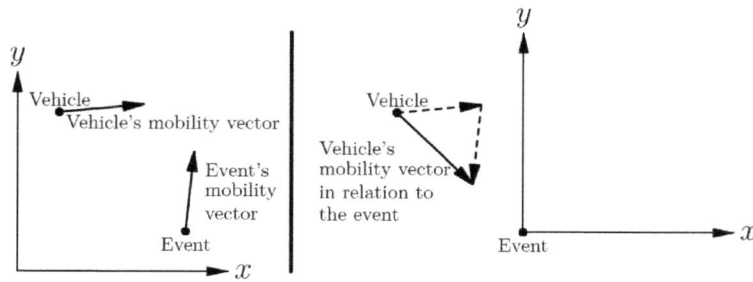

Fig. 3. Mobility vector of the vehicle in relation to the event: change of the frame of reference.

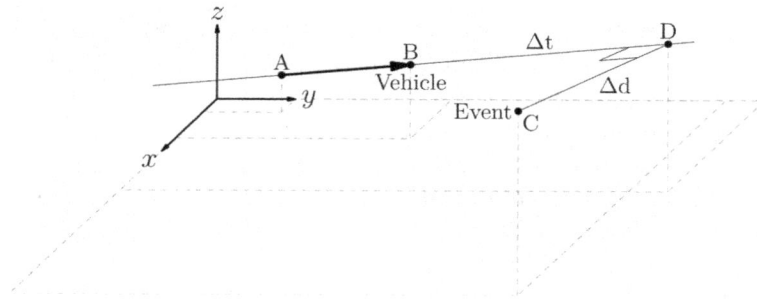

Fig. 4. Geometrical representation of Δd and Δt for a sample stationary event.

side. The change of frame of reference simplifies the computation of the EP by allowing a single vector for each couple <vehicle, event> to be managed, regardless of the type of event.

The mobility vector and direction vector of the vehicle in relation to the event are used to compute four elements (an example of the first two elements, Δd and Δt, is shown in Fig. 4, where B represents the position of the vehicle, C the position of the event, and $\overrightarrow{AB}$ is the mobility vector of the vehicle in relation to the event):

- The minimal geographical distance between the vehicle and the event over time (Δd).
- The difference between the current time and the time when the vehicle will be closest to the event (Δt).
- The difference between the time when the event is generated and the moment when the vehicle will be closest to the event (Δg).
- The angle between the direction vector of the vehicle and the direction vector of the event (represented by a colinearity coefficient c).

Once these values have been calculated, they are used to estimate the EP (a value in the range of 0% to 100%) between the vehicle and the event:

$$EP = \frac{100}{\alpha \times \Delta d + \beta \times \Delta t + \gamma \times \Delta g + \zeta \times c + 1}$$

where α, β, γ and ζ are *penalty coefficients* with values $\geqslant 0$. They are used to balance the relative importance of the Δd, Δt, Δg, and c values. The bigger the coefficient is, the more penalized the associated valued is when computing the EP. For example, the greater the α value, the shorter the spatial range where the event is considered relevant. β and γ are used so that only the most recent information and the information about events that will be encountered very rapidly is considered. Finally, ζ is used to weigh the importance of the colinearity coefficient. It should be noted that if the vehicle is moving away from the event, then Δt is 0 and Δd is the current distance to the event. Therefore, the computation of the EP makes sense even when an interesting event (e.g., a parking space) is behind the driver.

3.2.2. Use of the penalty coefficients

Considered individually, the penalty coefficients allow the definition of bounds on the relevance of events. For example, if the relevance threshold is set to 75% for the Encounter Probability, a value of $\alpha \geqslant \frac{1}{300}$ implies that if the minimum geographical distance between the vehicle and the event over time (Δd) is larger than 100 meters, then the event will be considered as not relevant whatever the values of the other parameters (i.e., Δt, Δg, and c):

$$75 \leqslant \frac{100}{(\alpha \times 100 + 1)} \Rightarrow \alpha \leqslant \frac{1}{300}$$

In the same way, β sets a maximum time interval between the current position of the vehicle and the position of the vehicle when it is expected to be at the closest location from the event; if this time interval is exceeded, the event is considered not relevant. For example, for values of $\beta \geqslant \frac{1}{900}$ an event will not be considered relevant if the time elapsed when the vehicle is at the closest distance from the event is estimated to be five minutes or more. Similarly, γ is used to penalize the relevance according to the age of the event. In practice, γ should be set according to the frequency used to generate new *versions* (see Section 2.2) of potentially long-term events (e.g., if this period is five minutes, then it is possible to set $\gamma = \frac{1}{900}$). Finally, ζ may induce a maximum tolerance on the angle formed by the direction vectors for direction-dependent events. For instance, when the vehicle is on the highway, the tolerance on the angle should be relatively low. Thus, if the tolerance is set to 45°, the value of ζ can be computed as follows:

$$75 \leqslant \frac{100}{\zeta \times 45 + 1} => \zeta \leqslant \frac{1}{135}$$

Naturally, the importance of the Δd, Δt, Δg and c parameters depends on the event considered (e.g., traffic congestion, parking space, emergency braking, etc.). For instance, a message describing a traffic congestion should be broadcasted several kilometers away from the place where it is located, for drivers to have the opportunity to change their itinerary. The penalty on Δt should so be very low. On the other hand, when dealing with parking spaces, the penalty on Δt should be more important because a driver is only interested in finding an available parking space if it can be reached quickly. In addition, for the same type of event, the penalty coefficients may have to be modified according to the current time and date. For example, when dealing with parking spaces in urban areas, the penalty on the age (i.e., the value of γ) should be more penalizing on Saturday afternoons than on Monday nights.

The different penalty coefficients can be fine-tuned by following different guidelines, similar to the ones described in this section. For more details about the computation of the Encounter Probability, along with an evaluation of its benefits, we refer the interesting reader to [13].

4. Second approach: Using digital road maps

In this section, we propose an alternative approach where, instead of relying on the computation of geographic (mobility and direction) vectors, the underlying road network topology is considered. Firstly,

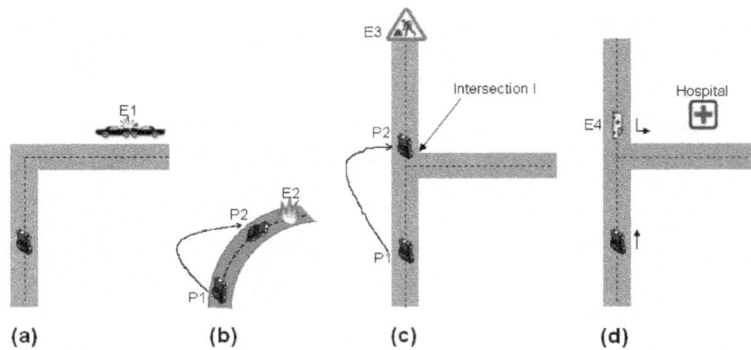

Fig. 5. Advantages of maps: sample scenarios.

in Section 4.1 we present some simple but illustrative situations where the use of digital road maps could be useful. Then, in Section 4.2 we describe how the EP is computed by exploiting the extra information provided by digital road maps.

4.1. Motivation for using digital road maps

A key idea of our general data management approach is that the relevance of an event should determine how the information about the event is disseminated and how a driver is alerted about such an event. The importance of considering the road network for this is justified by describing some simple situations where just considering the mobility and direction vectors, as described in Section 3, is not appropriate (see Fig. 5):

- In the situation shown in Fig. 5a, the car would consider the event *E1* (an accident) as not relevant since the values of Δd and c computed (see Section 3.2.1) would be too large. Intuitively, it is difficult to predict that the vehicle will meet the event unless we know that the road network will force a right turn.
- In Fig. 5b there is fire on a curve (represented by event *E2*). However, by just considering geographic vectors, *E2* is not relevant until the vehicle is at location *P2* (at *P1* the value of the colinearity coefficient c is too large and leads to a small EP computed). By then, alerting the driver is of little use (the driver herself/himself is able to see the fire), since it may be too late to react.
- Figure 5c shows a segment of a road that is difficult to traverse due to road works (event *E3*). Although the direction and mobility vectors of the vehicle computed at location *P1* suggest that the vehicle will meet the event, the value of Δd computed renders the event as not relevant yet (the vehicle is still far away from the event and so the situation may change in the future). When the event is considered relevant (at location *P2*) it is too late for the driver to take an alternative route, as s/he just missed the last possible intersection to avoid the event.
- A similar situation is shown in Fig. 5d, where there is a mobile event *E4*: an ambulance requesting the cooperation of nearby vehicles to get through. In this case, the value of the Δd computed is small enough and the event is considered relevant to the vehicle. However, the ambulance is about to turn left towards the hospital. As a consequence, the vehicle is never going to meet the ambulance. Therefore, alerting the driver would be an unnecessary disturbance. It should be noted that, in this scenario, information about the route of the mobile event would also be necessary to benefit from the use of digital road maps.

It is important to emphasize that, even though we have presented very simple examples for illustration purposes, similar situations can be observed in reality and considering real road maps. Despite the fact that the previous scenarios suggest the need of using digital road maps, we have obtained promising experimental results with the alternative approach (see Section 3) that uses geographic vectors [6,13]. Therefore, once we design an approach based on road maps, we will need to evaluate experimentally how the use of information contained in digital road maps can actually improve the accuracy of the relevance estimation.

It should be noted that, by using maps, it is trivial to estimate whether a vehicle will encounter an event or not if we assume that the route of the driver is known in advance (the route of the event is also needed if it is a mobile event, as exemplified before in relation to Fig. 5d). However, it is not possible to constantly ask the driver about her/his destination. This is for example required for route guidance with existing navigation systems, but these are used only occasionally, when the driver does not know her/his route. On the contrary, driver assistance systems should not be intrusive in order to facilitate their everyday use transparently. Indeed, even when the driver knows her/his route perfectly (e.g., driving to the office or back home) and s/he has not introduced the route's data in a navigation system, s/he may need to be informed about some interesting events (e.g., dangers) on the roads. Therefore, the solution that we propose to compute the EP with digital road maps does not assume that routes are known in advance.

4.2. EP with digital road maps

To benefit from digital road maps, we introduce one new field in the messages describing events exchanged between vehicles (see Section 2.2). Instead of the reference positions used for the EP computation based on geographic vectors (i.e., the *DirectionRefPosition* and *MobilityRefPosition* attributes, described in Section 3.2.1), we add here an *initialTTL* field which will be used to compute the EP. The *TTL* (*Time to Live*) of an event is an estimation of the time interval during which the event will continue being valid. If the event still exists after the TTL (e.g., the event may be a traffic congestion that has not disappeared yet when expected), a *new version* of the event (with a new TTL) has to be generated. The generator of an event sets the *initialTTL* of the event, and the current value of the TTL can be obtained by considering that value and the time elapsed since the creation of the event (the event's generation time is stored in the *CurrentPosition* field, as described in Section 2.2).

In the following, we present our solution to compute an Encounter Probability using digital road maps for both attraction and repulsion events (described in Section 2.1). Firstly, in Section 4.2.1 we define the concept of *Reachability Probability* for attraction events. Secondly, in Section 4.2.2 we define the concept of *Need to Escape Probability* for repulsion events. Finally, in Section 4.2.3, we summarize the computation of the EP based on the two previous concepts.

4.2.1. Dealing with attraction events: reachability probability

For attraction events (i.e., events that the driver would like to meet, such as parking spaces), the EP is computed as the *Reachability Probability* (*ReachP*):

$$ReachP = \begin{cases} 100 & \text{if } TTL > TTR \\ 0 & \text{otherwise} \end{cases}$$

where the TTR (*Time To Reach*) is the time needed for the vehicle to reach the event by taking the shortest path. Notice that $ReachP$ is either 0% or 100%, depending on whether it is estimated that the vehicle will be able to reach the event in time (i.e., before it disappears) or not. As there may be several attraction events relevant to the driver (i.e., with $ReachP = 100\%$), extra information is used to compute a *score* for each event and provide the driver with events of the same type ordered in a *ranked list*. For example, several reachable parking spaces are ranked according to different criteria, such as: 1) the distance to the vehicle (to minimize the time needed to reach the parking space, and therefore the probability that it becomes unavailable), 2) the probability of finding an available parking space it that area[1] (to minimize the impact of a situation where another vehicle occupies the space first), 3) the distance from that parking space to other alternative available parking spaces (to assign a higher score to a parking space if there are other events reporting available parking spaces near that space), or 4) the number of modifications in the route planned by the driver (if available) needed to meet the event.

Obviously, sharing resources such as parking spaces introduces some competition between vehicles. Indeed, only one vehicle will obtain the resource even if more than one receives a notification about it. To overcome such problems, we have proposed a complementary allocation protocol which aims at avoiding the competition between vehicles by electing, among the set of interested drivers, a single one [14,15].

4.2.2. Dealing with repulsion events: need to escape probability

In contrast to attraction events, repulsion events are events that the driver wants to avoid (e.g., traffic congestions, accidents, etc.). Dealing with such events, the EP is computed as the *Need to Escape Probability* (*NeedEsP*), which indicates the probability that the driver needs to perform some specific action to avoid the event:

$$NeedEsP = \begin{cases} 100 & \text{if } \text{TTL} > \text{TTE} \\ 0 & \text{otherwise} \end{cases}$$

where the *TTE* (*Time To Escape*) is the amount of time needed by the vehicle to reach the last intersection that offers the vehicle an alternative route to avoid the repulsion event (e.g., in Fig. 5c it would be the time needed to reach intersection *I*), or the *TTR* if there is no such intersection (i.e., if it is not possible to avoid the repulsion event). Therefore, by definition, it should be noted that the following always holds: $TTE \leqslant TTR$. The idea is that if the vehicle is not even able to reach that intersection before the end of the TTL of the event, then such an event is not relevant to the vehicle (it is expected to disappear before the vehicle reaches it). Otherwise, as commented before, a new version of the event will be generated if the TTL elapses and the event is still there; in this case, by having the last intersection to escape as a reference (instead of simply considering the TTL), the vehicle will still be able to avoid the event if necessary. Besides, a driver would not be alerted about a repulsion event if the route of the vehicle is known in advance and it does not pass through the event (unless the route changes in the future). However, even in this case the NeedEsP computed may be 100% (the event will be considered in this case relevant to the vehicle although not to the driver, similarly to what we discussed in Section 2.4), which implies that the vehicle will store and disseminate the event to inform other vehicles.

4.2.3. Computation of the EP by using digital road maps

According to the previous considerations, a vehicle will compute the EP for an event depending on whether the event is an attraction event or a repulsion event:

$$EP = \begin{cases} ReachP & \text{for an attraction event} \\ NeedEsP & \text{for a repulsion event} \end{cases}$$

---

[1]This knowledge can be extracted from aggregated information about events, as suggested in [11].

Fig. 6. GUI of VESPA for the GV-based approach.

By using digital road maps, the EP computed is thus either 0% or 100% (i.e., an event is either definitely relevant or irrelevant). Therefore, the specific values of the storage threshold, the relevance threshold, and the diffusion threshold (defined in Section 2.4), are not significant as long as these thresholds are higher than 0. Rankings of the attraction events (see Section 4.2.1) can be used to remove events with small scores in case of insufficient storage or to minimize the number of events disseminated. On the other hand, relevant repulsion events should be disseminated in any case, as they represent places that should be avoided and besides there will probably be a much smaller number of them.

5. Experimental evaluation

In this section, we evaluate and compare experimentally the two alternative approaches presented in this paper for data sharing in vehicular networks: using geographic vectors and exploiting the information stored in digital road maps, that we will call in the rest of this section the *GV-based approach* and the *Map-based approach*, respectively. We consider two different configurations for the approach based on geographic vectors: the GV-based configuration for both highway and urban scenarios, and the GV-based* configuration for highway scenarios. The *GV-based** configuration uses more selective penalty coefficients (e.g., the vehicle has to be closer to the event to consider it relevant), and it will be considered as an alternative to reduce the number of false warnings (i.e., irrelevant messages received by a vehicle) in a highway scenario. On the contrary, with the *GV-based* configuration the events will be disseminated far away from their location to provide drivers with more time to react (e.g., in the case of traffic congestions).

In the following, we present our experimental results. Firstly, in Section 5.1 we describe the experimental settings considered for the evaluation of our proposal. Then, we present experiments where we measure the amount of time available since a driver is notified about an event until the moment the driver meets the event (see Section 5.2), the number of irrelevant messages transmitted (see Section 5.3), and the performance of the system when dealing with mobile events (see Section 5.4).

5.1. Experimental settings

We have developed a prototype of VESPA (http://www.univ-valenciennes.fr/ROI/SID/tdelot/vespa/prototype.html), that works on mobile devices, and performed some experiments in real situations. Thus, Fig. 6 presents the graphical user interface (GUI) of VESPA for the *GV-based approach*

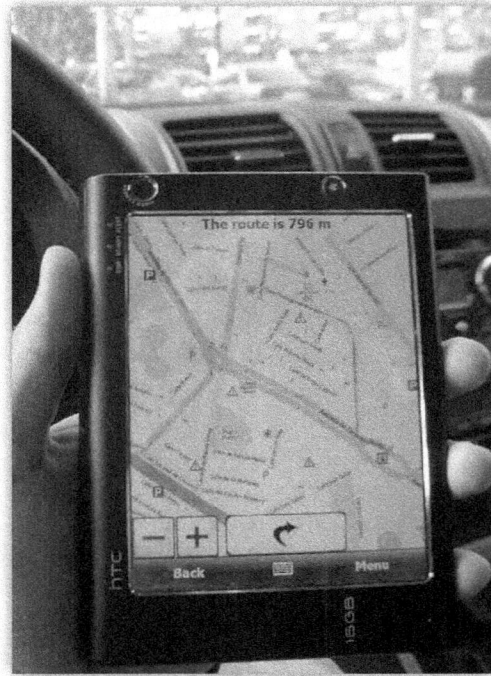

Fig. 7. GUI of VESPA for the Map-based approach.

considering an available parking space event. The GUI for the *Map-based approach* is different, as shown in Fig. 7, since the digital road maps used to evaluate the relevance of events are also exploited to show the events on a map, like navigation systems (e.g., TomTom Navigator, Navigon, Garmin, etc.) do, or to show the route needed to reach or avoid a specific event.

Despite the availability of this working prototype, due to obvious scalability reasons it is not possible to fully evaluate our approach through field tests. Thus, for example, it is difficult to obtain repeatable scenarios with a high number of vehicles in a real environment. Therefore, for experimental evaluation we use a vehicular network simulator that we have developed (http://www.univ-valenciennes.fr/ROI/SID/tdelot/vespa/simulator.html), which allows to simulate realistic contexts. Tests in a real environment are thus used mostly for verification and to calibrate our simulations.

During our experimentations, we have considered real road networks of an area of the city of Valenciennes (France) extracted from real digital road maps. The maps we used in our prototype and our simulator were provided either by *Tele Atlas* (http://www.teleatlas.com) or by *OpenStreetMap* (http://www.openstreetmap.org/). For the Map-based approach, these maps are transformed into a graph representation from which shortest paths can be computed by applying the Dijkstra algorithm [8, 41] (some additional heuristics are applied to speed up the performance of the computation of routes on the mobile device when large maps are managed). More precisely, we have considered two scenarios for evaluation:

- *Highway scenario*: a (15 Km long) segment of the highway between Valenciennes and Lille in the North of France; see Fig. 8, where different events are represented with their numbers: from *event1* to *event4*.

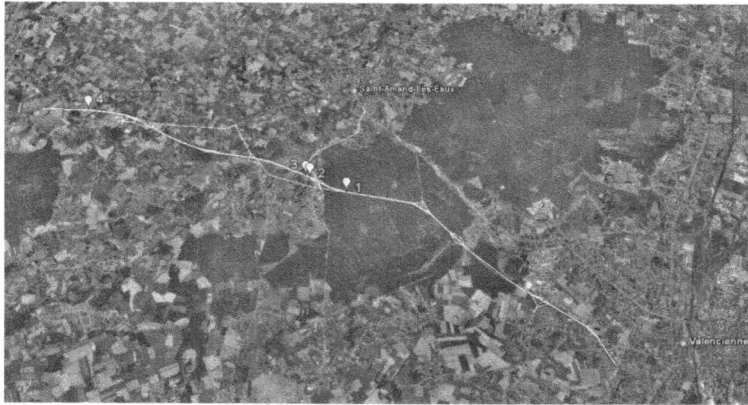

Fig. 8. Road network considered for the experimentation on the highway.

Fig. 9. Road network considered for the experimentation in an urban area.

– *Urban scenario*: the center of the city of Valenciennes in France; see Fig. 9, where a single event is marked.

On these road networks, we evaluate whether the two solutions proposed (based on geographic vectors vs. using digital road maps) allow the vehicles to exploit the information about attraction/repulsion events effectively (in order to avoid the repulsion events and meet the attraction events). The events considered for evaluation are those shown in the figures. Nevertheless, similar results have been obtained considering other maps and/or events placed at different locations.

In the simulation, vehicles are created every two seconds, with a random location, at one extreme of one of the roads considered in the scenario evaluated. Once created, vehicles follow a classical random mobility model through the road network. Their speed ranges from 20 Km/h to 130 Km/h with an average speed of 76 Km/h in the highway configuration, and varies from 15 to 60 Km/h with an average speed of 32.5 Km/h in the urban configuration. About 400 vehicles were considered during each simulation.

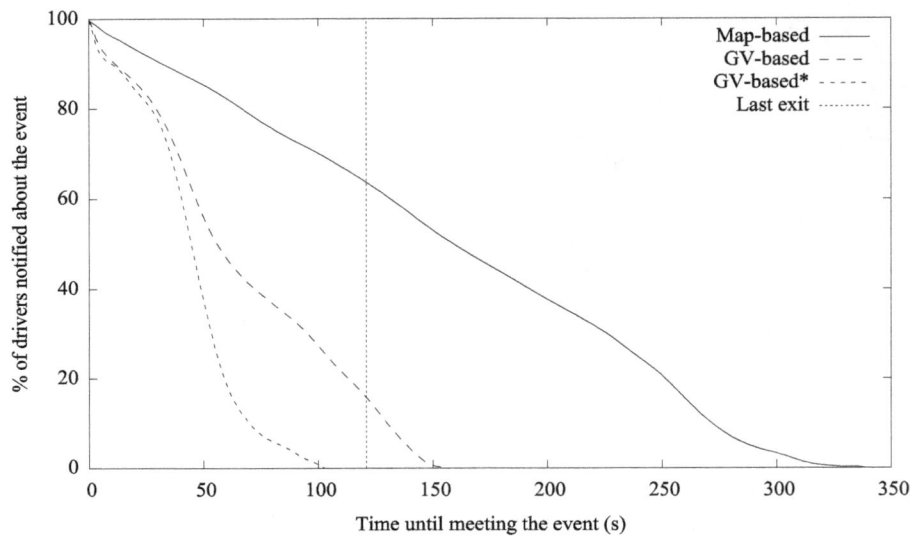

Fig. 10. Percentage of vehicles warned along time: stationary direction-dependent event on the highway.

The wireless communication range considered for each vehicle is 200 m. For each event, its time to live (TTL) is set to 500 s and the event is activated after 150 s of the start of the simulation. For the approach that computes the EP based on geographic vectors, mobility/direction vectors are obtained by using position statements performed every 500 m and every 30 m, respectively. Finally, the thresholds used to evaluate the incoming events in relation to their Encounter Probability (see Section 2.4) are set to 75%. The values used for the different parameters in the experiments, including the penalty coefficients in the GV-based approach, have been obtained through an extensive experimental evaluation; thus, we have selected values that behave well in a variety of different scenarios.

5.2. Evaluation of the reaction time available to drivers

During our simulations, we observed the vehicles which presented a warning to the driver before meeting events of different types (e.g., direction-dependent or not) located at different places on the road networks considered. More specifically, we measured the percentage of vehicles which presented a warning to the driver according to the time separating the moment when the vehicles received a warning and the moment when they would encounter the event. We expect that the greater the *accuracy* of the mechanism used to compute the Encounter Probability, the greater the amount of time available to drivers to react when events are received.

First, in Fig. 10 we focus on the event number *1* indicated in Fig. 8, that is, *event1*. This event is a repulsion, stationary, and direction-dependent event (e.g., a danger on the road or a traffic congestion) located on the highway. We first observe in Fig. 10 that 100% of the vehicles meeting *event1* inform their driver before meeting the event with both approaches. The Map-based approach significantly increases the time interval separating the moment when the vehicles warn their driver and the moment when they will actually encounter the event. By warning the drivers earlier, the probability to be able to exit the highway before meeting the event increases, which is particularly interesting for repulsion events that drivers need to avoid. For instance, the average travel time between *event1* and the last exit on the highway before *event1* is about 120 seconds; in this figure and in the upcoming figures, the *average*

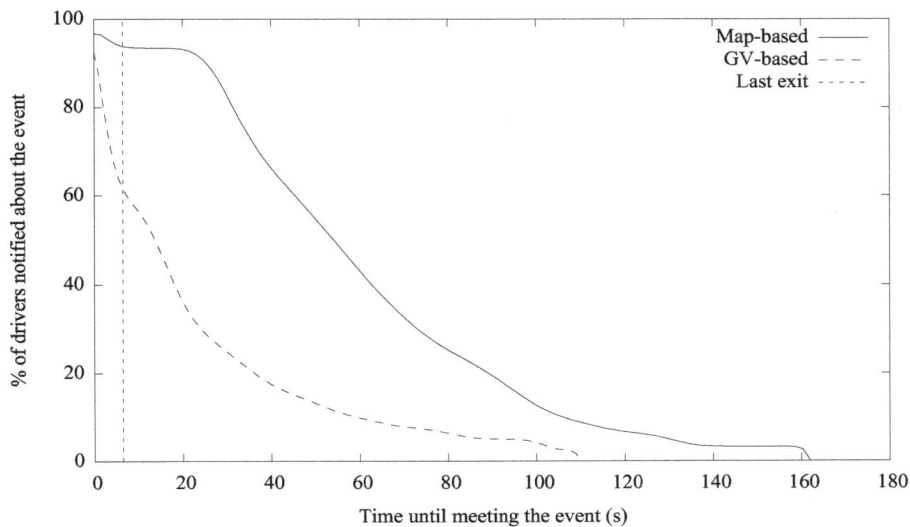

Fig. 11. Percentage of vehicles warned along time: stationary direction-dependent event in the urban scenario.

travel time from the last exit is represented by a vertical dashed line. So, the GV-based approach does not ensure that the messages describing this repulsion event are relayed far enough for drivers to have the opportunity to avoid that event. Indeed, the figure shows that only about 20% of the vehicles with the GV-based approach (and no vehicle with the GV-based* approach) had a chance to use the information received to avoid the event. Obviously, even with the Map-based approach, it is not possible to reach 100% of the vehicles informed 120 seconds before meeting the event, since some of them may be already very close to the event at the moment when the event is generated; according to our experimental results, with the Map-based approach about 70% of the vehicles received the information in time to avoid the event. Although in Fig. 10 we only show the results for *event1*, it is important to clarify that we obtained similar results for other types of events, located at different places on the highway (see Fig. 8).

During our tests, we compared the GV-based and Map-based approaches not only on highways but also in urban environments. Our goal with the evaluation in urban environments was to study the effects of frequent changes of the direction of vehicles. Figure 11 shows the moment when the vehicles are informed considering our different approaches applied on a stationary direction-dependent event located in the center of the city of Valenciennes (as shown in Fig. 9). In this scenario, the Map-based approach outperforms again the GV-based solution, although in this case the differences are reduced with respect to the highway scenario. Moreover, since the number of alternative roads is much higher here, both solutions allow a large percentage of the drivers to find an alternative route before encountering the event (about 95% with the Map-based approach and 60% with the GV-based approach), which is important in the case of repulsion events.

Finally, in Fig. 12 we present as another example the results for a stationary non-direction-dependent event in the urban scenario (located at the same location as the event considered in the previous experiment, see Fig. 9). The difference between the Map-based approach and the GV-based approach is again reduced in this case and we observe similar results for both approaches.

Before concluding this section, it is important to remind that we have evaluated also other types of events at different locations, in both urban and highway scenarios, obtaining similar results. Thus, we can conclude that the Map-based approach increases the reaction time available for drivers to avoid repulsion events, especially in scenarios where no many alternative roads exist.

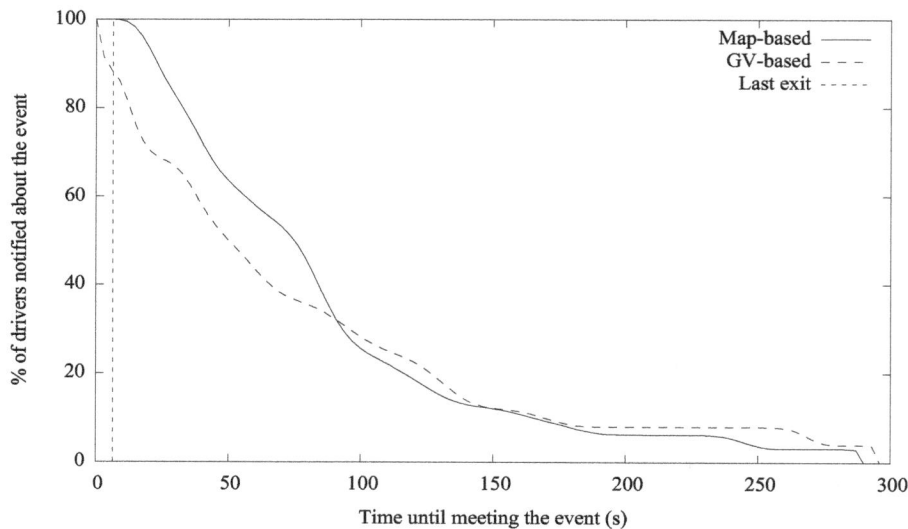

Fig. 12. Percentage of vehicles warned along time: stationary non-direction-dependent event in the urban scenario.

5.3. Evaluation of the number of irrelevant messages shown to the drivers

Obviously, the time interval available since a vehicle receives an event until it meets the event is a very important parameter, as it determines the reaction time provided to the driver by the system. However, it is not the only important one. The number of warnings about events communicated to drivers that these drivers finally did not encounter (called *false warnings* in the following) is interesting too. Indeed, while trying to communicate a piece of information to vehicles located far away from the event, it should be ensured that the information is delivered only to potentially interested drivers. Again, we expect that the greater the *accuracy* of the mechanism used to compute the Encounter Probability, the smaller the number of false warnings communicated to drivers. Of course, the existence of false warnings cannot be avoided when the final destination of the vehicles is not known. For instance, a vehicle can change its direction and finally not encounter an event that it previously estimated relevant (in that case the Continuous Query Processor, described in Section 2.3, will notify the driver that the event is not relevant anymore). Anyway, it is important not to disturb the driver with irrelevant information and the ratio of false warnings should remain limited.

Figure 13 shows the ratio of false warnings for different (direction-dependent) events considered during our tests on the highway (see Fig. 8). Since the ratio of false warnings observed strongly depends on the trajectory of the vehicles (e.g., it depends on the percentage of vehicles exiting the highway just before meeting a certain event), it is particularly important to ensure that the same trajectories are considered for the vehicles in the evaluation of all the approaches. We observe in Fig. 13 that the ratio of false warnings is quite limited with the Map-based approach whatever the location of the event. The GV-based* approach, using very penalizing coefficients, outperforms the Map-based approach for several events; however, as it was shown in Fig. 10, the GV-based* approach may leave little time for the reaction of the driver. As expected, the GV-based approach increases the number of false warnings, and the increase depends on the road configuration. For example, for *event1* we observe a very high percentage of false warnings because a high number of vehicles arriving on the highway using the road located between *event2* and *event3* in Fig. 8 consider this event relevant but finally do not meet it.

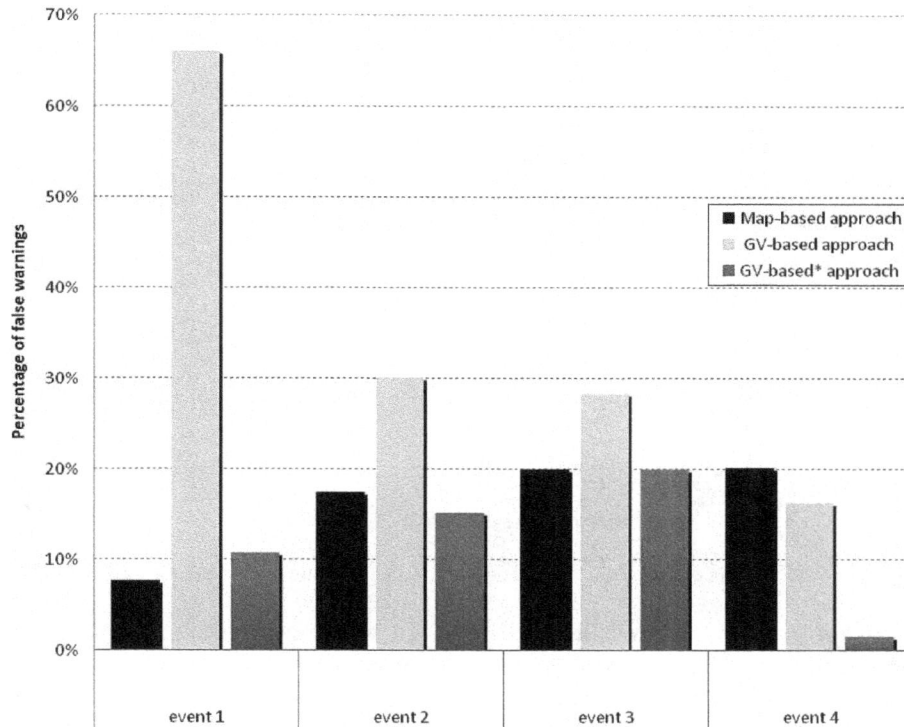

Fig. 13. Percentage of false warnings communicated to the drivers on the highway.

We also evaluated the percentage of false warnings generated for the different solutions in the urban environment. We omit the corresponding figure here because the general conclusions are similar to those observed in the previous case. Nevertheless, it should be emphasized that the ratio of false warnings in the urban scenario is always inevitably higher than in the highway configuration, since in an urban environment the vehicles keep changing their direction (due to the numerous turns and alternative routes available in urban roads) and often estimate an event as relevant at one point during their travel but it becomes irrelevant later. The percentage of false warnings about repulsion events shown to the driver is reduced with the Map-based approach, since it considers the Time to Escape to warn the driver only if s/he should react then to try to avoid the event.

5.4. Evaluation of the case of mobile events

Finally, we consider the evaluation of mobile events. Specifically, we present the reaction time available to drivers when receiving information about a mobile event corresponding to an emergency vehicle driving on the highway and asking the preceding vehicles to yield the right of way. The speed of the emergency vehicle is 163 Km/h. It moves over 9 Km, starting from the location of *event1* in Fig. 8, and generates a new version of the mobile event every two seconds to update its location. The TTL associated with each version is set to 25 seconds.

As shown in Fig. 14, similar results to those presented previously are observed. Thus, all the drivers reached by the emergency vehicle had been warned before. As in the previous cases, the vehicles are warned earlier with the Map-based approach. Anyway, with such mobile events, the GV-based approach

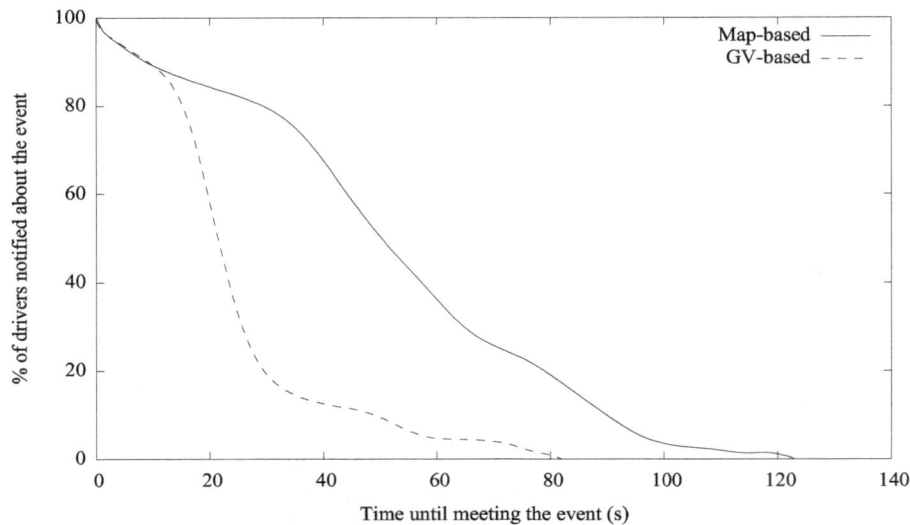

Fig. 14. Percentage of vehicles warned along time: mobile direction-dependent event on the highway.

can help to keep the diffusion area of the event limited, to avoid warning drivers too early (i.e, for vehicles still too far away from the mobile event), thus reducing the probability that the warning is unnecessary (e.g., the emergency vehicle may exit the highway before reaching the vehicle, as explained in Section 4.1).

6. Related work

Data dissemination in *Mobile Ad Hoc Networks* (*MANETs*) is an important research topic. In this area, several works have proposed broadcasting approaches to try to overcome the network congestion and other limitations of classical *flooding* (e.g., the *dynamic probabilistic flooding* approach presented in [20]). The special features of vehicular networks, which are highly dynamic MANETs, have attracted further research efforts. Thus, several previous works have addressed the problem of data management in vehicular networks [28], highlighting the importance of defining appropriate relevance measures for data exchange and query processing. As an example, we would like to highlight the following works:

– In the *Mobi-Dik* project (see, for example [39]), an *opportunistic exchange* mechanism, inspired by the field of epidemiology, is proposed for data sharing in vehicular networks. A vehicle with a certain piece of information acts as a disease carrier, and "contaminates" the nearby vehicles along its route. Once contaminated, these vehicles proceed to contaminate others. This dissemination principle is accompanied by mechanisms that monitor the relevance of the information (based on temporal and spatial criteria) in order to decide whether it should be stored in a local cache in the vehicle and/or broadcasted later on. Although this mechanism is well adapted for cars to share information about available parking spaces (which is the case study for Mobi-Dik), it has not been designed to deal with other types of events (e.g., to relay information about an accident or an emergency braking situation). Another work related to this approach is [27], which focuses on road hazards and proposes a data sharing strategy that is claimed to be similar to the opportunistic exchange proposed in Mobi-Dik. It is also interesting to mention the proposal in [19], which tackles the problem of content-based communication in a MANET based on the ideas of *opportunistic networking* (temporary contacts between mobile hosts are exploited to exchange documents) and *delay-tolerant networking* (a message can be stored temporarily in a host until forwarding it is possible).

- The importance of considering the relevance of data in a data management solution for vehicular networks is also emphasized in [9], which proposes the use of a *propagation function* to decide the route that a message has to follow in order to reach a target spatial area. The originator of a message defines an appropriate propagation function (e.g., by considering traffic conditions for the current time frame), which can be interpreted as a "gravitational field" where the message is attracted towards areas of minimum potential. The route traversed by the message is thus the result of evaluating the propagation function at each routing hop. On the basis of this propagation function, different dissemination approaches (both deterministic and probabilistic) are proposed and compared. However, how to define appropriate data propagation functions for different scenarios, which is a key element for the dissemination strategy, is not studied in such paper.
- Finally, the importance of considering the relevance of events (called the *expected benefit* in [2,17]) for data sharing in VANETs, especially when the bandwidth is scarce, is also emphasized in works such as [1,3]. For example, in [1], a relevance-based dissemination approach is proposed, although it is not detailed how the values of the different relevance parameters are set. In [5], the focus is on road accidents and a zone-of-relevance is also defined.

Thus, all these previous works agree on the importance of evaluating the spatio-temporal relevance of data. However, they mostly focus on the dissemination of events, whereas our approach considers a relevance evaluation mechanism for both data sharing among vehicles and data management on the vehicles (data storage, dissemination, and reporting of events to drivers). Moreover, our approach is a general proposal for all the types of events relevant on the roads. Finally, the previous works do not exploit the information available about road networks for the evaluation of the relevance of the data.

Although the importance of considering the underlying road network has been highlighted in several works related to moving objects and location-based services (e.g., [7,16,18,36]), as far as we know the road information has not been used so far for data sharing and data management in vehicular networks. There are only some works that use road map information but focus on multi-hop data delivery [22,38, 42]. Thus [22], presents a position-based routing approach that makes use of the navigational systems of vehicles to route a message following the shortest path to a target node. In [38], the *MDDV* approach is presented, based on the observation that taking the path with the shortest distance from the source is not necessarily the best approach, as road segments with a high traffic density lead to a faster propagation. Finally [42], proposes several Vehicle-Assisted Data Delivery (*VADD*) protocols, that benefit from the predicted vehicle mobility and digital road maps to improve the basic *carry-and-forward* solutions [10].

Summing up, although the previous works show the importance of the topic studied in this paper, none of them considers the use of road maps to estimate the relevance of events or uses this relevance to decide an appropriate dissemination strategy and a suitable driver alert mechanism for different types of events. On the contrary, we propose a complete and general system for data management and sharing in vehicular networks, that is able to benefit from the information stored in digital road maps when it is available.

7. Conclusions and future work

In this paper, we have presented a general and complete data management solution for intelligent vehicles, and we have studied and evaluated two different approaches to compute an *Encounter Probability* in order to estimate the relevance of events exchanged in a vehicular ad hoc network. The first approach is based on the management of geographic vectors whereas the second one exploits digital road maps. Whatever the mechanism used to compute it, the Encounter Probability is used to decide when to alert the driver and whether the event should be rediffused or not. Up to the authors' knowledge, this is the first work that proposes to benefit from information stored in digital road maps for data management and sharing in vehicular networks.

We have implemented a working prototype of the proposed system and tested it in a real environment. Moreover, we have evaluated our approach in a larger scale by using simulations. Our experimental evaluation proves that digital road maps can increase the accuracy of the system considerably, especially when dealing with repulsion events that should be diffused far away from their initial location. However, the other alternative also provides good results and we believe that both approaches are rather complementary. Indeed, digital road maps may be unavailable (e.g., they can be expensive to buy or accessible only through the Internet when a 3G connection is available), or some elements may be missing in existing digital road maps (e.g., roads in parking lots, entrances of parking lots, etc.). So, the approach based on geographic vectors could be the only one that could be used in some situations.

As future work, we will analyze whether using information about the planned trajectories could be useful, especially in the case of a mobile event (whose expected trajectory could be stored as an attribute of the event). Besides, our current research involves studying the applicability of mobile agent technology [26,33] in vehicular networks. Thus, for example, we have proposed encapsulating environment monitoring tasks in mobile agents that hop from car to car as needed [34,35]. We are also considering using mobile agents to implement a pull-based query processing approach that supports location-dependent queries [21,24] in vehicular networks. Finally, it could be interesting to study agent-based data dissemination approaches, as proposed in [25].

Acknowledgements

The present research work has been supported by the International Campus on Safety and Intermodality in Transportation, the Nord-Pas-de-Calais region, the European Union, the Ministry of Higher Education and Research and by the CICYT project TIN2007-68091-C02-02. The authors gratefully acknowledge the support of these institutions.

References

[1]   C. Adler, Information dissemination in vehicular ad hoc networks. Master's thesis, University of Munich (Germany), April 2006.

[2]   C. Adler, S. Eichler, T. Kosch, C. Schroth and M. Strassberger, Self-organized and context-adaptive information diffusion in vehicular ad hoc networks. In *Third International Symposium on Wireless Communication Systems (ISWCS'06)*, pages 307–311. IEEE Computer Society, September 2006.

[3]   C. Adler and M. Strassberger, Putting together the pieces – a comprehensive view on cooperative local danger warning. In *13th ITS World Congress and Exhibition on Intelligent Transport Systems and Services (ITS'06)*, October 2006.

[4]   K. Ahmadian, M.L. Gavrilova and D. Taniar, Multi-criteria optimization in GIS: Continuous K-nearest neighbor search in mobile navigation. In *International Conference on Computational Science and its Applications (ICCSA 2010)*, volume 6016 of *Lecture Notes in Computer Science (LNCS)*, pages 574–589. Springer, 2010.

[5]   L. Briesemeister, L. Schäfers and G. Hommel, Disseminating messages among highly mobile hosts based on inter-vehicle communication. In *Intelligent Vehicles Symposium (IV'00)*, pages 522–527. IEEE Computer Society, October 2000.

[6]   N. Cenerario, T. Delot and S. Ilarri, Dissemination of information in inter-vehicle ad hoc networks. In *IEEE Intelligent Vehicles Symposium (IV'08)*, pages 763–768. IEEE Computer Society, June 2008.

[7] A. Civilis, C.S. Jensen and S. Pakalnis, Techniques for efficient road-network-based tracking of moving objects. *IEEE Transactions on Knowledge and Data Engineering* **17**(5) (May 2005) 698–712.

[8] T.H. Cormen, C.E. Leiserson, R.L. Rivest and C. Stein, *Introduction to Algorithms*, The MIT Press, 2009. Third Edition.

[9] P. Costa, D. Frey, M. Migliavacca and L. Mottola, Towards lightweight information dissemination in inter-vehicular networks. In *Third International Workshop on Vehicular Ad Hoc Networks (VANET'06)*, pages 20–29. ACM Press, September 2006.

[10] J.A. Davis, A.H. Fagg and B.N. Levine, Wearable computers as packet transport mechanisms in highly-partitioned ad-hoc networks. In *Fifth IEEE International Symposium on Wearable Computers (ISWC'01)*, pages 141–148. IEEE Computer Society, October 2001.

[11] B. Defude, T. Delot, S. Ilarri, J.L. Zechinelli and N. Cenerario, Data aggregation in VANETs: The VESPA approach. In *MobiQuitous First International Workshop on Computational Transportation Science (IWCTS'08)*. ICST (Institute for Computer Sciences, Social-Informatics and Telecommunications Engineering), July 2008. 6 pages.

[12] T. Delot, N. Cenerario and S. Ilarri, Estimating the relevance of information in inter-vehicle ad hoc networks. In *MDM Workshop on Sensor Network Technologies for Information Explosion Era (SeNTIE'08)*, pages 151–158. IEEE Computer Society, April 2008.

[13] T. Delot, N. Cenerario and S. Ilarri, Vehicular event sharing with a mobile peer-to-peer architecture, *Transportation Research Part C: Emerging Technologies* **18**(4) (2010), 584–598.

[14] T. Delot, N. Cenerario, S. Ilarri and S. Lecomte, Cooperative parking space allocation in vehicular ad hoc networks, In *Sixth Annual International Conference on Mobile and Ubiquitous Systems: Networking & Services (MobiQuitous'09)*, pages 1–2. IEEE Computer Society, July 2009.

[15] T. Delot, N. Cenerario, S. Ilarri and S. Lecomte, A cooperative reservation protocol for parking spaces in vehicular ad hoc networks, In *Sixth International Conference on Mobile Technology, Applications and Systems (ACM Mobility Conference 2009)*. ACM Press, September 2009. 8 pages. Best paper award.

[16] Z. Ding and R.H. Güting, Managing moving objects on dynamic transportation networks. In *16th International Conference on Scientific and Statistical Database Management (SSDBM'04)*, pages 287–296. IEEE Computer Society, June 2004.

[17] S. Eichler, C. Schroth, T. Kosch and M. Strassberger, Strategies for context-adaptive message dissemination in vehicular ad hoc networks. In *Second MobiQuitous International Workshop on Vehicle-to-Vehicle Communications (V2VCOM'06)*. IEEE Computer Society, July 2006. 9 pages.

[18] H. Güting, V.T. Almeida and Z. Ding, Modeling and querying moving objects in networks, *The VLDB Journal* **15**(2) (June 2006), 165–190.

[19] J. Haillot and F. Guidec, A protocol for content-based communication in disconnected mobile ad hoc networks, *Mobile Information Systems* **6**(2) (2010), 123–154.

[20] A.M. Hanashi, I. Awan and M. Woodward, Performance evaluation with different mobility models for dynamic probabilistic flooding in MANETs, *Mobile Information Systems* **5**(1) (2009), 65–80.

[21] S. Ilarri, E. Mena and A. Illarramendi, Location-dependent query processing: Where we are and where we are heading, *ACM Computing Surveys* **42**(3) (March 2010), 12:1–12:73.

[22] C. Lochert, H. Hartenstein, J. Tian, H. Füßler, D. Hermann and M. Mauve, A routing strategy for vehicular ad hoc networks in city environments. In *Intelligent Vehicles Symposium (IV'03)*, pages 156–161. IEEE Computer Society, June 2003.

[23] J. Luo and J.P. Hubaux, *Embedded Security in Cars*, chapter "A Survey of Research in Inter-Vehicle Communications", pages 111–122. Springer, 2006.

[24] Z. Mammeri, F. Morvan, A. Hameurlain and N. Marsit, Location-dependent query processing under soft real-time constraints, *Mobile Information Systems* **5**(3) (2009), 205–232

[25] S.S. Manvi, M.S. Kakkasageri and J. Pitt, Multiagent based information dissemination in vehicular ad hoc networks, *Mobile Information Systems* **5**(4) (2009), 363–389.

[26] D. Milojicic, F. Douglis and R. Wheeler, *Mobility: processes, computers, and agents*, Addison-Wesley Professional, April 1999.

[27] S. Nittel, M. Duckham and L. Kulik, Information dissemination in mobile ad-hoc geosensor networks. In *Third International Conference on Geographic Information Science (GIScience'04)*, volume 3234 of *Lecture Notes in Computer Science (LNCS)*, pages 206–222. Springer, October 2004.

[28] S. Olariu and M.C. Weigle, editors. *Vehicular Networks: From Theory to Practice*. Chapman & Hall/CRC, 2009.

[29] B. Ostermaier, Analysis and improvement of inter vehicle communication security by simulation of attacks. Master's thesis, Technische Universität München, Institut für Informatik, Munich (Germany), 2005.

[30] V. Pham, E. Larsen, Ø. Kure and P. Engelstad, Routing of internal MANET traffic over external networks, *Mobile Information Systems* **5**(3) (2009), 291–311.

[31] M. Safar, K nearest neighbor search in navigation systems, *Mobile Information Systems* **1**(3) (2005), 207–224.

[32] M.L. Sichitiu and M. Kihl, Inter-vehicle communication systems: A survey, *IEEE Communications Surveys and Tutorials* **10**(2) (2008), 88–105.

[33] R. Trillo, S. Ilarri and E. Mena, Comparison and performance evaluation of mobile agent platforms. In *Third International Conference on Autonomic and Autonomous Systems (ICAS'07)*, pages 41–46. IEEE Computer Society, June 2007.

[34] O. Urra, S. Ilarri, T. Delot and E. Mena, Mobile agents in vehicular networks: Taking a first ride. In *Eight International Conference on Practical Applications of Agents and Multi-Agent Systems (PAAMS 2010)*, volume 70 of *Advances in Intelligent and Soft Computing*, pages 118–124. Springer, April 2010.

[35] O. Urra, S. Ilarri, E. Mena and T. Delot, Using hitchhiker mobile agents for environment monitoring. In *Seventh International Conference on Practical Applications of Agents and Multi-Agent Systems (PAAMS'09)*, volume 55 of *Advances in Intelligent and Soft Computing*, pages 557–566. Springer, March 2009.

[36] M. Vazirgiannis and O. Wolfson, A spatiotemporal model and language for moving objects on road networks. In *Seventh International Symposium on Advances in Spatial and Temporal Databases (SSTD'01)*, volume 2121 of *Lecture Notes in Computer Science (LNCS)*, pages 20–35. Springer, July 2001.

[37] T.L. Willke, P. Tientrakool and N. F. Maxemchuk, A survey of inter-vehicle communication protocols and their applications, *IEEE Communications Surveys and Tutorials* **11**(2) (2009), 3–20.

[38] H. Wu, R. Fujimoto, R. Guensler and M. Hunter, MDDV: A mobility-centric data dissemination algorithm for vehicular networks, In *MobiCom Workshop on Vehicular Ad Hoc Networks (VANET'04)*, pages 47–56. ACM Press, October 2004.

[39] B. Xu, A.M. Ouksel and O. Wolfson, Opportunistic resource exchange in inter-vehicle ad-hoc networks. In *Fifth International Conference on Mobile Data Management (MDM'04)*, pages 4–12. IEEE Computer Society, January 2004.

[40] K. Xuan, G. Zhao, D. Taniar and B. Srinivasan, Continuous range search query processing in mobile navigation. In *14th International Conference on Parallel and Distributed Systems (ICPADS'08)*, pages 361–368. IEEE Computer Society, 2008.

[41] N. Zhang, Shortest path queries in very large spatial databases. Master's thesis, University of Waterloo (Canada), 2001.

[42] J. Zhao and G. Cao, VADD: Vehicle-assisted data delivery in vehicular ad hoc networks. In *25th IEEE International Conference on Computer Communications (INFOCOM'06)*. IEEE Computer Society, April 2006. 12 pages.

Thierry Delot (http://www.univ-valenciennes.fr/ROI/SID/tdelot/) is an associate professor at the University of Valenciennes since 2002. He is a member of the LAMIH laboratory (FRE CNRS 3304). He got a PhD in Computer Science at the university of Versailles in 2001. His research interests mainly concern mobile data management and query processing. Since 2007, Thierry is particularly interested in vehicular ad hoc networks.

Sergio Ilarri received his B.S. and his PhD in Computer Science from the University of Zaragoza in 2001 and 2006, respectively. Now, he is an Assistant Professor in the Department of Computer Science and Systems Engineering. For a year he was a visiting researcher in the Mobile Computing Laboratory at the Department of Computer Science at the University of Illinois in Chicago, and he has also cooperated (through several research stays) with the University of Valenciennes and with IRIT in Toulouse. His research interests include data management issues for mobile computing, vehicular networks, mobile agents, and semantic web.

Nicolas Cenerario earned his Research Master and Ph.D. degrees in Computer Science at the University of Valenciennes, France, in 2006 and 2010, respectively. His research interests concern information sharing in inter-vehicular communication networks. In this context, he is particularly interested in information dissemination, continuous query processing and resource allocation.

Thomas Hien is the holder of a Master's degree in Computer Science from the University of Valenciennes, France, obtained in 2009. Since then, he is a member of the LAMIH laboratory (FRE CNRS 3304) devoted to the VESPA project as a research engineer. His work mainly concerns the evaluation of inter-vehicle communication systems in realistic environments.

9

Context-aware mobile service adaptation via a Co-evolution eXtended Classifier System in mobile network environments

Shangguang Wang[a,*], Zibin Zheng[b], Zhengping Wu[c], Qibo Sun[a], Hua Zou[a] and Fangchun Yang[a]

[a]*State Key Laboratory of Networking and Switching Technology, Beijing University of Posts and Telecommunications, Beijing, China*
[b]*Shenzhen Research Institute, The Chinese University of Hong Kong, Hong Kong, China*
[c]*Department of Computer Science and Engineering, University of Bridgeport, Bridgeport, CT, USA*

Abstract. With the popularity of mobile services, an effective context-aware mobile service adaptation is becoming more and more important for operators. In this paper, we propose a Co-evolution eXtended Classifier System (CXCS) to perform context-aware mobile service adaptation. Our key idea is to learn user context, match adaptation rule, and provide the best suitable mobile services for users. Different from previous adaptation schemes, our proposed CXCS can produce a new user's initial classifier population to quicken its converging speed. Moreover, it can make the current user to predict which service should be selected, corresponding to an uncovered context. We compare CXCS based on a common mobile service adaptation scenario with other five adaptation schemes. The results show the adaptation accuracy of CXCS is higher than 70% on average, and outperforms other schemes.

Keywords: Mobile service, context-aware, service adaptation, learning classifier system, eXtended Classifier System, Co-evolution

1. Introduction

With the development of mobile internet, mobile terminals[1] are networked via heterogeneous access technologies.[2] Obviously, these mobile terminals have entered almost every part of life. For instance, in UK, Mobile Squared[3] forecasts, by 2015, smartphones will number 63.83 million- approximately 100% population penetration. Despite the overall rise in smartphone penetration and data usage, however, text messaging still dominates mobile operator non-voice revenues worldwide.

Unfortunately, from the survey of Mobile Squared, a total of three-quarters of all operators surveyed thought that Instant Messaging (IM) tools[4] on smartphones do pose a threat to traditional operator-based services around SMS and voice. Specifically, 63% of respondents agreed with the statement,

---

*Corresponding author: Shangguang Wang, State Key Laboratory of Networking and Switching Technology, Beijing University of Posts and Telecommunications, Box 187, 10 No., Road Xitucheng, Beijing, China. E-mail: sguang.wang@gmail.com.

[1]Such as smartphones, PDAs, table PCs and so on.
[2]Such as Internet, WiFi, WiMAX, 3G and 4G.
[3]www.mavenir.com/RCS-survey-release.htm.
[4]Such as Skype, iMessage, Google Talk, Facebook Messages, WhatsApp.

while a further 17% strongly agreed. Hence, how to cope with more and more IM tools from Internet company, is becoming more and more urgent for the major operators groups.[5] We find if Service Delivery Platform (SDP) can adapt to end users' context and adaptively provide suitable mobile services, may be a good solution. The viewpoint is similar to context-sensitive SDP.[6] A context-sensitive SDP provides a user interface (UI) that exhibits some capability to be aware of the context and to react to changes of this context in a continuous way. As a result such a UI will be adapted to a person's devices, tasks, preferences, and abilities, thus improving people's satisfaction and performance compared to traditional SDP based on manually designed UIs.

The context-sensitive SDP is complex and enormous. We can not implement its whole function. In this paper, we only focus on mobile service[7] adaptation according to different end user (device) context such as Location,[8] Activity[9] Illumination[10] and Acceleration[11] by using Event-Condition-Action rule [1] of database system and machine learning theory. In our proposed adaptation scheme, we map each mobile service into an action and map each (end) user context (device) into a condition. The condition is a logical test that, if satisfied or evaluates to true, causes the action to be carried out. The action consists of updates or invocations on the local data or condition. Our scheme can perform the best action (mobile service) according to the given condition (user context) by using a Co-evolution eXtended Classifier System. In the paper, our proposed adaptation system is deployed in an Adaptation Server of SDP of mobile network.

Different traditional adaptation system, in mobile service adaptation process, different users may have different preferred adaptive actions [2]. For example, if the available bandwidth is narrowing, John may want to turn a video call into an audio call, while Sam may be less sensitive to the video distortion and maintain the video call. Hence, some previous schemes such as utility function-based method [3], work-flows scheme [4] and colored Petri-net scheme [5] cannot be used in SDP. The main reason is that these schemes rely on the off-line analysis of conditions, they can not adaptively provide suitable action (mobile service) according to different condition (user context). Hence, SDP need an effective context-aware adaptation to enhance the quality of the experience of mobile services and improve communication service adhesiveness. Fortunately, some research findings such as machine learning theory in the filed of artificial intelligence, may be able to support the adaptation problem. Then, some schemes from machine learning theory have been proposed to support the adaptation, such as the Granular Computing [6,7], Rough Set [8,9] and Decision Tree models [10,11]. These schemes can give adaptation rules automatically by monitoring user interaction, but we are disappointed to find that they cannot support context-aware online preference learning. Obviously, they cannot be immediately used as the mobile service adaption schemes in SDP. Fortunately, an Event-Condition-Action rule-based system with machine learning theory can perform the mobile service adaptation. For instance, The Learning Classifier System (LCS) can manipulate the rule to decide which mobile service (action) will take place in a given context (condition) [12].

The (LCS) [13], as an adaptive rule-based system, can build a rule set automatically. It is also a machine learning technique combining evolutionary computing and reinforcement learning. Wilson's

---

[5] Such as AI&T, Verizon, Deutsche Telekom, Mepro PCS, China Mobile and so on.
[6] http://www.serenoa-fp7.eu/.
[7] Such as SMS, MMS, Voice Mail, Voice Call and Video Call.
[8] Home, Road, Car, Company/Meeting.
[9] Walking, Eating, Running, Sleeping.
[10] Very Weak (AW), Weak (W), Middle (M), Strong (S) and Very Strong (VS).
[11] AW, W, M, S, VS.

eXtended Classifier System (XCS) [14] is a type of LCS, in which a classifier's fitness is determined by the measure of its prediction's accuracy. XCS, the most successful LCS, solved a number of well-known problems, such as classification, planning tasks, function approximation and general prediction [15]. However, XCS can not been applied to perform mobile service adaptation. The reasons are as follows. First, XCS needs a long time to train itself to a reasonable accuracy because of the deficiency of user interactions as new-users enter a mobile service adaptation system. The slow converging speed fail in adapting users' mobile service requirements quickly according to user context. Second, the reinforcement credit assignment mechanism of XCS requires many training records to cover the total learning space. However, when the training records are incomplete, XCS cannot gvie any classifiers to match the current context. The uncovered-context leads to inaccurate mobile service adaptation.

To address above weaknesses, we propose a novel LCS based on co-evolution XCS (called CXCS) to perform context-aware mobile service adaptation. In this paper, the user context can be mapped to the classifier conditions, and mobile services can be mapped to classifier actions. Then the generation of the adaptation rule will be transformed into the learning of the classifier population. Once the learning process is completed, it can provide each user with mobile service adaptation through the matching and competition of classifiers. The contributions of this paper are as follows:

1. We survey why some traditional adaptation scheme cannot be used in mobile service adaption. We also point out an Event-Condition-Action rule-based system with machine learning theory can be employed to perform context-aware mobile service adaption for SDP. Then we proposed a context-aware mobile service adaptation scheme via a Co-evolution eXtended Classifier System.

2. To lessen the computation time of the mobile service adaptation, we introduce other users' evolutionary information to "inject new blood" into the classifier population and propose a new-user initialization algorithm to quicken the convergence speed.

3. To improve the accuracy of the mobile service adaptation, we employ an uncovered-context algorithm and an action prediction algorithm to predict which mobile service should be selected when there is a failure in covering the user context.

4. We conducted an experiment based on a mobile service adaptation dataset. The user context contains Location, Activity, Illumination and Acceleration. The mobile services contain Mail, SMS, MMS, video and Audio. The experimental results show the adaptation accuracy of CXCS is better than other five mobile service adaptation schemes.

The rest of this paper is organized as follows. Section 1 reviews XCS. Section 2 describes the proposed CXCS, which contains new-user initialization and uncovered-context prediction. Section 3 describes the experiments conducted to evaluate our proposed scheme. Section 4 introduces related work, and Section 5 ends the paper.

2. XCS review

To sketch our proposed CXCS, we first review XCS. Figure 1 shows the framework of XCS. XCS is a rule-based system, in which each rule has a condition and a set of parameters: prediction, fitness, accuracy, experience and so on. The complete set of rules forms the Classifier Population. Based on the two interface modules, a Detector (Sensor) and an Effector, the procedure of XCS contains three phases, as follows:

- First, the detector receives an input from the environment. The rules whose conditions match the input data form a Match Set. In the Match Set, each rule has an action. An action selection mechanism is used to select an action where the same action forms the Action Set.

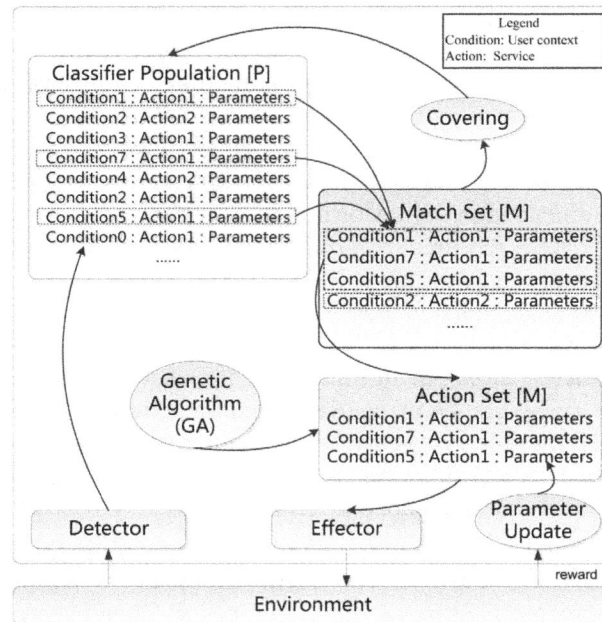

Fig. 1. XCS framework.

- Second, the Effector outputs the action to the environment, and a reward is received. Reinforcement learning is applied to the Action Set to update the parameters contained in the rules according to the rewards.
- Finally, the processes are included in a learning cycle. A genetic algorithm (GA) runs on the Action Set to guide the search for better rules, so XCS learns repeatedly to evolve the rule population. The learned rules are the solution to a given problem (context-aware mobile service adaptation).

For further illustration, based on Algorithm 1, we will introduce the running process of XCS by the following four components: finite classifier population, performance component, reinforcement component and discovery component, in the following sections.

Algorithm 1 XCS

Initiate the maximum iteration times N;
while $i + + < N$ **do**
    Get the environmental state from the Detector;
    All the classifiers of [P] that match the current state are selected to form the Match Set [M];
    Classify the classifier of [M] into several subgroups according to the classifier action, and prepare the prediction array;
    $xp = random[0,1]$;
    if $xp < P_{\exp}$ **then**
        Do the exploration scheme, choose a random action A;
        Update all the classifiers of [A], and apply GA on [A] with probability θ_{GA};
    else
        Do the exploitation scheme;
        Choose the action [A] with the largest fitness weighted prediction;
    end if
    Send [A] to the Effector to execute, and receive the reward from the environment;
end while

Table 1
Classifier parameters

Param	Usages
P	The payoff XCS receives after the execution of the chosen action.
ε	The prediction error of the classifier.
k	The accuracy of the classifier.
F	The fitness of the classifier which is the inverse of k.
exp	The matching time of the classifier.
num	The copy numbers of the classifier.
as	The average size of the Action Set.

2.1. Finite classifier population

The finite classifier population (Classifier Population [P]) is a finite classifier set. It represents the current knowledge. Each classifier is composed of three parts: the condition, the action and the classifier parameters.

A classifier is activated only if its condition matches the environment state. Once a classifier is activated, its action will be selected to execute. The classifier condition uses a ternary encoding scheme (i.e., it is represented by a string of characters using a ternary alphabet {0, 1, #}, where # is a "don't care" symbol), which means that the condition will match the state whether this position is 0 or 1. As shown in Table 1, there are 7 types of classifier parameters, which are P, ε, k, F, exp, num, as.

2.2. Performance component

All the classifiers of XCS that match the current context state will be selected to form the Match Set [M], and then one action will be selected as the final output of XCS. All the classifiers will be classified into subgroups according to their actions.

During the training phase, the ε-greedy action selection will be used. There will be two action selection schemes, i.e., the exploration and the exploitation, where their execution probabilities are $P_{\exp}$ and $1 - P_{\exp}$, respectively. In the exploration scheme, a random action will be selected; in the exploitation scheme, the action whose subgroup has the highest fitness weighted prediction will be selected. Then, the subgroup with the chosen action will be selected as the Action Set [A], and its action will be sent to the Effector to carry out.

2.3. Reinforcement component

The reinforcement component is also called the credit assignment component. It takes charge of the assignment of the environmental reward. Once the Effector has executed the selected action, XCS will receive the reward from the environment. This reward will be used to update all the classifiers of [A]. More information about the following parameter can be found in Table 1.

First, the experience time of the classifier, exp, will be increased as follows:

$$exp = exp + 1. \tag{1}$$

Then the prediction p and prediction error ε of the classifier will be calculated as follows:

$$p = \begin{cases} p + (P - p)/\exp, & \exp < /\beta \\ p + \beta \cdot (P - p), & otherwise \end{cases} \tag{2}$$

$$\varepsilon = \begin{cases} \varepsilon + (|P - p| - \varepsilon)/\exp, \exp < 1/\beta \\ \varepsilon + \beta \cdot (|P - p| - \varepsilon), \quad otherwise \end{cases} \tag{3}$$

where $\beta(0 < \beta \leqslant 1)$ denotes the learning rate constant, which is used to control the updating speed.

Afterwards, the accuracy and relative accuracy of each classifier will updated for the calculation of its fitness value as follows:

$$k = \begin{cases} 1, & \varepsilon < \varepsilon_0 \\ \alpha \cdot (\varepsilon/\varepsilon_0)^{-u}, & otherwise \end{cases} \tag{4}$$

where $\varepsilon_0(\varepsilon_0 > 0)$ is used to determine the threshold error when a classifier is considered to be accurate, $\alpha(0 < \alpha < 1)$ and $v(v > 0)$ are used to control the degree of the decline in accuracy, respectively.

Finally, once the classifier accuracy is updated, its fitness valued, F can be computed by the following:

$$F = F + \beta \cdot (k' - F) \tag{5}$$

with

$$k' = k \times n \bigg/ \sum_{i=1}^{|[A]|} k_i \times n_i.$$

2.4. Discovery component

The discovery component of XCS includes the genetic algorithm (GA) based on the evolution of the classifier population, the cover mechanism and the deleting mechanism.

The GA process will only occur in the exploration scheme with a possibility of θ_{GA}. Once the GA is invoked, XCS will choose two classifiers from [A] as parents with the probability proportional to the fitness, and make a copy of them. These two cloned classifiers will mutate or crossover with a separate probability of p_x and p_m. The newly generated child or produced offspring will be injected back into [P] with the probability p_{sel} by the following:

$$p_{sel} = F_i \bigg/ \sum_{j=1}^{|[A]|} F_j. \tag{6}$$

Whenever there are no classifiers in [P] to match the current state, the covering mechanism will be executed. The classifier condition will be created to match the current state, but with a probability P# of adding the # character. The classifier action will be randomly selected, and its fitness and prediction value will be initiated.

During the GA or covering process, the population will be expanded. However, if the size of [P] exceeds the population limit, the deleting mechanism will be activated. The classifiers with a low fitness will be deleted to keep the population size with a constant upper bound, and the deleting probability p_{del} will be calculated as follows:

$$p_{del} = (F_i/num_i) \bigg/ \sum_{j=1}^{|[P]|} (F_j/num_j). \tag{7}$$

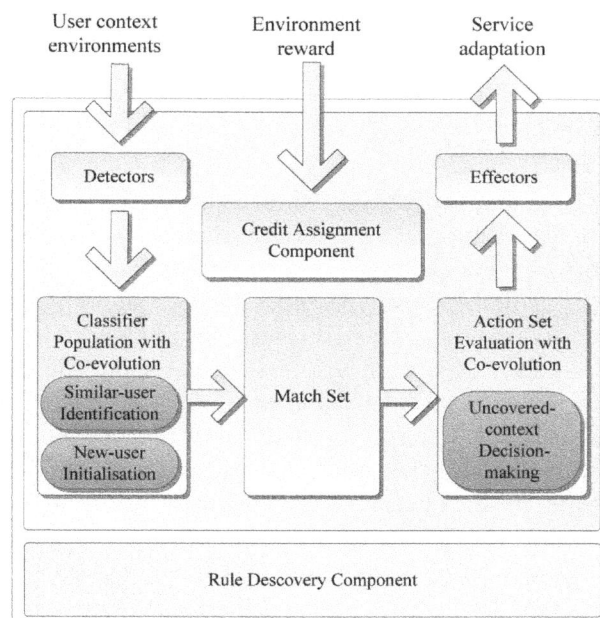

Fig. 2. Procedure of CXCS for context-aware mobile service adaptation.

3. Proposed CXCS for service adaptation

Because the high cost and low accuracy of XCS obstruct its application in mobile service adaptation systems, we propose CXCS to cope with these problems. The idea of CXCS is to add other users' evolutionary information into the current classifier population by introducing the co-evolution mechanism [16] into XCS and to utilize similar-users to help predict which mobile service is invoked for the current user.

As shown in Fig. 2, the proposed CXCS is a rule-based system, which contains three main components for context-aware mobile service adaptation, i.e., similar-user identification, new-user initialization and uncovered-context prediction. To illustrate, we first describe the concept of the co-evolution mechanism.

3.1. Co-evolution mechanism

The term co-evolution is derived from biology. It means "change in a trait of individuals of one population in response to a trait of individuals of a second population, followed by an evolutionary response of the second population to a change in the first". For example, the common influence of herbivore and plant will produce stronger and better generations of both over time.

There are two main classes of co-evolutionary approaches, competitive co-evolution and cooperative co-evolution. Just as the names imply, in competitive co-evolution, populations evolve simultaneously through their competition with each other, while cooperative co-evolution is implemented by the cooperation of the populations.

For competitive co-evolution, two main subclasses can be classified: competition and amensalism. The species of the former type inhibit each other, which means that the success of one species will result in the failure of the other species. In the latter type, the inhibition of species is single direction. There are three identified subclasses of cooperative co-evolution. In mutualism, every species benefits from the improvements of the other species. In commensalism, only one of the species benefits, while the other is not affected. In parasitism, only one of the species benefits, and the other is harmed.

3.2. Similar-user identification

First, the cooperative co-evolution needs to identify all the similar-users. The reward for the adaptive action under a specified context state indicates the user's preference. Hence, by calculating the deviation of the users' preference shown for every contextual adaptive action, the similarity between users can be measured. The smaller the preference deviation is, the more similar the users are. The similarity between users includes two parts, the static part and the dynamic part. The measurement of the static similarity uses the static user profile information, such as gender, age, and occupation. The calculation of the dynamic similarity can use all the adaptation records for the user. Hence, the identification of similar-users can be divided into two types of cases. The first type is when a user uses the mobile service for the first time, and there are no training samples at all. This similarity measurement can only use the users' static user profile information. The greater the difference between the users' profiles is, the smaller their similarity. a user u_i's user profile can be defined as $UP_i = (Item_1, \ldots, Item_I)(1 \leqslant i \leqslant I)$. Another user u_j's user profile can be defined as $UP_j = (Item_1, \ldots, Item_I)(1 \leqslant j \leqslant I)$, and the similarity value, σ between u_i and u_j can be obtained as follows:

$$\sigma(u_i, u_j) = \sigma_{static}(u_i, u_j) = \sum_{k=1}^{I} \sqrt{\frac{\lambda_k^2}{I}}(1 \leqslant k \leqslant I) \tag{8}$$

with

$$\lambda_k = \begin{cases} 1, \textit{if} \quad u_i.Item_k == u_j.Item_k. \\ 0, \textit{else} \end{cases}$$

where $\sigma_{static}(u_i, u_j)$ denotes the static similarity value.

The second type of circumstance is when a user has used the service for some time, in which case the similarity value calculation should use both the static user profile and the experience records, as follows:

$$\sigma(u_i, u_j) = R(t) \cdot \sigma_{static}(u_i, u_j) + (1 - R(t)) \cdot \sigma_{dynamic}(u_i, u_j) \tag{9}$$

with

$$R(t) = e^{-t},$$

where t denotes a time sample.

where $\sigma_{dynamic}(u_i, u_j)$ denotes the dynamic similarity value and it can be obtained according to the training rules of CXCS as show Fig. 3, $R(t)$ is added for the weight adjustment over time. The longer a user has used this service, the more familiar the user will be with the service, and the greater the weight of the dynamic part.

As shown in Fig. 3, every training rule is composed of context state C, adaptive action A_j and reward R, which represents that if, under a given context state, CXCS selected an adaptive action, it will receive a reward from the user. It can be defined as $S = \{(C, A_j, R)|1 \leqslant j \leqslant Q\}$ where Q denotes the number of adaptive action. The reward indicates the user's preference degree for this rule. Every context state C has n context attributes, $C = (C_1, \ldots, C_n)$. Each context attribute C_i can choose one value from

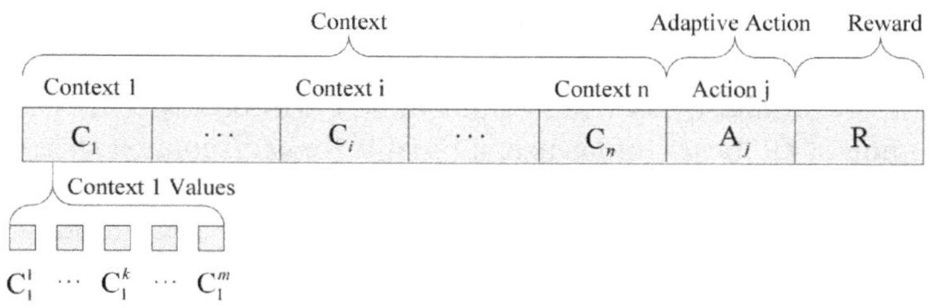

Fig. 3. Training rule.

$\{C_i^1, C_i^2, \ldots, C_i^{m-1}, C_i^m\}$ where m denotes the number of training sample. Then the user's preference for A_k under C_i^j can be obtained as follows:

$$P(C_i^j, A_k) = \frac{1}{B} \sum_{m=1}^{M} f(C_i^j, A_k, S_m) R_m - \frac{1}{M} \sum_{m=1}^{M} R_m \tag{10}$$

with

$$B = \sum_{m=1}^{M} f(C_i^j, A_k, S_m).$$

where M is the training set size, Q is the number of adaptive actions. If in training sample S_m, the value of C_i is C_i^j and the adaptive action is A_k, the value of $f(C_i^j, A_k, S_m)$ will be 1; otherwise, the value of $f(C_i^j, A_k, S_m)$ should be 0.

Once $P(C_i^j, A_k)$ is determined, the user's preference for the contextual adaptive action can be computed as follows:

$$P = \begin{pmatrix} P_1 \\ \ldots \\ p_Q \end{pmatrix} = \begin{pmatrix} P(C_1^1, A_1) \ldots P(C_1^{|C_1|}, A_1) \ldots P(C_n^1, A_1) \ldots P(C_n^{|C_n|}, A_1) \\ \ldots \quad \ldots \quad \ldots \quad \ldots \quad \ldots \quad \ldots \quad \ldots \\ P(C_1^1, A_Q) \ldots P(C_1^{|C_1|}, A_Q) \ldots P(C_n^1, A_Q) \ldots P(C_n^{|C_n|}, A_Q) \end{pmatrix}$$

Then, the dynamic user similarity can be obtained as follows:

$$\sigma_{dynamic}(u_1, u_2) = \sum_{i=1}^{Q} \sigma_{dynamic}(P_i^{u_1}, P_i^{u_2}) =$$

$$\sum_{i=1}^{Q} \sum_{j=1}^{n} \sqrt{\frac{1}{|C_j|} \sum_{k=1}^{|C_j|} \left(P_{u_1}\left(C_j^k, A_i\right) - P_{u_2}\left(C_j^k, A_i\right) \right)^2}. \tag{11}$$

3.3. New-user initialization

For the new-user, the administrators or developers can initialize some preference rules. These rules can help users configure their mobile service settings or parameters, however, it is difficult to perform mobile service adaptation. The reason is that the adaptation action preferences of each user are different. To solve this problem, we use some similar-users to initialize their preference rules and propose a new-user initialization algorithm as shown in Algorithm 2. Because there are no training samples at all, the users' similarity measurements only use their static user profiles. Once the similar-users are identified, all these users will cooperate to initialize the new-user.

As shown in Algorithm 2, first, the similar-users within a threshold ς will be selected. Then, the algorithm obtains the context state union of all these similar-users. Finally, for each context state obtained, all these similar-users will cooperate to produce a new classifier and inject it into the new-user's classifier population. Each similar-user who has experienced the current state will vote for the predicted action. When a ratio of λ users have agreed on the maximum vote action, a new classifier with this voted action will be inserted into the current user's classifier population. Then the prediction value of the action will be set as the average of all the similar-users with the correct voted action.

Algorithm 2 New-user initialization

Input: {new-user u_c, candidate actions A, similar-user ratio threshold λ, similarity degree threshold ς}
similarList = Null;
for u_e in U **do**
    if calculate Similarity $(u_c, u_e) < \varsigma$ **then**
        similarList.insert(u_e);
    end if
end for
S = Null;
for u_e in similarList **do**
    for t in u_e.*trainingSamples* **do**
        if notS.hasState(t.state) **then**
            S.insert(t.state);
        end if
    end for
end for
for s in S **do**
    actionVoteList = [0 0];
    predictionList = [0.0 0.0];
    for u_e in similarList **do**
        if u_e.XCS.coverState(s) **then**
            action = u_e.XCS.getActionByState(s);
            prediction = u_e.XCS.getPredictionByState(s);
            actionVoteList[action] + = 1;
            predictionList[action] + = prediction;
        end if
    end for
    maxNum = max(actionVoteList);
    if maxNum>int(len(similarList) $\times \lambda$ **then**
        for act In A **do**
            if actionVoteList[act] == maxNum **then**
                cl = new Classifier();
                cl.action = act;
                cl.prediction = predictionList [act]/ float(actionVoteList[act]);
                u_c.XCS.insertClassifier(cl);
            end if
        end for
    end if
end for

3.4. Uncovered-context prediction

When introducing the reinforcement credit assignment mechanism, the training process of XCS always takes a long time and often needs a large number of training samples to cover the entire problem space. Because some context states have been experienced by other similar-users, it is highly possible that similar-users may make the same decisions. When these new decisions can be "borrowed" from similar-users, the learning speed might be greatly speeded up and the uncovered prediction accuracy might be increased. Hence, we propose an uncovered-context prediction algorithm for the CXCS, as shown in Algorithm 3.

By using Algorithm 3, when an action needs to be covered, but the covered classifier cannot be generated by the original mechanism, our CXCS produce the covered classifier by extracting the relevant information from all similar-users.

Algorithm 3 Uncover-context prediction algorithm

Input: {uncovered-context state s, similarity threshold ς}
Output: {generated cover classifier by similar-users}
actionVoteList = [0 0];
predictionList = [0.0 0.0];
for u_e in U **do**
    if calculateSimilarity$(u_c, u_e) < \varsigma$ **then**
        if u_e.XCS.coverState(s) **then**
            action = u_e.XCS.getActionByState(s);
            prediction = u_e.XCS.getPredictionByState(s);
            actionVoteList[action] + = 1;
            predictionList[action] + = prediction;
        end if
    end if
end for
for act In A **do**
    if actionVoteList[act] > 0 **then**
        cl = new Classifier();
        cl.action = act;
        cl.prediction = predictionList [act]/float(actionVoteList[act]);
        u_c.XCS.insertClassifier(cl);
    end if
end for

In the testing phase (or working phase) of CXCS, when an uncovered-context state is encountered, the final predicted action is voted on by all the similar-users through Algorithm 3. However, because the new classifiers generated by all possible decisions of all users flood the current user populations, the classifier fails to update. To overcome this problem, we propose an action prediction algorithm. As shown in Algorithm 4, we only take the action with the maximum number of votes as the output of the current user. If the maximum number of votes for multiple actions is the same, an action should be selected randomly as the input.

4. Experiments

In this section, we look into the performance of CXCS and compare it with other adaptation schemes base on a common mobile (communication) service adaptation scenario.[12] In experiments, we do not fulfil the whole scenario. We only focus on how to provide the following five mobile services adaptively, i.e., Mail, SMS, MMS, Audio and Video according the following four context, i.e., Location, Activity, Illumination and Acceleration.

---

[12]http://cordis.europa.eu/fp7/home_en.html.

Algorithm 4 Action prediction algorithm

Input: {uncovered-context state s, similarity threshold ς}
Output: {generated cover classifier by similar-users}
similarList = Null;
for u_e in U **do**
    if calculateSimilarity$(u_c, u_e) < \varsigma$ **then**
        if u_e.XCS.coverState(s) **then**
            action = u_e.XCS.getActionByState(s);
            prediction = u_e.XCS.getPredictionByState(s);
            actionVoteList[action] $+ = 1$;
            predictionList[action] $+ = $ prediction;
        end if
    end if
end for
maxNum=max(actionVoteList)
actionList=Null;
for act In A **do**
    if actionVoteList[act]=maxNum **then**
        actionList.insert(act);
    end if
end for
Return random (actionList)

4.1. Experiment setup

As shown in Fig. 4, a scenario of mobile service adaptation is used to evaluate our proposed CXCS. The scenario revolves around a user called Julie. Generally, Julie gets up at 7 in the morning. One day, her boss (John) called her at 6. Because Julie set the adaptation system on the adaptation server before she went to sleep, the mobile service is automatically switched to Voice Mail. As soon as she gets up, Julie finds a unread Voice Mail informing her that she must go into town to the company for an unusual meeting at 9 in the morning. While driving, Julie receives a Video Call from her mother. Because she is driving at a high speed, the mobile service automatically changes to an Audio Call. After 15 minutes, she reaches the company and immediately attends the meeting. During the meeting, she receives a Video Call from her daughter. Because Julie is at a company meeting, the mobile service is changed to SMS.

Based on this scenario, we employ a mobile service adaptation system to validate the performance of CXCS. The mobile service uses four types of context attributes, *location, activity, illumination, and acceleration*, to identify the user's preferred mobile service. For example, the parameters "Illumination" and "Acceleration", which belong to [1,100], have 5 linguistic variables, Very Weak (AW), Weak (W), Middle (M), Strong (S) and Very Strong (VS). The adaptive action is the mobile service. The detailed setup is shown in Table 2. The membership function of the fuzzy linguistic variables of illumination and acceleration is the same, as depicted in Fig. 5.

In the experiments, we invite 10 users to collect our datasets.[13] First, they are required to provide some adaptation rules for the mobile service (MS), as shown in Table 3. Second, for each user, the adaptation rules are applied to a randomly generated set of the records using fuzzy inference. Finally, these datasets are encoded into binary format as the training samples. Some examples of these datasets are shown in Table 4. Note that the illumination and acceleration are both divided into 16 discrete values so that they can be easily encoded in the algorithm. For each user, the encoded datasets are split into two parts, i.e., the training set and the test set.

---

[13] http://sguangwang.com/Source%20code/XCSNew_Coevolution.zip.

Table 2
Experiment configuration

Parameter	Value
Location	{Home, Road, Company, Car}$\pi < P_{\exp}$
Activity	{Walking, Running, Eating, Sleeping}
Illumination	{VW, W, M, S, VS}
Acceleration	{VW, W, M, S, VS}
Communication service	{Mail, SMS, MMS, Audio, Video}

Table 3
Adaptation rule

Location	Activity	Illumination	Acceleration	Communication mode
Home	Sleeping	♮	♮	Mail
♮	♮	♮	♮	SMS
♮	Walking	Strong/Very Strong	♮	Audio
...	...	...	...	...

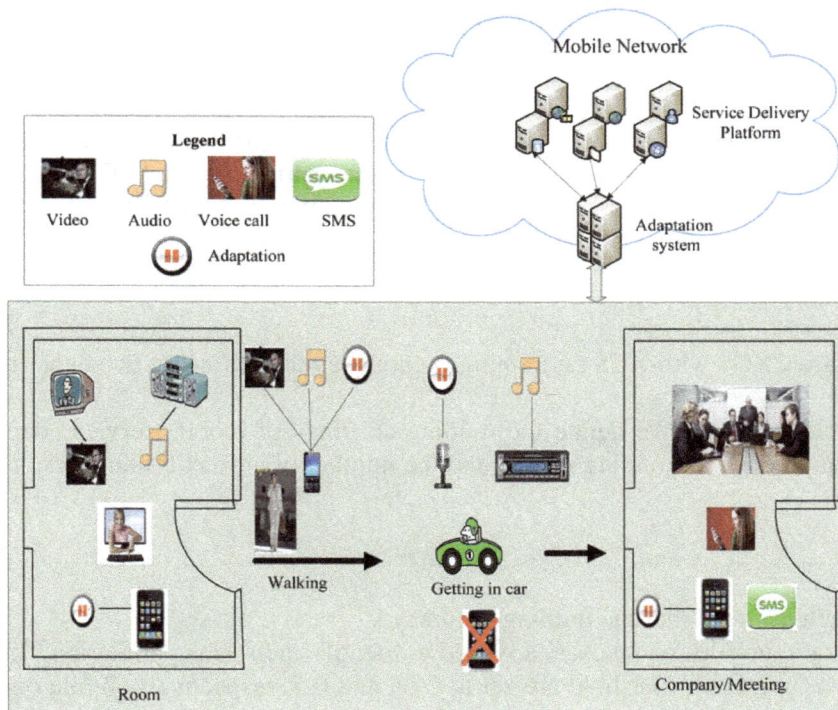

Fig. 4. Service adaptation scenario.

In experiments, all parameter are set prudently. The aim is to find a good tradeoff among all parameters, obtain the best performance of all schemes, and evaluate them objectively. Based on lots of experimental results, the XCS and CXCS configuration parameters are set with a tradeoff as follows: *iterationNum* = 10000, *popNum* = 5000, $\alpha = 0.1$, $\beta = 0.2$, $\varepsilon_0 = 0.25$, $p_0 = 10.0$, fitness$_0 = 0.01$, $\theta_{GA} = 25$, $p_x = 0.8$, $p_m = 0.04$, $\theta_{sub} = \theta_{exp} = \theta_{del} = 20$. The payoff is set to 1000 if the correct action is taken; otherwise, it is 0.

Table 4
Experimental dataset

No.	Location	Activity	Illumination	Acceleration	Communication mode
1	00	01	1001	0001	0
2	01	11	0000	0101	3
3	11	01	1111	1011	2
...	...	...	...	...	...

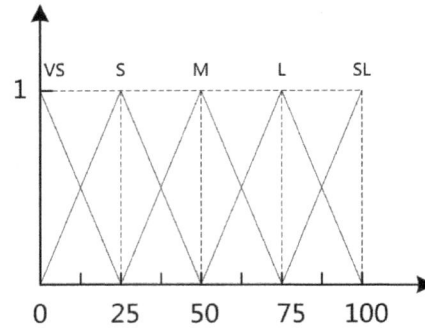

Fig. 5. The membership function.

All experiments are conducted on three PCs with an Intel Core2 2.8 GHz processor, 2.0 GB of RAM, Windows XP SP3, and Python 3.2.2. Two PCs is used to simulate smartphone users. One PC that is used to deploy our adaptation system, is used to simulate the Adaptation Server of SDP. All results are reported below as averages.

4.2. Experimental results on rraining accuracy

In this section, we compare CXCS with XCS on training accuracy by the following two experiments.

Definition 1. Training Accuracy (TA): We define the training accuracy of mobile service adaptation system as the ratio of the number of all training results and the number of correct training results with the following:

$$TA = 100\% \times count_{correct-train} / count_{train} \tag{12}$$

where the better the approach is, the higher the training accuracy is.

In first experiment, the static profiles of 10 users are used to identify their static similarity. The user similarity degree and similar-user ratio threshold are set to 0.85 and 0.5, respectively. While one user is selected as the current user, the other users are requested to initialize this user's classifier population. The average training accuracy is compared with XCS in Fig. 6.

The second experiment is conducted to validate co-evolutionary XCS's cooperative covering and predicting functions. First, all the training records are used to calculate the user similarities. Then, for each user, the similar-users assisted with the covering and predicting process. Finally, the average training accuracy is shown in Fig. 7.

From the comparison results, the training accuracy of our proposed CXCS is higher than XCS. Obviously, CXCS deployed on adaptation server, can support mobile service adaptation more effectively than XCS in mobile network.

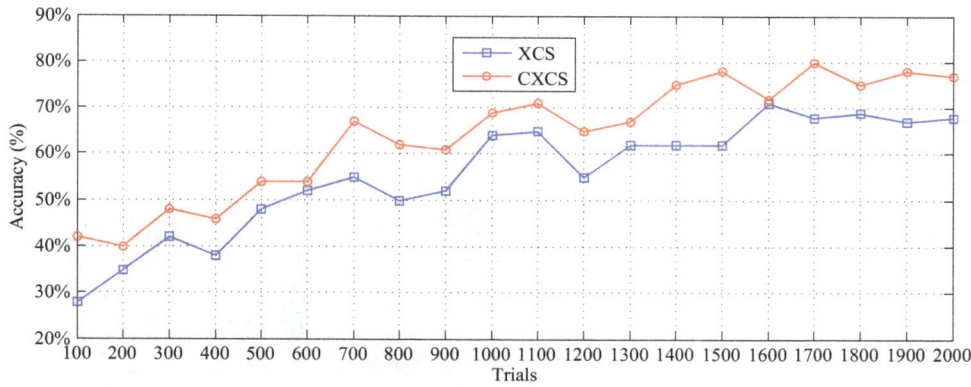

Fig. 6. Training accuracy using cooperative initialization.

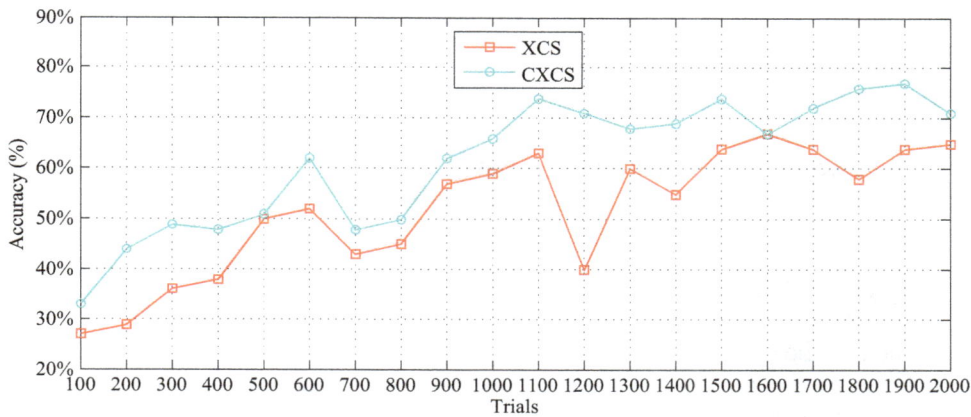

Fig. 7. Training accuracy using cooperative cover and predicting.

4.3. Experimental results on convergence speed

In this experiment, we evaluate the convergence speed of CXCS.

Definition 2. Convergence Speed (CS): We define the convergence speed of mobile service adaptation system as the ratio the fitness and the convergence fitness with the same Trails by the following:

$$CS = \frac{Convergence\text{-}Fitness_i}{Fitness_i}(1 \leqslant i \leqslant Trails) \tag{13}$$

where the better the approach is, the higher the convergence speed is.

As shown in Table 5, it can be concluded that CXCS converges more quickly than XCS because of its cooperative co-evolution mechanism.

4.4. Experimental results on adaptation accuracy

In this experiment, we compare our proposed CXCS with other five schemes such as XCS, Bayesian, SVM, C4.5 and Decision tree, in term of the adaptation accuracy.

Table 5
Comparison results on convergence speed

Trails	Convergence speed	
	XCS	CXCS
200	0.320	**0.420**
400	0.380	**0.470**
600	0.520	**0.580**
800	0.475	**0.560**
1000	0.615	**0.675**
1200	0.475	**0.680**
1400	0.585	**0.720**
1600	0.660	**0.760**
1800	0.655	**0.775**
2000	0.665	**0.780**

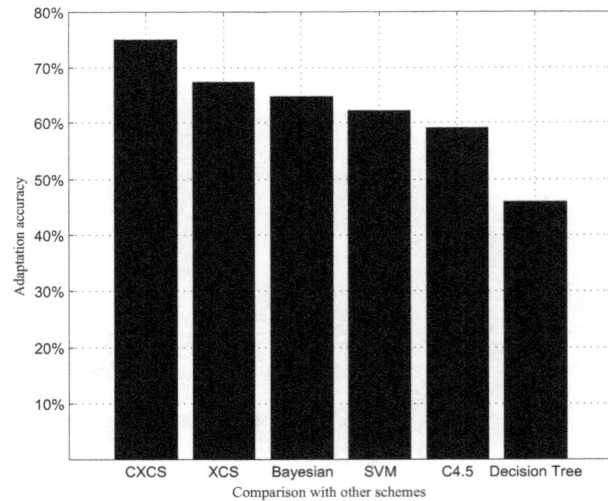

Fig. 8. Comparison with other schemes on adaptation accuracy.

Definition 3. Adaptation Accuracy (AA): We define the adaptation accuracy of mobile service adaptation system as the ratio of the number of all test results and the number of correct test results with the following:

$$AA = 100\% \times count_{correct-test}/count_{test} \tag{14}$$

where the better the approach is, the higher the adaptation accuracy is.

The experiment is run 15 times in testing process and the results are obtained, on average, as shown in Fig. 8. According to the comparison results, the adaptation accuracy of CXCS is higher 70%. Obviously, it outperforms other mobile servcie adaptation schemes on the adaptation accuracy for mobile services in mobile network environments. The main reason is that new users can be initialized accurately and all context can be covered effectively.

5. Related work

Many approaches [17–21] have previously been proposed for learning the rules for context-aware service adaptation. Nearchos Paspallis et al. [3] proposed to use the utility function-based method, but its fatal weakness is that the weight in the utility function is very difficult to determine during real applications. Choi et al. [4] applied work-flows to implement the adaptation, and Miraoui et al. [5] used a coloured Petri-net for the adaptive decisions. However, most of these previous approaches can only run in an off-line way, and the adaptation ability is still very limited.

In contrast to these previous schemes, our study focuses not only on an on-line context-aware approach but also on effective mobile service adaptation ability. LCS offers good performance as well as excellent understandability, and it has wide applications in data mining, automatic robots, traffic controlling, automatic fighter control, pattern recognition and complex adaptive systems. However, due to its complexity, it is too difficult to implement for practical applications. Wilson proposed ZCS [14] as a simplified version of LCS. It maintained most of the LCS framework, but increased the understandability and performance significantly. One characteristic of ZCS is that it used a strength-based mechanism to represent the classifier payoff. However, it also introduces a shortcoming in distinguishing the classifier accuracy. Therefore, Wilson presented another improved version called XCS [13], which used a fitness-based accuracy, niche-based GA, modified Q learning credit assignment mechanism. Thus far, XCS is the most successful variety of LCS, and it has been widely used in many areas.

However, because of the new-user problem and the context-cover problem, XCS cannot effectively perform context-aware mobile service adaptation. Hence, in this paper, CXCS extends XCS by adding a cooperative co-evolution mechanism [16], modifying the rule structure and working process, and adding a user-interactive feature. Research on co-evolutionary LCS is still very rare. Dumitrescu [22] applied co-operative and competitive co-evolutionary mechanisms to the classifiers of the same species. ETANI [23] proposed a cooperative co-evolutionary architecture consisting of a collection of LCS, where each one attempts to evolve its subcomponents (agents), and the assembling of all these subcomponents represents the complete solution. Although the two studies produced good results, they still failed to solve the new-user problem and the context-cover problem in context-aware mobile service adaptation. Shang Gao et al. [24] reported on the development of a survey instrument designed to measure user perception on mobile services acceptance. Junho Ahn and Richard Han [25] focused on recognizing mobile users' daily behavior patterns and unusual event detection. They built a general unusual event classification model to analyze the activity, location, and audio sensor data collected from mobile phone users to identify these users' personalized normal daily behavior patterns. The model may be able to provide valuable information for mobile service adaptation systems.

6. Summary and future work

In this paper, we incorporate a cooperative co-evolutionary mechanism into XCS, proposing CSC, to perform context-aware mobile service adaptation. In CXCS, similar-users are used to initialize the classifier population of new-users. Moreover, the current user's uncovered-context states are also predicted. Experimental results show that CXCS can quicken the convergence speed and improve the training accuracy. Moreover, CXCS is also better than five adaptation schemes in terms of the adaptation accuracy in mobile network environments. Hence, CXCS can effectively perform context-aware mobile service adaptation in mobile network environments.

In the future, we will focus on implementing more adaptation function of CXCS. Besides, more evaluation for CXCS in a real mobile service environment is also our future work.

Acknowledgements

The work presented in this study is supported by the NSFC(61202435); NSFC(61272521); Natural Science Foundation of Beijing under Grant No.4132048; Specialized Research Fund for the Doctoral Program of Higher Education under Grant No. 20110005130001; Program for New Century Excellent Talents in University of China under Grant No.NCET-10-0263; and Innovative Research Groups of the National Natural Science Foundation under Grant No.61121061. We also thank Dr. Guoqiang Li of Amazon China.

References

[1]   Klaus R. Dittrich, S. Gatziu and A. Geppert, The Active Database Management System Manifesto: A Rulebase of ADBMS Features, *Lecture Notes in Computer Science 985*, Berlin, Germany, Springer (1995), 3–20.
      A. Attou and K. Moessner, Context-aware service adaptation management IEEE International Symposium on Personal, *Proceedings of the IEEE International Symposium on Personal, Indoor and Mobile Radio Communications (PIMRC 2007)*, (2007), 1–5.

[2]   G. Gehlen, F. Aijaz, Y. Zhu and B. Walke, Mobile P2P Web Services using SIP, *Mobile Information Systems* **3** (2007), 165–185.

[3]   N. Paspallis, K. Kakousis and G.A. Papadopoulos, A multi-dimensional model enabling autonomic reasoning for context-aware pervasive applications, *Proceedings of the 5th Annual International Conference on Mobile and Ubiquitous Systems: Computing, Networking, and Services (MobiQuitous 2008)*, (2008), 1–6.

[4]   J. Choi, Y. Cho, K. Shin and J. Choi, A context-aware workflow system for dynamic service adaptation Lecture Notes in Computer Science (including subseries Lecture Notes in Artificial Intelligence and Lecture Notes in Bioinformatics), *Proceedings of the 2007 international conference on Computational science and its applications (ICCSA 2007)*, (2007), 335–345.

[5]   M. Miraoui, C. Tadj and C.B. Amar, Dynamic context-aware service adaptation in a pervasive computing system, *Proceedings of the 3rd International Conference on Mobile Ubiquitous Computing, Systems, Services, and Technologies (UBICOMM 2009)*, (2009), 77–82.

[6]   A.B. Kocaballi and A. Kocyigit, Granular Best Match Algorithm for Context-Aware Computing Systems, *Proceedings of the ACS/IEEE International Conference on Pervasive Services (ICPS 2006)*, (2006), 143–149.

[7]   G. Koutrika and Y. Ioannidis, Personalizing queries based on networks of composite preferences, *ACM Transactions on Database Systems* **2** (2010), 1735886–173589.

[8]   Z. Huang, X. Lu and H. Duan, Context-aware recommendation using rough set model and collaborative filtering, *Artificial Intelligence Review* **1**, (2011), 85–99.

[9]   R. Yasdi, Learning classification rules from database in the context of knowledge acquisition and representation, *IEEE Transactions on Knowledge and Data Engineering* **3** (1991), 293–306.

[10] K. Chrysostomou, S.Y. Chen and X. Liu, Identifying user preferences with Wrapper-based Decision Trees, *Expert Systems with Applications* **4** (2011), 3294–3303.

[11] M.-W. Tong, Z.-K. Yang and Q.-T. Liu, A novel model of adaptation decision-taking engine in multimedia adaptation, *Journal of Network and Computer Applications* **1** (2010), 43–49.

[12] A. Shankar and S. Louis, Learning classifier systems for user context learning, *Proceedings of the IEEE Congress on Evolutionary Computation (CEC 2005)*, (2005), 2069–2075.

[13] J.H. Holland, Adaptation in natural and artificial systems, *MIT Press*, Cambridge, MA, USA, 1992.

[14] S.W. Wilson, Classifier fitness based on accuracy, *Evolutionary Computation* **2** (1995), 149–175.

[15] L.D. Shi, Y.H. Shi, Y. Gao, L. Shang and Y.B. Yang, XCSc: A novel approach to clustering with extended classifier system, *International Journal of Neural Systems* **1** (2011), 79–93.

[16] M.A. Potter and K.A.D. Jong, Cooperative Coevolution: An Architecture for Evolving Coadapted Subcomponents, *Evolutionary Computation* **1** (2000), 1–29.

[17] K. Henricksen, J. Indulska and A. Rakotonirainy, Using context and preferences to implement self-adapting pervasive computing applications, *Software – Practice and Experience* **11–12**, (2006), 1307–1330.

[18] J. Xiao and R. Boutaba, QoS-aware service composition and adaptation in autonomic communication, *IEEE Journal on Selected Areas in Communications* **12** (2005), 2344–2360.

[19] C. Jacob, D. Linner, I. Radusch and S. Steglich, Context-aware Data Dissemination and Service Adaptation, *Proceedings of the 16th IST Mobile and Wireless Communications Summit (IST 2007)*, (2007), 1–5.

[20] C. Baladron, J. Aguiar, B. Carro, L. Calavia, A. Cadenas and A. Sanchez-Esguevillas, Framework for intelligent service adaptation to user's context in next generation networks, *IEEE Communications Magazine* **3** (2012), 18–25.

[21] W.Y. Lum and F. Lau, A context-aware decision engine for content adaptation, *IEEE Pervasive Computing* **3** (2002), 41–49.

[22] D. Dumitrescu, M. Preuss, C. Stoean and R. Stoean, Coevolution for classification, *Technical report, No. CI-239/08.* Collaborative Research Center on Computational Intelligence University of Dortmund, Germany, 2008.

[23] E. N, Modeling Agent-Based Coevolution with Learning Classifier System, *IEIC Technical Report* **535** (2002), 49–53.

[24] S. Gao, J. Krogstie and K. Siau, Developing an instrument to measure the adoption of mobile services, *Mobile Information Systems* **1** (2011), 45–67.

[25] J. Ahn and R. Han, Personalized behavior pattern recognition and unusual event detection for mobile users, *Mobile Information Systems* **2** (2013), 99–122.

Shangguang Wang is an assistant professor at Beijing University of Posts and Telecommunications. He received his Ph.D. degree at Beijing University of Posts and Telecommunications in 2011. His research interests include service computing and cloud computing. He has served as reviewers for numerous journals, including IEEE Internet Computing, Mobile Networks and Applications, IET Software, Computer Journal, Service Oriented Computing and Applications, Security and Communication Networks, Journal of Network and Computer Applications, International Journal of Systems Science, Entropy, Sensor Letters, etc. He is an IEEE member, ACM member as well as a CCF senior member. Homepage: http://www.sguangwang.com/.

Zibin Zheng is an associate research fellow at Shenzhen Research Institute, The Chinese University of Hong Kong. He is also a guest research member at State key Laboratory of Networking and Switching Technology, Beijing University of Posts and Telecommunications, Beijing, China. He received Outstanding Ph.D. Thesis Award of The Chinese University of Hong Kong at 2012, ACM SIGSOFT Distinguished Paper Award at ICSE2010, Best Student Paper Award at ICWS2010, and IBM Ph.D. Fellowship Award at 2010. He served as PC member of CLOUD2009, SCC2011, SCC2012, ICSOC2012, etc and reviewers for journals including IEEE Transactions on Software Engineering, IEEE Transactions on Parallel and Distributed System, ACM Transactions on the Web, IEEE Transactions on Service Computing, etc. His research interests include service computing and cloud computing. Homepage: http://www.zibinzheng.com/.

Zhengping Wu is an assistant professor of Computer Science and Engineering at the University of Bridgeport in USA. He received his Ph.D. in Computer Science from the University of Virginia in 2008. He has authored or co-authored eleven book chapters and over forty peer-refereed papers. His research interests include services computing, cloud computing, network security, distributed systems, wireless networks, operating systems, and medical informatics. He has served as program committee members and reviewers for numerous conferences and journals, including IEEE Transactions on Services Computing, IEEE Transactions on Systems, Man, and Cybernetics (A), IEEE Transactions on Fuzzy Systems, IEEE Transactions on Control Systems Technology, IEEE Transactions on Industrial Electronics. Homepage: http://www1bpt.bridgeport.edu/~zhengpiw/.

Qibo Sun received his PhD degree in communication and electronic system from the Beijing University of Posts and Telecommunication in 2002. He is currently an associate professor at the Beijing University of Posts and Telecommunication in China. He is a member of the China computer federation. His current research interests include services computing, internet of things, and network security.

Hua Zou received her PhD degree in communication and electronic system from the Beijing University of Posts and Telecommunication in 2010. She is currently a professor at the Beijing University of Posts and Telecommunication, China. Her current research interests include network intelligence and services computing.

Fangchun Yang received his PhD degree in communication and electronic system from the Beijing University of Posts and Telecommunication in 1990. He is currently a professor at the Beijing University of Posts and Telecommunication, China. His current research interests include network intelligence and services computing. He is a fellow of the IET.

10

GeoVanet: A routing protocol for query processing in vehicular networks

Thierry Delot[a,*], Nathalie Mitton[b], Sergio Ilarri[c] and Thomas Hien[a]

[a]*University Lille North of France, LAMIH FRE CNRS, Valenciennes, France*
[b]*INRIA Lille-Nord Europe, Villeneuve, France*
[c]*IIS Department, University of Zaragoza, Zaragoza, Spain*

Abstract. In a vehicular ad hoc network (VANET), cars can exchange information by using short-range wireless communications. Along with the opportunities offered by vehicular networks, a number of challenges also arise. In particular, most works so far have focused on a push model, where potentially useful data are pushed towards vehicles. The use of pull models, that would allow users to send queries to a set of cars in order to find the desired information, has not been studied in depth.

The main challenge for pull models is the difficulty to route the different results towards the query originator in a highly dynamic network where the nodes move very quickly. To solve this issue, we propose GeoVanet, an anonymous and non-intrusive geographic routing protocol which ensures that the sender of a query can get a consistent answer. Our goal is to ensure that the user will be able to retrieve the query results within a bounded time. To prove the effectiveness of GeoVanet, an extensive experimental evaluation has been performed, that proves the interest of the proposal for both rural and urban areas. It shows that up to 80% of the available query results are delivered to the user.

Keywords: vehicular networks, mobile computing, query result routing

1. Introduction

Developing data management techniques that can be useful to build information services for drivers is a hot research topic (e.g., see [13,49]). A key driving force is that, in the last decade, a number of small-sized wireless devices (e.g., PDAs or laptops) with increasing computing capabilities have appeared in the market at very affordable costs. These devices have started to be embedded into modern cars in the form of on-board computers, GPS navigators, or even multimedia centers. This has lead to the emergence of vehicular ad hoc networks (VANETs) [36]. In this kind of networks, cars traveling along a road can exchange information with other nearby cars. The lack of a fixed communication infrastructure, characteristic of ad hoc networks, implies that vehicles usually communicate with one another by using short-range wireless communications. Nevertheless, a piece of information can be disseminated and reach a far distance by using moving cars as intermediates, following multi-hop routing protocols [27].

To provide drivers with useful information, accessing data in vehicular networks has become a major issue. As we briefly describe in the following, push and pull approaches can be considered. The basic idea of these approaches is similar to the traditional push and pull approaches considered in the field of distributed and mobile databases [37,45]. The main difference is their application to a scenario

---

*Corresponding author: Thierry Delot, University Lille North of France, UVHC/LAMIH FRE CNRS 3304, Le Mont Houy, 59313 Valenciennes, France. E-mail: Thierry.Delot@univ-valenciennes.fr.

of vehicular networks, where there is no single server (several vehicles can act as servers for a given query or for certain data items) and each vehicle can communicate directly only with others within its communication range.

Some works rely on a *push model*. With such an approach, each vehicle receives information (e.g., about an emergency braking, a traffic congestion, an available parking space, etc.) from its neighbors and has to decide whether that information is relevant enough to be transmitted to the driver or not. The major difficulty for these solutions is how to disseminate data in the vehicular network so that vehicles receive the relevant information efficiently (timely and without unneeded overheads such as duplicate packets or irrelevant data) [11].

Nevertheless, with such a push model, we cannot imagine that every data item will be communicated to every vehicle, as this would consume too much bandwidth and lead to high communication and processing efforts on the vehicles. On the contrary, only data about events that are potentially interesting for a large set of vehicles (e.g., an emergency braking or a traffic congestion) are diffused among the vehicles that the system estimates as potentially interested. Information about other events will not be disseminated, and so it is impossible to share information among a small set of interested vehicles, for example to build vehicular social networks.

Several works also use a *pull model*, where a query is actually communicated to other vehicles in the vehicular network. This provides more flexibility in terms of the types of queries that can be considered, as opposed to the approaches based on a push model, since a query could in principle be diffused far away to retrieve remote data. This implies that vehicles should be able to understand, route, and process those queries. The basic idea, inspired by traditional Peer-to-Peer (P2P) systems [1], consists of diffusing the queries to different data sources, either directly or by using multi-hop relaying techniques [27]. Then, each node can compute a partial query result based on its local data and then deliver it to the destination node. However, since no fixed data server or any kind of infrastructure is necessarily available in vehicular ad hoc networks, new techniques to access data are needed. Indeed, the mobility of nodes makes the management of an indexing structure, used in traditional P2P systems to decide how to route queries, impossible (as indicated in works such as [19]). These works must also face the problem of routing the query results back to the query originator. This is indeed a challenge because the vehicle that issued the query can move in the meanwhile, and so routing the results based on simple geographic criteria may not be enough (it is difficult to know where the originator is currently located). Furthermore, since the vehicles keep moving, it is not even possible to ensure that there is at that moment a communication path to the originator node.

To overcome the issues related to the high mobility of nodes, the use of mobile telephony networks (as an alternative or a complement of a VANET) has been investigated [16,26]. The queries can then be evaluated on a central server storing information sent by the vehicles. An example of a remote service for drivers is the *Waze* application (http://world.waze.com/), which is a social mobile application available on smartphones. Waze users can publish and consume real-time maps and traffic information. Maps can be provided to mobile users and traffic information is retrieved using a mobile telephony network. Anyway, the use of a central server leads to obvious scalability issues. The types of data shared are so limited to those interesting for a majority of the users. This approach can also cause confidentiality issues, since a driver may not accept to send his/her location for it to be stored by an untrusted peer.

To summarize, the different access techniques for vehicular networks mentioned previously are really interesting and have lead to the design of real systems. Nevertheless, they also impose severe limits. Indeed, whereas they are well-suited to the dissemination of information useful for a large set of users, they are not adapted to information-sharing in vehicular networks. Existing techniques only allow diffusing a small subset of the data shared by each vehicle (i.e., the data interesting for the majority of the vehicles nearby) to neighboring vehicles through the network. Different objectives justify such choices (e.g., bandwidth saving, difficulty to identify the set of recipients, etc.).

In this paper, we present an anonymous and non-intrusive solution for drivers to share data in vehicular networks, by describing a geographic mailbox-based routing protocol called GeoVanet. This paper extends and improves the initial proposal of GeoVanet presented in [14], including an extensive experimental evaluation, a discussion of the use of maiboxes, and more detailed explanations. Our goal with GeoVanet is to allow drivers to query information shared by other vehicles. Therefore, a query has to be disseminated in the network in a bounded time. Then, once computed on remote nodes in the VANET, the results have to be delivered to the node that issued the query. We present the algorithms needed to solve this routing problem. We use a model which identifies a fixed geographical location where a mailbox is dedicated to the query to allow the user to retrieve his/her results in a bounded time.

The rest of the paper is organized as follows. Section 2 presents some related works. Our assumptions and our model are introduced in Section 3. Our proposal is detailed in Section 4. Section 5 presents our experimental evaluation. Finally, we conclude and present the perspectives of this work in Section 6.

2. Related work

Numerous recent research works have addressed the problem of information gathering in vehicular networks. In this section, we consider works focused on information dissemination, works that concern about query processing, and finally other works that are related to ours due to the similarity of some of the techniques proposed.

2.1. Works on data dissemination (push-based approach)

In [29], the authors focus on urban environments. They explain that the network connectivity is a limiting factor for information dissemination, since chains of vehicles are needed for broadcasting and a low traffic density may become a problem. The authors so make a clear distinction between data transportation via locomotion (i.e., vehicles carry data to areas where they can be disseminated) and via wireless communications, emphasizing the problem of lack of network connectivity that may occur depending on the density of equipped vehicles in an area. The main idea is that *Stationary Supporting Units (SSUs)* are needed to alleviate this problem, and different heuristics are proposed to decide the best locations to place them.

In [10], the use of a *propagation function* is proposed to decide the route that a message has to follow in order to reach a target spatial area. The originator of a message defines an appropriate propagation function (e.g., by considering traffic conditions for the current time frame), which can be interpreted as a "gravitational field" where the message is attracted towards areas of minimum potential. The route traversed by the message is thus the result of evaluating the propagation function at each routing hop. On the basis of this propagation function, different dissemination approaches (both deterministic and probabilistic) are proposed and compared. The metrics used for performance evaluation are: *message delivery* (ratio between the nodes receiving the message and the total number of nodes, where *delivery-IN* –within the target area– should be closed to 100% and *delivery-OUT* –outside the target area– should be minimized) and *network traffic* (total number of messages transmitted).

As a final example of push-based approach for vehicular networks, *VESPA (Vehicular Event Sharing with a mobile P2P Architecture)* [9,11,12] is a system developed to support sharing information about different types of events in inter-vehicle ad hoc networks. One of the interesting features of VESPA is that it has not been developed with a specific type of event or application scenario in mind, and therefore can deal with many types of events in a VANET (e.g., information about available parking spaces, accidents, emergency brakings, obstacles in the road, real-time traffic information, information relative to the coordination of vehicles in emergency situations, etc.). VESPA is based on the concept of *Encounter Probability* (*EP*) between a vehicle and an event (measure of the likelihood that the vehicle will meet the event) for both alerting the driver about important events [11] and for diffusion decisions in the proposed dissemination protocol [9]. EP values can be computed based on maps, which can increase the precision of the system in some situations [12].

2.2. Works on query processing (pull-based approach)

PeopleNet [33] is an infrastructure-based proposal for information exchange in a mobile environment. However, it relies on the existence of a fixed network infrastructure to send a query to an area that may contain relevant information. Even though, once the query has arrived in the target area, it uses epidemic query dissemination through short-range communications within the area (to save economic costs), the answer to the query is communicated again to the originator using the fixed network (e.g., by sending an SMS or an email). Thus, problems related to query routing and result routing do not appear in this context. Moreover, this proposal does not focus on vehicular networks.

FleaNet [25] is a virtual market organized over a vehicular network. It proposes a *mobility assisted query dissemination* where the node that submitted the query periodically advertises it only to its one-hop neighbors, which will see if they can provide some answers from information stored on their local caches. With this approach the query spreads only due to the motion of the vehicle that submitted the query. This avoids overloading the network with many messages. It also solves the problem of routing back the results to the query originator, since it is only at one-hop distance. However, it is not general enough to process some kinds of queries, since the vehicle that submits the query must move near the vehicles which store the information needed.

Roadcast [51] is a content sharing scheme for VANETs. As in FleaNet, a vehicle can only query other vehicles that it encounters on the way. So, the problems of query and result routing do not arise either. In this case, keyword-based queries are submitted by the users and the scheme proposed tries to return the most popular content relevant to the query, as this content is likely to be useful for more vehicles in the future. Thus, not necessarily the most relevant data are returned for a query, as the popularity of the data is also taken into account.

In [47], a combination of pull and push is considered for *in-network query processing in mobile P2P databases*. When two mobile nodes encounter each other, they first exchange queries and results (pull phase). Then, they broadcast other popular data items that may help the other peer to improve its capabilities as source of relevant data in the future (push phase). Once again, multi-hop query/result routing is not considered.

In [13], the authors focus on multi-scale query processing in vehicular networks. They consider both the pull and push models and claim that different types of data sources (e.g., the local data cache containing data diffused by neighboring vehicles, remote web services, etc.) can be interesting for the evaluation of a single query. The authors explain that these data sources provide complementary information that increases the probability to answer the users' queries. However, no specific solution is proposed for the routing problem tackled in this paper.

In [48] the problem of searching documents in a vehicular network is considered. In this approach, the authors adapt the concept of *Distributed Hash Table* (*DHT*) [1] to a mobile environment and propose a *Hybrid Retrieval* (*HR*) approach which, based on the expected costs, adapts itself to choose between a flooding scheme and a DHT scheme for indexing and searching. The geographic space is divided into regions, such that each region must keep certain data. So, when a vehicle leaves a region it must transfer the documents belonging to the region to some other vehicle within the region. Thus, the purpose and approach of this work is quite different from those traditionally found in the context of vehicular networks. Besides, the problem of routing the results to the query originator is not discussed.

2.3. Other interesting related works

The problem of broadcasting/flooding has been widely studied (e.g., [28]), not only for the specific case of wireless ad hoc networks but for mobile environments in general (e.g., [46]). In [41], different routing protocols for vehicular networks are compared. Works such as [30,40] have studied the problem of routing a message towards a certain location.

As far as we know, only a few works have considered the problem of routing a message to a moving vehicle using multi-hop routing [22,24,44]. [24] tackles the problem of geographic routing in MANETs (mobile ad hoc networks) in scenarios where the communicating nodes are moving. In [22], the trajectories of the vehicles are considered for data delivery from infrastructure nodes to moving vehicles; however, the approach proposed requires knowledge about trajectories. In works such as [44], the problem of routing the results to a moving vehicle is also under study, but again some knowledge about the trajectories is needed. In this work, we apply some geographical routing but, contrary to related works, we consider mobile nodes and at the same time we want to preserve anonymity and avoid the need to know and manage information about trajectories. Using the mailbox-based approach presented in this paper we do not incur the overhead of managing trajectories and we are able to deliver the results more efficiently towards a static destination.

Finally, it is also interesting to mention that a mailbox approach has also been proposed in the context of mobile agent technology [43], in works such as [6]. Mobile agents are programs that can move from one computer to another and resume their execution on the target computer. Given the mobility of mobile agents, achieving an efficient communication among the agents is a challenge, as some mechanism is needed in order to route a communication to the location where an agent is currently executing. Moreover, the communication should eventually succeed even if the agent moves during the communication. Thus, the goal of the mailbox-based communication schemes is to increase the efficiency and reliability of the communication protocols by decoupling agent migrations and message delivery. However, although there is some relation, it should be noticed that the application context is totally different from the one considered in this work (distributed environments with mobile agents vs. vehicular networks). Thus, our proposal focuses on query processing in vehicular networks, it preserves the anonymity of the user, and it is able to work independently of whether a fixed support infrastructure is available or not.

3. Preliminaries

In this work, we consider a vehicular ad-hoc network (VANET) in which vehicles (nodes) are aware of their geographical position and moving direction. VANETs are highly mobile networks whose nodes are vehicles traveling along a road or a highway, and which communicate among themselves using short-range (e.g., 100–200 meters) wireless communication devices (such as Wi-Fi or UWB). Using these *vehicle-to-vehicle* (*V2V*) communications provides some interesting advantages. The most important benefit is probably that there is no need to deploy an expensive and wide-coverage dedicated support infrastructure. Moreover, the users will not be charged by the use of such a network. Finally, there are applications that require a quick and direct exchange of data (without intermediate proxies or routers) between two vehicles within range of each other, for example in the area of safety.

3.1. Disseminating queries in a VANET

In this paper, we assume that users of the VANET may need to spread queries in the network for which the answer is needed in a bounded time (but not necessarily immediately). For instance, let us consider a tourist visiting a city or a region and searching for information about the interesting places to visit. Some of the vehicles receiving the query broadcasted by this user may belong to tourists sharing information about the sites they have visited. If the shared information matches the user's needs, it has to be delivered to him/her. In this scenario, the user does not require an immediate answer as long as enough information is gathered before visiting the city/region. As opposed to traditional query processing techniques, whose objective is to deliver the query result as quickly as possible, our goal here is to guarantee that the maximum amount of results will be delivered in a bounded time (e.g., with a deadline set to tomorrow). This allows spending more time searching for relevant information, since it is here distributed over a large set of nodes. Our contribution in this paper resides in an anonymous geographic routing protocol, called GeoVanet, which ensures the routing of query results in highly mobile and dynamic networks. Our protocol is suitable:

- both when an infrastructure is available and when it is not. Indeed, we assume that cities can be easily equipped (at each crossroad, for instance) with reliable infrastructure nodes, whereas it would be much more costly to reach a sufficient coverage in the countryside.
- both when some road maps are available and when they are not.

3.2. Ad hoc communications vs. 3G wide-area communications

At this stage, the reader may wonder why we do not consider the use of mobile telephony networks (e.g., 3G) to share information between drivers but only a partial infrastructure when it is available. We indeed believe that the use of such mobile telephony networks presents several drawbacks:

1. First of all, not all devices are able to communicate using mobile telephony networks (e.g., connection not available everywhere, no subscription to a mobile telephony operator, etc.).
2. As opposed to short-range communication networks, mobile telephony networks are not free. The users may then not agree to be charged to provide others useful information. So, the fees related to the use of mobile telephony networks do not match with the cooperative applications considered in this paper.
3. As concerns privacy preservation, users may not accept that the personal data that they share with other users in their vicinity (e.g., information about the places they visited with the date of the visit, personal comments or pictures they may appear on) may be stored in a remote repository. However, the use of mobile telephony networks implies the use of such a repository where shared data will be stored.
4. Finally, bandwidth limitations and scalability issues may arise when considering a high number of drivers sharing an important amount of data with multimedia contents (e.g., pictures or videos of places, audio descriptions, etc.).

Moreover, new short-range communication protocols dedicated for vehicular applications are under standardization, such as the Wireless Access in Vehicular Environments (WAVE) protocol [32]. They will provide vehicles and pedestrians with the ability to communicate with each other and with the road-side infrastructure with a bandwidth up to 27 Mbps. This will also ease the exchange of important quantities of data between mobile units, for example to share multimedia contents.

Hence, in the rest of this paper we consider the use of ad hoc communications between vehicles and, only if available, with infrastructure nodes.

3.3. Use of mailboxes

With the proposed protocol, when a driver submits a query, a *mailbox* is assigned to the query. Several mailboxes will be available, which could communicate with each other (if necessary) by using a fixed network. The mailbox acts as a collector for the results of the query. Thus, it allows an indirect routing. The idea is that the answers will not be directly routed to the requester but to the mailbox assigned to his/her query, which is identified by a field inserted into the request. Different mailboxes can be set up as part of the fixed support infrastructure (thus, mailboxes could be considered as *infrastructure nodes*). The density of mailboxes in an area could affect the performance, as each mailbox will collect some of the query results and there may be a certain routing overhead to deliver the results to the mailbox and to retrieve the results from it. Thus, the location and distribution of mailboxes could be planned by following design principles inspired by those used in cellular telecommunication networks [39].

It should be noted that an alternative approach would be to consider that a mailbox is not a physical entity but just refers to an area. In such a case, vehicles in that area (or close to it, if there is no vehicle inside) would be responsible for maintaining the results available in the area. If the vehicles holding the results are going to leave the area, they would need to relay the information to other vehicles in the area. This alternative option would require a deep study on how to maintain the results alive in the area until they are picked up by the vehicle that submitted the query. As it is complementary to our approach (based on physical fixed entities called mailboxes) and the problem of keeping information alive in an area has been studied in other works [7], we do not study it further in this paper.

The use of the mailbox avoids the great difficulty that would arise if we intended to route the results directly back to the moving querying vehicle, as its location would be changing constantly and for privacy reasons no information about the route of the vehicle (if known) should be disseminated with the query. Indeed, even if the estimated route is available, routing the results would still be challenging, as it would be difficult to estimate the precise location of the destination vehicle at that moment. So, the advantage is that the mailbox is static, it has a fixed location and therefore routing the results towards it does not involve the aforementioned problem, especially the privacy issue. In this way, our scheme preserves the identity of the user.

Of course, due to this indirect routing, when the initiator of the request wishes to retrieve the answers to his/her query, he/she will either need to drive towards the mailbox (until the vehicle is within the communication range of the mailbox) to obtain the results using ad hoc communications (querying the mailbox can be considered a form of *V2I* –*vehicle-to-infrastructure*– communication [3]) or ask the mailbox remotely about the answers. In this latter case, the collecting mailbox will send the results by using the proposed routing algorithm (described in Section 4). The target of this routing process can be directly the user (assuming that he/she will not move during the process, as otherwise this would not be the most appropriate approach, due to the challenges involved in routing to a moving node), who in this case needs to provide his/her location, or another mailbox chosen by the user based on his/her planned trajectory. Even if the results are routed directly towards the user, the collecting mailbox can first send the results (through the fixed network) to another mailbox that is located near the location of the user, in order to reduce the efforts invested in V2V communications (this routing process is transparent to the user).

It should be noted that the user could indicate with the query that he/she wants the mailbox to automatically send the results after the request expires (*push-based retrieval of results*). Alternatively, he/she could explicitly ask about the results when he/she is near any other mailbox (*pull-based retrieval of results*); since mailboxes are connected, this other mailbox will be able to contact the one which has collected the results and to retrieve them.

We assume that the available mailboxes and their locations are known by the vehicles, as this information can be easily obtained through a simple initial registration process when the software module implementing the proposed routing protocol is installed on the vehicles. Now the question is: how is an appropriate mailbox assigned to a query? In the following, we analyze three possible alternatives.

Option 1: Selection of mailbox by the user

The driver himself/herself could select a target mailbox among those available. For example, if he/she knows quite well his/her driving plans, he/she could choose a mailbox that is near his/her destination in order to facilitate the retrieval of the results (the vehicle could just query the mailbox when entering its communication range). Alternatively, he/she could choose a mailbox along his/her planned route, in order to pick up the results on his/her way to the final destination.

The main advantages of this strategy are its simplicity and the fact that the user is in control of the mailbox that is assigned for his/her query. However, it has also important disadvantages. First, the user has the responsibility of selecting an appropriate mailbox. Second, this strategy does not ensure load-balancing, as there may be popular mailboxes that are frequently selected as collectors of results of queries; this will not only increase the load at the mailbox itself but also the routing overhead near it. Third, this strategy may compromise the privacy of the querying vehicle. For example, based on the mailboxes selected for queries it may be possible to infer information about locations frequently visited by that vehicle. Another privacy concern is that if the mailbox is in an area with not many vehicles it may be easier to infer which vehicle submitted a certain query, as it will be more difficult to preserve a *k-anonymity* [42] with a high k (in this context, k-anonymity would imply that the issuer of a query could be any among at least k vehicles). Besides, it could happen that the driver does not know exactly where he/she is going or does not want to bother himself/herself with selecting an appropriate mailbox. In these cases, the system should automatically determine a mailbox for the query by using one of the other two options that we discuss in the following.

Option 2: Automatic assignment of mailbox based on geographic criteria

With this strategy, the idea is to exploit some information about the contents of the query in order to decide a suitable mailbox. In particular, it can be reasonably assumed that most queries of interest in the context of mobile users are associated to a certain *spatial scope*, that delimits the range of the area that is interesting for the query (e.g., in the so-called *range queries* and their variants) [21]. The scope of a query could be defined by indicating a simple rectangular or circular range or based on the concept of *location granules* proposed in [20]. So, for example, if a driver is approaching the city center and wants to retrieve information about available parking spaces, then the scope of the query will be the city center and its surroundings. If the user is asking about that area, it is likely that he/she is going towards it. So, allocating a mailbox located in that area seems a good idea.

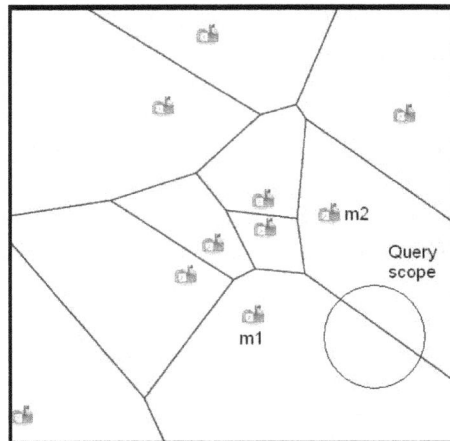

Fig. 1. Example of map of mailboxes and their Voronoi cells.

As there could be several mailboxes in a certain area, the one assigned to the query is selected based on the overlapping between that area and a set of Voronoi cells [2,50] defined by the locations of the mailboxes (see Fig. 1). For each location within each Voronoi cell there is no closer mailbox than the one defining that cell. So, the intersection between the spatial scope associated to a query and the Voronoi cells is obtained. For each intersecting cell, the percentage of overlapping is computed and finally the mailbox in the cell with the largest overlapping is selected (ties are solved randomly); for example, for the spatial scope indicated in Fig. 1, mailbox *m1* would be selected. The Voronoi diagram would be computed off-line. For simplicity, we can assume that a single Voronoi diagram is managed. Nevertheless, it could be partitioned to reduce the space needed to store it in the vehicles. On the one hand, a vehicle routing the results only needs to know the identifier of the target mailbox and its location, which can be embedded in the query.[1] On the other hand, a vehicle submitting a query needs to compute the target mailbox. For this, it could download from a server (if it does not have this information already locally available) only the part of the complete Voronoi diagram corresponding to the spatial scope associated to the query. In practice, a vehicle will eventually store locally fragments of the Voronoi diagram corresponding to the areas it frequently queries about, which will avoid the need to download this information from a server.

Of course, the procedure described here for an automatic assignment is just a heuristic approach and it may fail in assigning the best mailbox in some cases. For example, the user could want to query something about an area in order to decide whether he/she would like to go there or not; in this case, he/she will need to retrieve the query results before going there, and therefore assigning a mailbox within the area may not be the optimal choice. On the other hand, we consider that this strategy is flexible in the sense that different heuristics can be incorporated in order to decide the most appropriate mailbox in a given area (e.g., the decision can also be based on the current load of the candidate mailboxes).

The main disadvantage of this approach is that some mailboxes in popular areas could get easily overloaded, although (as mentioned above) there is some flexibility in the assignment of mailboxes and so load-balancing criteria could be considered. Another problem with this strategy is that privacy could

---

[1]Having information about nearby Voronoi cells would also be useful, as the vehicle could route the results to the target mailbox through a nearby mailbox, but for simplicity we do not consider this possibility here.

Table 1
Strategies for assigning a mailbox for a query: summary

Strategy	Load balancing	Privacy	Automatic assignment	Flexibility
1) The user selects	$\emptyset$	$\emptyset$	$\emptyset$	$\checkmark$
2) Geographic-based	$\approx$	$\emptyset$	$\checkmark$	$\approx$
3) Hash-based	$\checkmark$	$\checkmark$	$\checkmark$	$\emptyset$

be more easily violated because there exists a relation between the query and the mailbox assigned, and therefore probably also between the destination of the driver and the mailbox. The density of mailboxes and their Voronoi cells may have an impact on performance. Anyway, as commented later, we advocate a different approach for the assignment of mailboxes.

Option 3: Automatic assignment of mailbox based on hashing

Another solution is to use a hash function applied on a query identifier (a query identifier could be assigned by the vehicle launching the query, for example by concatenating the GPS current time and the location of the vehicle) and use the result returned by the hash function to allocate a mailbox randomly to the query. So, the idea would be to exploit a *Distributed Hash Table (DHT)* [1].

The main problem of this approach could be that the mailbox assigned can be located anywhere. Thus, the use of a uniform hash function will destroy the spatial locality information [52] (at least, unless some kind of *proximity-aware or locality-aware hash function* [38] is used). However, this can also be seen as an advantage, as the random assignment of mailboxes makes this approach less susceptible to privacy attacks (there is no apparent relation between a query/driver and the mailbox assigned, as the hash function may select a far away mailbox, a nearby one, or one in between). Besides, the use of a uniform random hash function will provide load-balancing. This approach leaves no flexibility for choosing a mailbox, as the one determined by applying the hash function must be selected. If the number of mailboxes available changes, the hashing must be re-adapted to the new situation.

It is interesting to mention that other works have proposed the use of distributed hash tables in mobile ad hoc networks and vehicular networks (e.g., [15,23]).

Conclusions

As described above, different strategies can be used to allocate mailboxes. The first strategy relies on the decision of the driver, whereas the two other strategies try to automatically locate a suitable mailbox. The first two strategies try to minimize the effort required by a vehicle to retrieve the results from the mailbox, whereas the last strategy attempts to achieve a good load-balancing and preserve the privacy of the user. A summary of the comparison among the three strategies considered is shown in Table 1, where $\emptyset$ represents a missing feature, $\checkmark$ represents an existing feature, and $\approx$ represents a feature partially supported.

In order to maximize the privacy of the user and offer a good load-balancing, we advocate the third option, which is based on hashing. Thus, a *key* field (the query identifier) is attached to each query message, which can be mapped to a mailbox identifier by applying a hash function. However, it is important to emphasize that the strategy used to choose the mailboxes does not have an impact on the routing process of GeoVanet.

GeoVanet preserves the sender's anonymity, since the driver does not need to provide his/her identity or intended route during the process. This is important because privacy protection is a major concern in vehicular networks [5,8] and in any kind of location-based system [17].

3.4. No need of digital road maps

To conclude this section, we would like to emphasize that, even though we consider the use of digital road maps for routing purposes with one of the possible implementations of the protocol that we propose (see Section 5.3.2), we do not assume that the destination of a vehicle submitting a query is known. The knowledge of the destination is for example required for route guidance when using existing navigation devices, but these are used only occasionally, when the driver does not know his/her route. In our case, the system is anonymous, non-intrusive and works without requiring information about the destination of the cars.

4. Proposed routing protocol

In this section, we present GeoVanet, a geographic routing protocol which spreads efficiently a request in a vehicular network in such a way that the sender can easily and quickly get a consistent answer within a bounded time. The use of mailboxes allows an indirect routing, as explained in the previous section. The routing protocol proposed consists of four steps:

1. Query broadcasting: the query is spread over the network.
2. Query processing: the query is processed locally on the vehicles that receive it.
3. Reply delivery: the answers are routed from the vehicles to the mailbox.
4. Data retrieval: the request initiator retrieves the results to his/her query.

In the rest of this section, we first introduce our notation and the two types of messages considered. Then, we detail how GeoVanet works for every step of the process, particularly the query broadcasting, the result delivery to the mailbox, and the retrieval of the results from the mailbox by the requester.

4.1. Basic notation

We denote by $N(u)$ the set of vehicles a vehicle u can directly communicate with. Since the nodes are aware of their location, direction and speed, they are able to determine whether they are aiming or not towards mailbox d. We denote $dir_{\to d}(u)$ the variable representing this relative movement of u towards d. This relative movement can be evaluated either in the Euclidean plane (if no maps are available) or based on maps, if available. Indeed, a vehicle following a road may get closer to a point based on maps and get farther based on Euclidean distances if the road makes some detours. $dir_{\to d}(u) > 0$ means that u is getting closer to d while $dir_{\to d}(u) < 0$ means that u is getting further from d. $dir_{\to d}(u) = 0$ means that u is static. In this latter case, u may be an infrastructure node.

4.2. Types of messages

We consider different kinds of messages: Query messages and Reply messages. On the one hand, a Query message represents the request issued by a vehicle. This message is as follows: Query=[request, exp-date, key], in which:

 – 'request' is the core of the request i.e., 'what are the interesting sites to visit in Paris?'.
 – 'exp-date' is the date by which the answer is expected.
 – 'key' (query identifier) is used to determine the location where the answer should be sent and retrieved (i.e., the mailbox).

On the other hand, the Reply message is created by a vehicle which is able to answer the query and has to send its answers to the mailbox. This message is as follows: Reply = [request, data, exp-date, key], in which:

- 'data' is the core of the answer, e.g., 'Visit the Eiffel Tower'.
- 'exp-date' is the expiration date of the request to be answered. It is used to stop the routing in case the date expires before the reply reaches the mailbox.
- 'key' (query identifier) is used to determine the location of the destination of the reply. It is the same as in the Query message.

In the following, we describe the four steps of the proposed protocol.

4.3. Step 1: Query broadcasting

When a driver needs some information, he/she issues a Query message that is spread over the network. Several solutions may be considered to spread the query expressed by a user over the vehicular ad hoc network. More precisely, the following strategies may be considered:

- *Flooding*, where each vehicle receiving a message (i.e., the query to disseminate) relays it to its neighbors, without any limitation on the number of diffusions.
- *Contention-based forwarding*, where the principle "the farthest broadcasts" is applied to restrict the number of vehicles relaying the message. In our case, in order to diffuse the query in every direction, the number of relaying vehicles should not be limited to one, but only the farthest vehicle in each direction should broadcast, as done in [9].
- *Dissemination using a fixed infrastructure*. The use of fixed relays (usually called *Stationary Supporting Units –SSUs–* [29], *Road-Side Units –RSUs–* [31], or stationary gateways [34]) can help disseminate the query and keep it alive in some areas where the density of vehicles may not be high enough to ensure its propagation.

In each of the above mentioned strategies, the diffusion stops by the expiration date associated to the query. Any of these strategies, with their particular advantages and disadvantages, could be used for query broadcasting. Thus, works on data/query dissemination are complementary to our approach. We will compare these strategies in the context of GeoVanet in Section 5.2.

4.4. Step 2: Processing of the query

In this paper, we do not concern about the query language used to formulate the query. Indeed, our proposal is independent from it. Numerous candidates are therefore available (e.g., SQL, XQuery, etc.). We also assume that the time needed to process the query on the vehicles holding interesting information is not significant when compared to the time needed to route the query result towards its destination.

4.5. Step 3: Delivery of the query results

Upon reception of a Query message, a vehicle u first checks the expiration date. If the message is still valid, u forwards it as needed (for query broadcasting) and checks whether it has information to provide to the Query sender. If so, u retrieves the location of the destination d that should gather every Reply message R. To do so, it uses the *key* of the Query.

The procedure then used by the vehicles to forward the Reply message R to the destination node d is described in Algorithm 1. In Algorithm 1, R can either be issued by the node u or by another node.

In this latter case, the node u is a forwarder of the original query. Algorithm 1 runs as follows. Every Δt seconds, while the Reply message R is still valid (i.e., while the expiration date that was set in the query and associated to R is valid), u checks its direction. If it aims at d ($dir_{\rightarrow d}(u) > 0$), it keeps the message as long as it either reaches the mailbox d itself or it ends up getting further from the mailbox. In the former case, u delivers directly the reply message to d. In the latter case, u has to pass R to a more promising vehicle. It then chooses the closest one among its neighbors v ($v \in N(u)$) that are either infrastructure nodes ($dir_{\rightarrow d}(v) = 0$) or aiming towards d ($dir_{\rightarrow d}(v) > 0$), which will run the same algorithm, and so on until either R is delivered to d or it expires.

The choice of the closest neighbor driving in the direction of the mailbox aims at maximizing the probability that the connection time between the two vehicles exchanging the query result is long enough to allow the complete transfer. To determine whether a vehicle is driving towards a mailbox or not, either simple geographic computations or computations based on road maps could be considered (evaluated in Sections 5.3.1 and 5.3.2, respectively).

Algorithm 1 GeoVanet(u, d, R) – Run at vehicle u holding a Reply R to deliver to d

---

```
 1:   ToBeDelivered:=TRUE
 2:   while ToBeDelivered == TRUE do
 3:      Every Δt seconds
 4:      if R.exp-date ≤ Current-date then
 5:         ToBeDelivered := FALSE {R has expired.}
 6:      else
 7:         if dir→d(u) > 0 then
 8:            {Vehicle u is moving towards the target position. It keeps the message.}
 9:            if d ∈ N(u) then
10:               Send R to d
11:               ToBeDelivered := FALSE {Destination is within the range of u. u successfully delivers its Reply.}
12:            else
13:               ToBeDelivered := TRUE {u keeps on holding R.}
14:            end if
15:         else
16:            {Node u is getting further from d or it does not move, so it has to release R.}
17:            N' = {v} such that v ∈ N(u) and dir(v)→d ≥ 0
18:            {u selects the closest neighbor which is either an infrastructure node or moving towards d.}
19:            Send R to node v ∈ N' which minimizes distance(u, v)
20:            ToBeDelivered := FALSE
21:         end if
22:      end if
23:   end while
24:   RETURN;
```

Notice that some variants to this basic scheme could be considered to try to maximize the performance of the protocol. For example, it is possible to consider several carriers simultaneously to route a message towards its destination (this will be evaluated in Section 5.3.3). As another example, even if the current vehicle is driving towards the mailbox, it could be more efficient to relay the message to another vehicle that could approach it faster. This last optimization would require extra knowledge about the status of the neighbors of a vehicle and so it is not further explored in this paper.

4.6. Step 4: Retrieval of the results

When the query has expired, the node S that issued the query Q will contact the mailbox d to retrieve every answer that has been collected. As explained in Section 3.3, S can either move towards d to query it directly or it can send a message to d asking for the results. However, in this latter case, it would have to provide its location to allow the result to be routed to it (and it should remain at the same location until the results have arrived); the same algorithm (Algorithm 1) would be used first to route the message asking for the results from S to d and then to route the results from d to S.

Notice that, in some cases, it may be even possible for the vehicle that submitted the query to contact the mailbox directly by using a 3G connection. However, even if this connection may also be available for other vehicles to route their results to the mailbox, using ad hoc communications during step 3 is convenient and economical, which will also encourage the participation of the vehicles. Thus, our approach does not assume the existence of a support infrastructure (other than mailboxes).

5. Experimental evaluation

In order to evaluate our solution with an important number of vehicles over a significant period of time, we have tested it using a simulator. We needed a testing system that could simulate realistic vehicles' movements and wireless exchanges and support the implementation of both the query dissemination strategies and GeoVanet to route the set of the computed query results. Therefore, we chose to reuse the simulator we developed in the context of the VESPA project (see http://www.univ-valenciennes.fr/ROI/SID/tdelot/vespa/), which is a simulator dedicated to the evaluation of any data management system designed for vehicular networks. We have successfully used this simulator in several works before (e.g., [9,11,12,14]).

In this section, we first describe the experimental settings and then different experiments that we have performed to evaluate our approach. It should be noted that in a highly dynamic distributed environment like the one presented in this paper it is not possible to guarantee that a complete answer has been obtained (the traditional *closed world assumption* gives way to an *open world assumption* [4]), and therefore an interesting metric to consider for evaluation will be the amount or percentage of interesting results retrieved.

5.1. Experimental settings

During the simulations, the vehicles drive from a random departure location to a random destination location through roads defined according to real maps. The choice of roads used for each vehicle to reach its destination is computed by using the Dijkstra's shortest path algorithm. To calibrate our simulations (e.g., communication times, communication ranges, etc.), we extracted values from tests performed in real environments using our VESPA prototype.

As shown in Fig. 2, we have considered during our experimentations a real road network of two different areas by using digital maps provided by Tele Atlas (http://www.teleatlas.com). The first one (Fig. 2(a)) is an urban area, specifically the area surrounding the city of Lille (France) while the second one (Fig. 2(b)) is a rural area, specifically a countryside area in the North of France. We chose to present results obtained for such areas because they clearly correspond to two different scenarios for our GeoVanet algorithm. Indeed, on the one hand, in an urban area vehicles keep changing direction all the time, which considerably increases the probability that useless changes of the carrier of a message arise. On the other hand, in a rural area vehicles keep the same direction for a long time but the density of vehicles is much lower than in a city and thus next forwarders are rarer.

For each area considered, GeoVanet is evaluated by measuring the quality of the query dissemination and the retrieval of results. In the latter case, GeoVanet is run both when nodes are aware of existing road maps and when they are not.

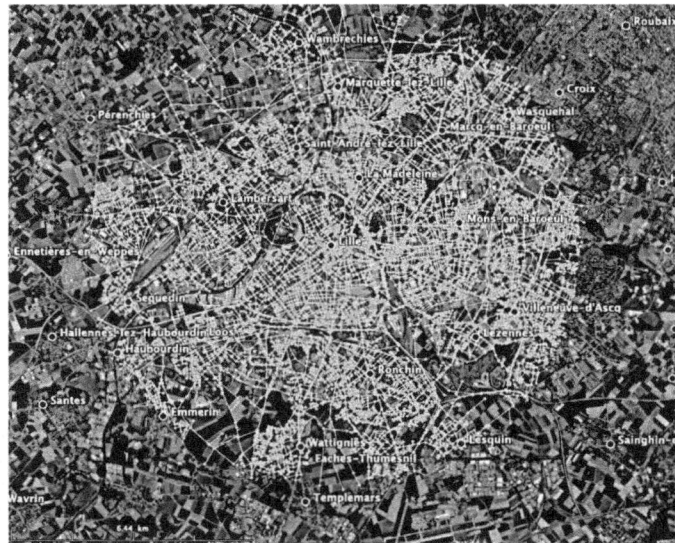

(a) Urban area

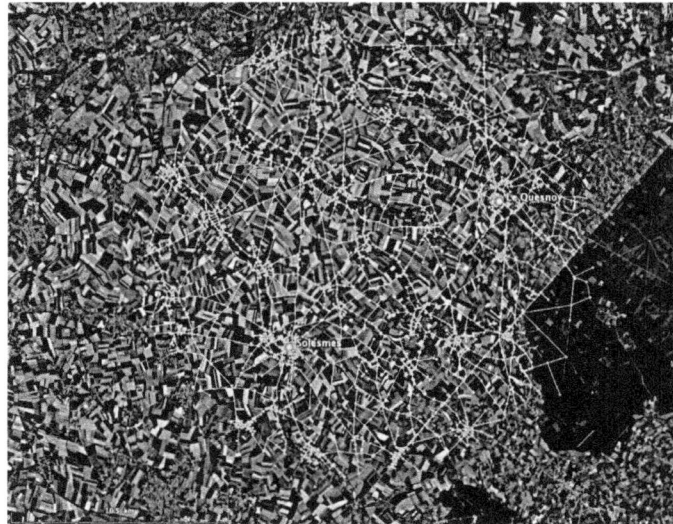

(b) Rural area

Fig. 2. Geographic areas considered during the experimentations.

Each simulation is 16000 seconds long and includes about 25000 vehicles. The evolution of the number of vehicles during the simulations is presented in Fig. 5(a). We tried to model a realistic flow of vehicles with several phases where the traffic increases or decreases. After a warm-up phase, one of the vehicles starts disseminating a query in the vehicular network. Each time a query reaches a vehicle holding a query result, this result is routed towards the mailbox by using GeoVanet, so that it can be eventually delivered to the user. We considered that only 2% of the vehicles carry a query result (we maintain constant this percentage during the whole simulation). Finally, the communication range used by the vehicles to exchange data is set to 200 meters.

5.2. Evaluation of the query dissemination

In this paper, we do not focus on the effectiveness of the query dissemination process, as this is complementary to the GeoVanet protocol. Nevertheless, we present in this section some results obtained for the query dissemination. Our objective here is to show that the percentage of vehicles contacted among those holding interesting information for a particular query is high enough for our approach to collect a suitable number of results. More precisely, we consider four dissemination strategies based on the ideas introduced in Section 4.3:

- *Flooding.* With this strategy, each vehicle receiving the query from one of its neighbors relays it immediately.
- *Contention-based forwarding.* This strategy aims at avoiding problems like the "broadcast storm" [18,35], which may happen with the previous flooding strategy. Therefore, we apply the contention-based forwarding principle described in Section 4.3 to limit the number of vehicles relaying the query in the network. In order to better adapt the process to the environment, we apply two different contention-based forwarding strategies depending on whether the vehicle density is high or not. In an urban environment, where a message is more likely to be received by another vehicle when sent, the message is sent once (*one-time contention-based forwarding*). In a rural environment, the message is regularly sent till it has been received (*periodic contention-based forwarding*).
- *Flooding using a fixed infrastructure.* In this configuration, for urban settings we considered 4 fixed hotspots around the city of Lille. Once these hotspots have received a query they regularly broadcast it in the network. For rural settings, a few hotspots are set in the center of the town or village (in main places like the town hall or the church), in order to facilitate the dissemination when the density of vehicles is low. Each vehicle receiving the query diffused by one of these access points then relays it.
- *Contention-based forwarding using an infrastructure.* This strategy relies on the same infrastructure than the previous one, but in this case the contention-based forwarding principle is applied when the vehicles receiving the query from the hotspot relay it.

When applicable, the broadcast frequency is set to 400 seconds. During our experimentations, we evaluated the percentage of vehicles that held a result and received the query. Some videos illustrating the dissemination of the query in the network for the different strategies considered can be seen at the following address: http://www.univ-valenciennes.fr/ROI/SID/tdelot/GeoVanet/videos/. The experimental results are presented in Fig. 3. All the strategies behave quite well in the urban area, since the percentage of relevant vehicles receiving the query is between 60% and 70% whatever the strategy used. In such an environment, every strategy acts similarly. On the contrary, in a rural environment, the classical flooding performs very poorly compared to other strategies that reach between 70% and 90% of the vehicles owning information of interest. This is due to the low density of vehicles. With the flooding strategy, once a message is relayed, if there is no vehicle within communication range to receive it, the message is lost. By using the periodic contention-based strategy, a message is periodically sent till being reached by another vehicle, which improves the performance of the diffusion. In the same way, when there are some fixed hotspots relaying the message when a vehicle is within range, the results are better.

We also measured, for each strategy, the number of messages diffused. The results are depicted in Fig. 4. The more messages are sent, the more vehicles are expected to be reached. However, the more messages are sent, the more expensive is the strategy in terms of energy consumption and traffic overhead. Therefore, there is a trade-off between the cost and the number of vehicles reached, that must be considered when choosing the query broadcasting strategy. As expected, we observe that the network

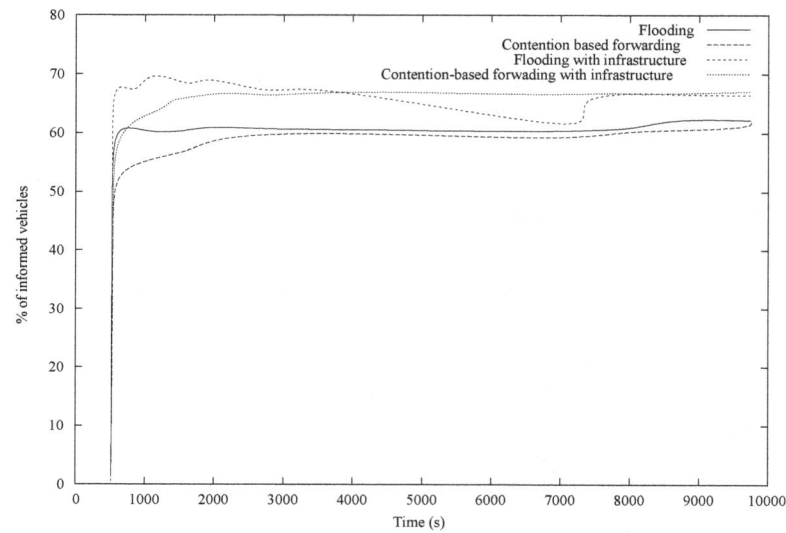

(a) Urban area

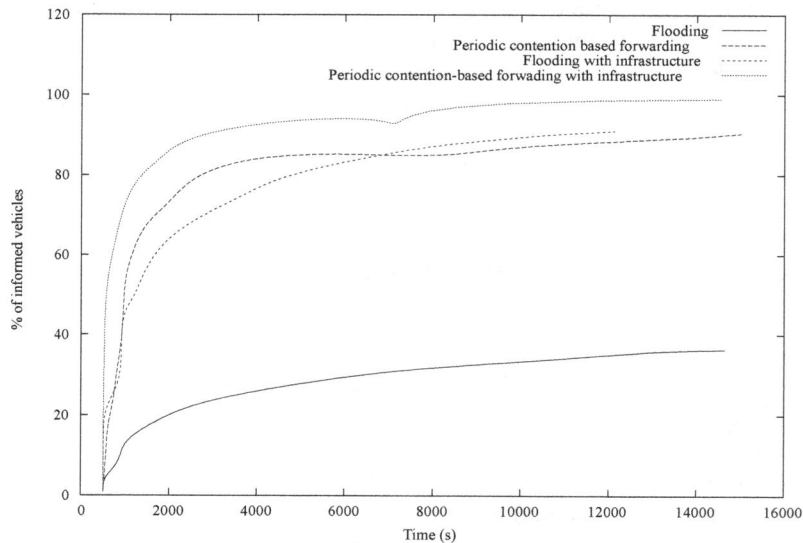

(b) Rural area

Fig. 3. Evolution of the percentage of vehicles holding results that have received the query over time.

traffic load for the strategies based on flooding is higher in an urban environment, since the density of vehicles is high. In a rural environment there are less diffusions with the plain flooding strategy because the message is quickly lost. Another interesting point is the difference in scale of the number of diffusions between both environments (compare the Y-scale of Figs 4(a) and 4(b)). This can be explained by the fact that in a rural environment a periodic contention-based forwarding is needed, unlike in an urban environment, to alleviate the problems derived from the low density of vehicles. Thus, in a rural environment, a vehicle will keep sending regularly a message till reaching a new vehicle not already aware of it.

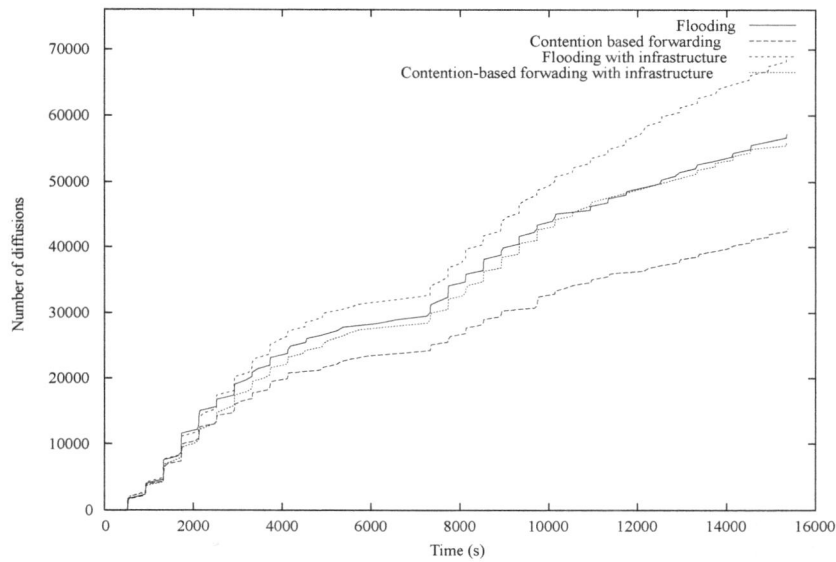

(a) Urban area

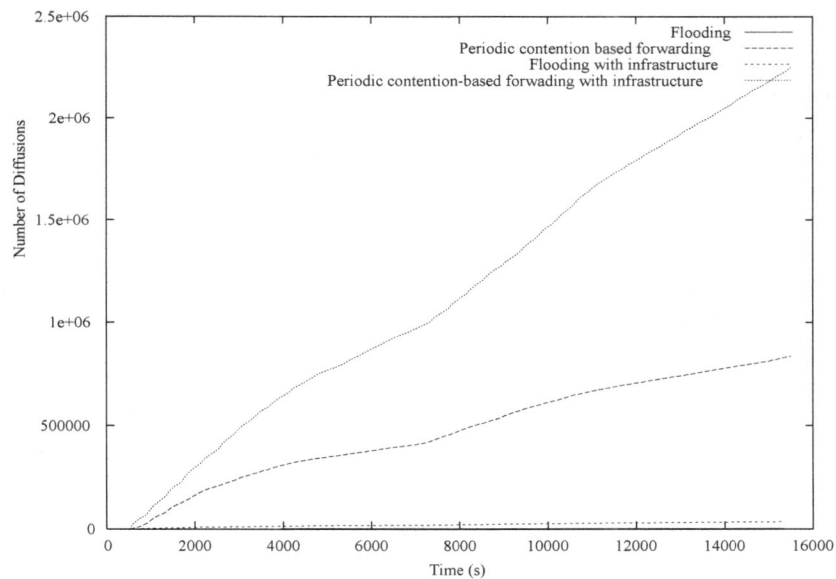

(b) Rural area

Fig. 4. Number of messages diffused for the dissemination strategies considered.

5.3. Evaluation of the delivery of the results using GeoVanet

In this section, we present the results obtained for the evaluations performed using GeoVanet to route the query results towards a fixed mailbox. Once spread in the vehicular network, the query generates different results which are routed towards the corresponding mailbox. For this evaluation, we focus on

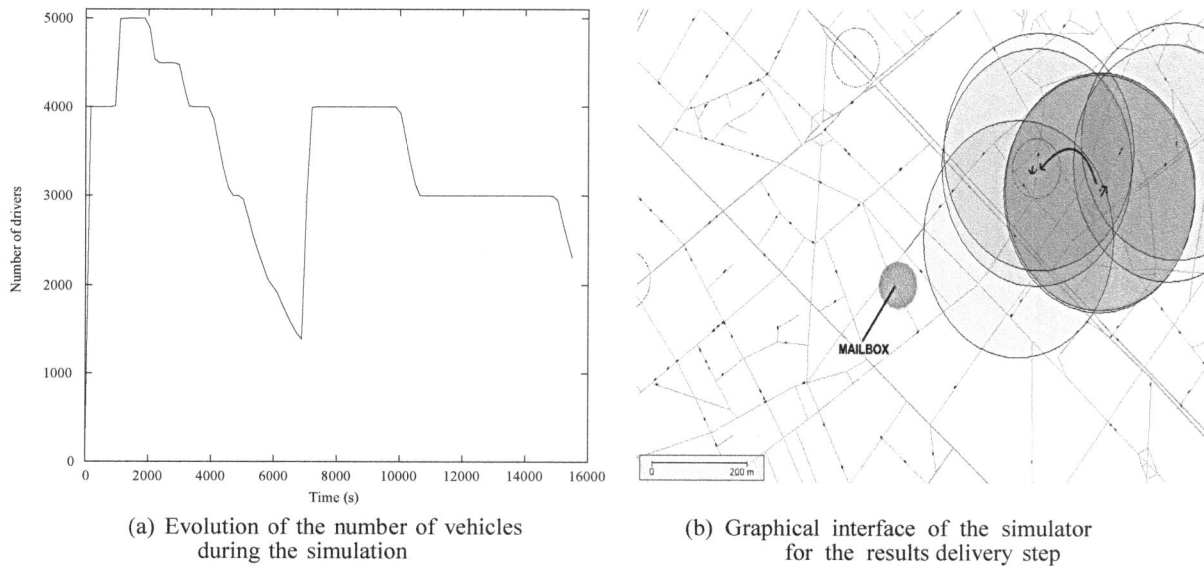

(a) Evolution of the number of vehicles   (b) Graphical interface of the simulator
during the simulation   for the results delivery step

Fig. 5. Simulation settings.

the case where no infrastructure (other than the target mailbox) is available, which can be considered a worst-case situation. In Fig. 5(b), we present the graphical interface provided by our simulator during the routing phase of the query result. The small rectangles correspond to vehicles driving on the roads. The rectangles surrounded by a dotted circle represent vehicles holding interesting information for the query considered. Finally, the larger circles correspond to the different messages exchanged between vehicles when the carrier of a result has to be changed (e.g., discovery of the neighbors' driving direction, diffusion of the result, etc.).

As explained in Section 4.6, we assume that the mobile user who formulated the query will be able to download the set of results collected in the mailbox after the deadline set for the query expiration. In the following, we consider two different configurations to evaluate our GeoVanet algorithm. In the first one, GeoVanet is implemented on top of geographic vectors (vehicles are not aware of road maps), whereas the second one exploits information about the road network provided by digital road maps. In the following, we consider a value of Δt (see Algorithm 1) equal to 10 seconds. We evaluated also several other values (even if the results are not presented due to space constraints) but no important changes were observed with the variation of this parameter.

5.3.1. Implementation of GeoVanet on top of geographic vectors

First, we rely on geographic vectors to estimate the direction of the vehicles and determine whether they are carrying the results towards the target (i.e., the mailbox) or not. Hence, we assume that vehicles are equipped with GPS receivers supporting the storage of previous locations to try to predict future ones. Obviously, the distance between the former (or reference) position and the current one plays a critical role. As illustrated in Fig. 6, a close reference position can be used to generate a vector indicating the instantaneous direction of the vehicle (called *direction vector* in Fig. 6), whereas a more distant one can be used to represent a vector that gives an indication of the global trajectory of the vehicle (called *mobility vector* in Fig. 6).

To illustrate the efficiency of our GeoVanet algorithm, we performed numerous experimentations. In the following, we first study the impact of the size of the vector. Thus, Fig. 7 shows the evolution of the

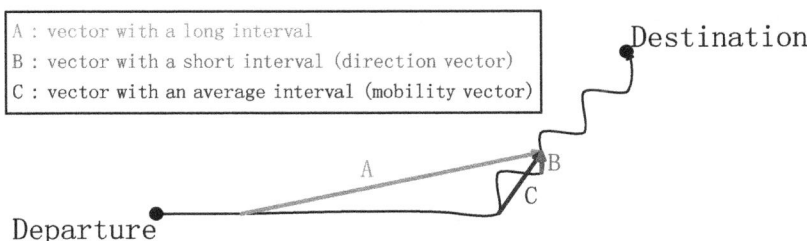

Fig. 6. Influence of the reference position on the geographic vector considered.

percentage of results collected in the mailbox over time according to the reference position considered to generate the vector. In both environments, it appears that the use of a vector providing a good indication of the vehicle's direction (i.e., a direction vector) is the best choice to decide when the carrier of the information has to be changed. Moreover, we observe that we are able to successfully route more than 60% of the results in the urban area and more than 40% in the rural area. Besides, let us note that the time needed to collect the majority of the results is very limited since most results are already available after one hour. This is satisfactory and proves that GeoVanet can be used even in scenarios where the time limit could be much smaller than the one we considered for the touristic application we introduced in Section 3.

Obviously, the choice of the reference position (i.e., the size of the vector) also has an incidence on the network traffic. Indeed, the shorter the vector is, the higher the probability to decide to change the carrier, whatever the environment. Nevertheless, as shown in Fig. 8, the accumulated number of diffusions does not significantly increase for the shorter vectors, that provide a precise indication of the direction of the vehicle.

As depicted in Fig. 7, we do not collect all the query results transmitted by vehicles reached with the query. One reason why we lose results using geographic vectors is due to the presence of loops. Due to the road topology, useless changes of carrier may arise because a car is temporarily driving in the wrong direction. More precisely, we observe in Fig. 9 that a vehicle (i.e., the rectangle surrounded by a circle) may decide at one moment that it is not driving in the direction of the mailbox and choose to pass the result to carry to another vehicle. However, the first vehicle may be driving in the wrong direction only for a short period of time, and so deciding to change the carrier may be a bad decision. Moreover, the carrier may even choose a vehicle following exactly the same path and only temporarily driving towards the mailbox. Such loops may lead to delays or even to a loss of results which may never reach the target. This may also generate an overhead in terms of the number of changes of carrier and hops needed to reach the mailbox.

During our experimentations, we also observed the number of hops (i.e., changes of carrier) needed for a result to reach the mailbox between the moment when the vehicle carrying the result receives the user query and the moment when the result joins the mailbox. Figure 10 presents the percentage of information collected according to the number of hops. Whatever the size of the vector, the number of hops needed to reach the mailbox is quite high (i.e., more than 30 hops to navigate at most 7 km for 50% of the results collected) and similar in both environments. In the urban environment, these results can be explained by the facts that the vehicles frequently change direction and the traffic is dense, and thus a message is very likely to reach a vehicle heading in the right direction. In the rural environment, this can be explained by the low density of vehicles and the use of distance vectors. Because of the low density of vehicles, as soon as a vehicle apparently heading towards the right direction is passed by, it

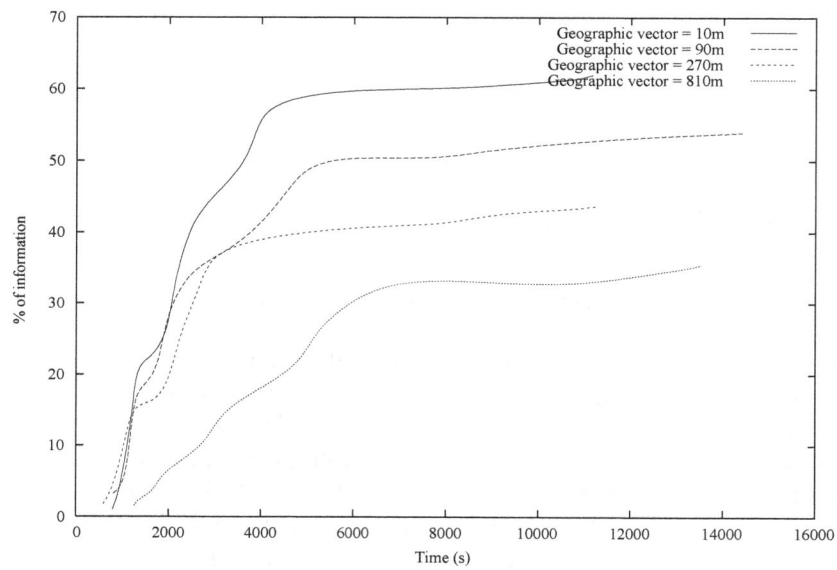

(a) Urban area

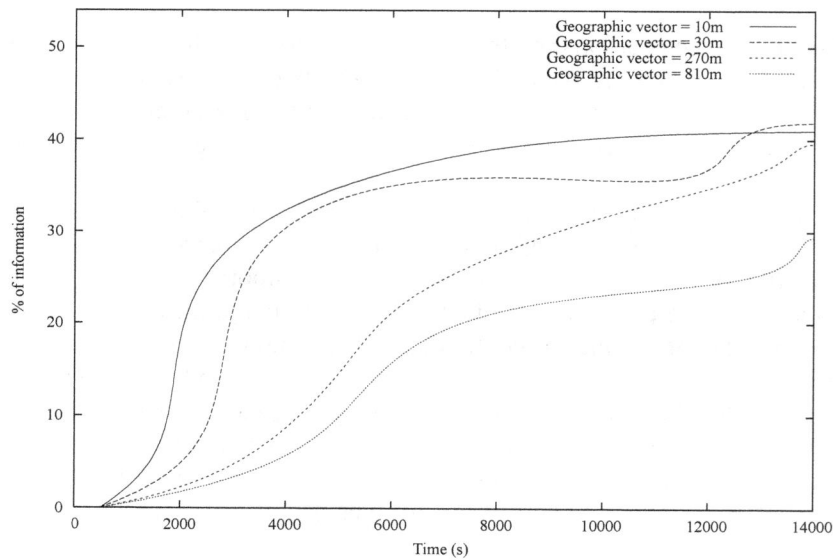

(b) Rural area

Fig. 7. Percentage of information collected over time using variable length vectors.

takes charge of the message. However, since the routes are longer, the route may actually head towards the wrong direction. Since the density is low, a large amount of time may pass before crossing another vehicle.

5.3.2. Implementation of GeoVanet on top of digital road maps

Using maps, each vehicle detects every $\Delta t = 20s$ that it is not driving in the right direction if the shor-

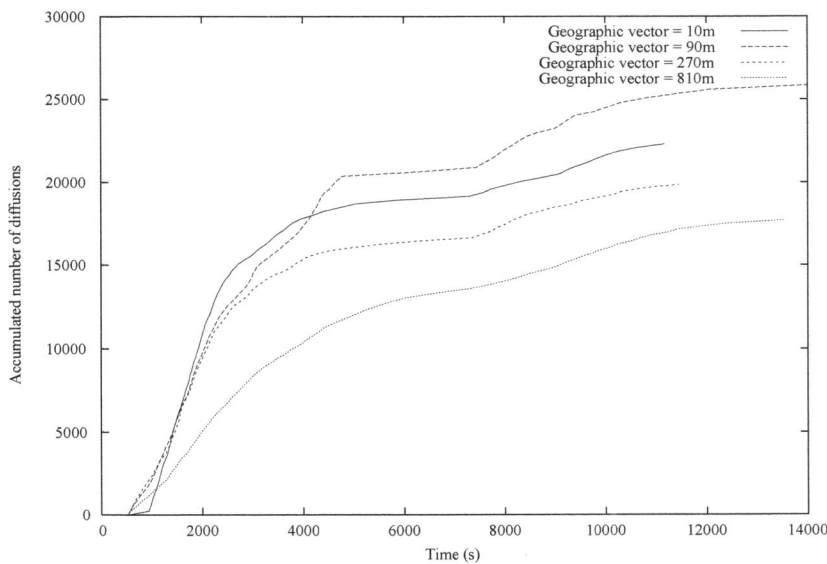

(a) Urban area

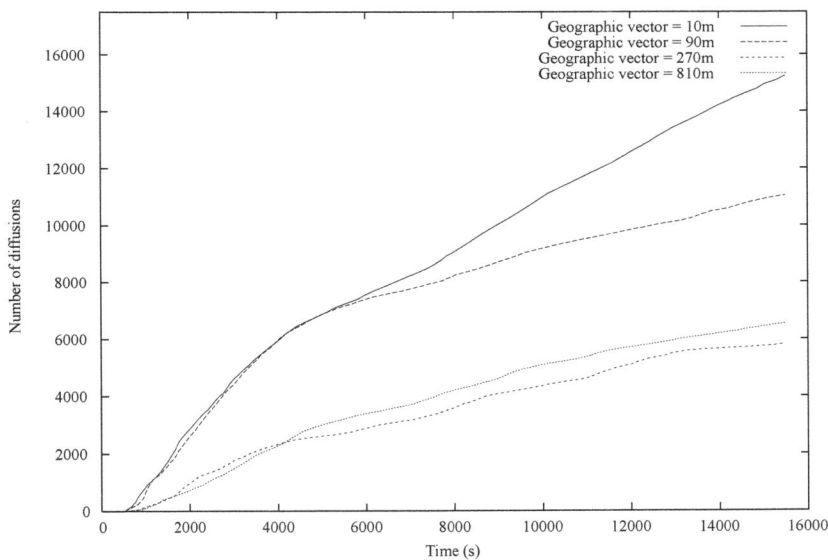

(b) Rural area

Fig. 8. Evolution of the total number of diffusions over time.

test path between this vehicle and the target (i.e., the mailbox) increases. In that case, GeoVanet advocates a change of carrier in order to reach the target. Even if we consider here digital maps, geographic vectors are also necessary to determine the closest vehicle driving towards the mailbox, which should become the new carrier.

Figure 11 shows the evolution of the results collected in the mailbox for different thresholds. A threshold of 0% means that the carrier of the result is changed as soon as an increase of the shortest path

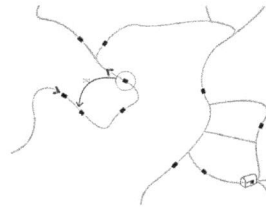

Fig. 9. Illustration of possible loops.

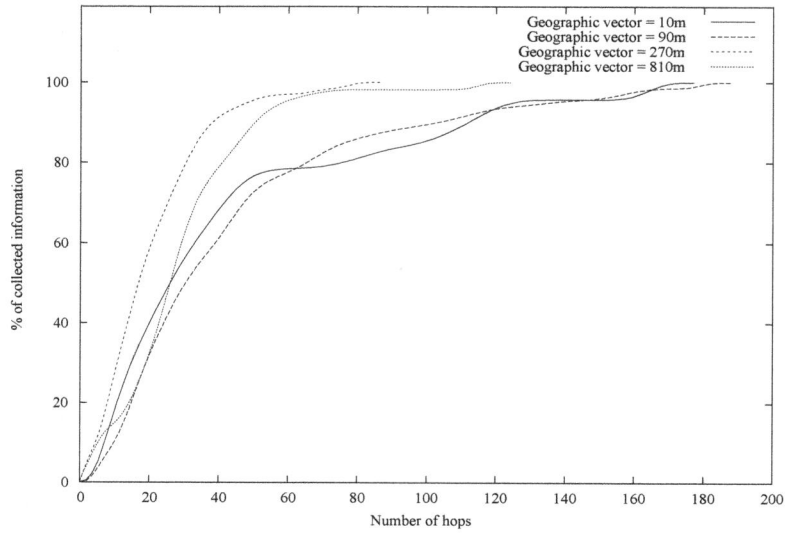

(a) Urban area

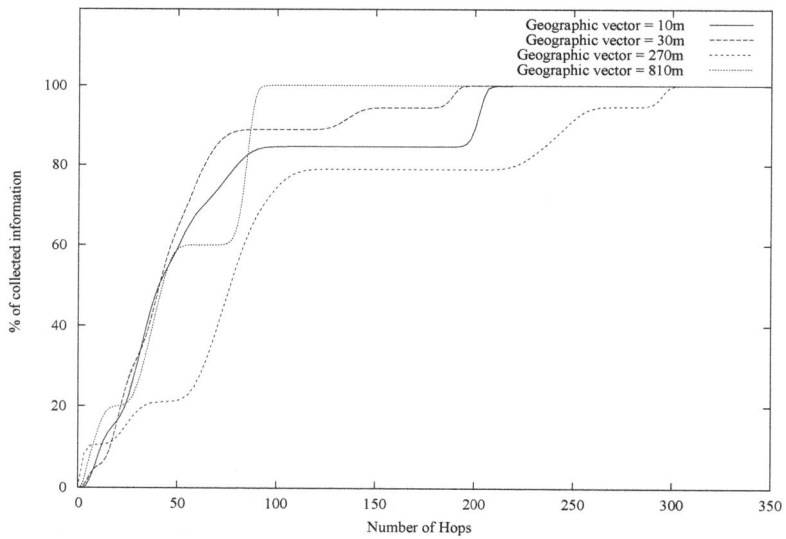

(b) Rural area

Fig. 10. Percentage of recovered information according to the number of hops.

(a) Urban area

(b) Rural area

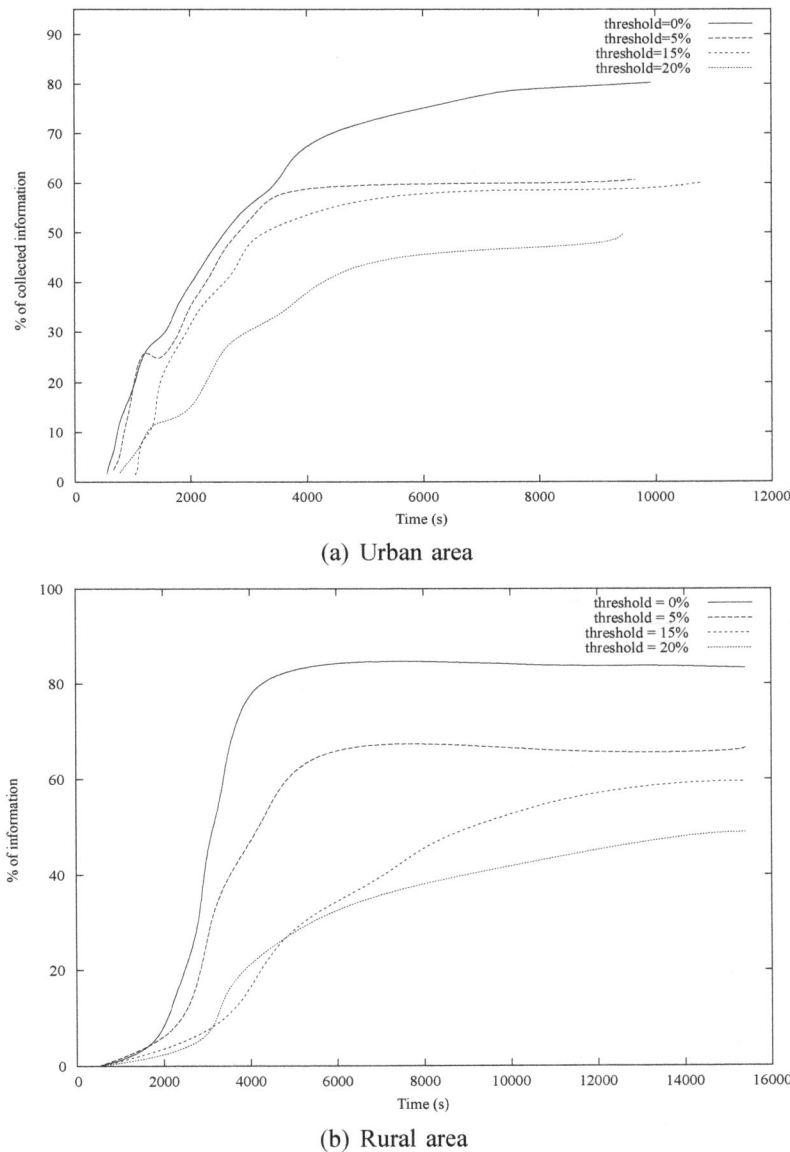

Fig. 11. Percentage of information collected over time using digital maps.

is detected. More generally, with a threshold set to $x\%$, the change is initiated only if the increase is greater than $x\%$. We observe in Fig. 11 that the percentage of results collected is higher if the change is initiated immediately when the shortest path increases, as this leads to the fastest reaction possible when a wrong route is being taken. Using digital maps, the quantity of results collected is higher than using geographic vectors. Indeed, the percentage of results collected is close to 80% in the best case. The time needed is similar to the case using geographic vectors. So, most of the results are available at most one hour after the first dissemination of the query.

As we did for the implementation of GeoVanet on top of geographic vectors, we measured the number

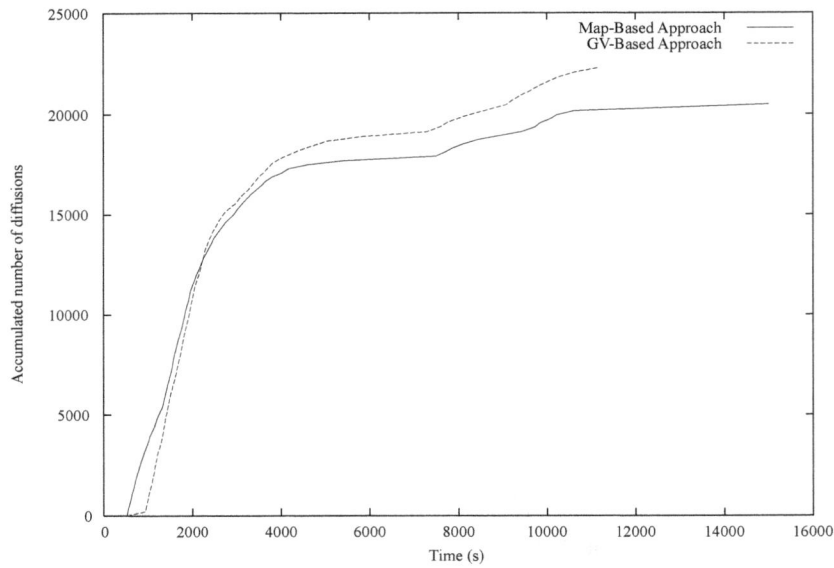

(a) Urban area

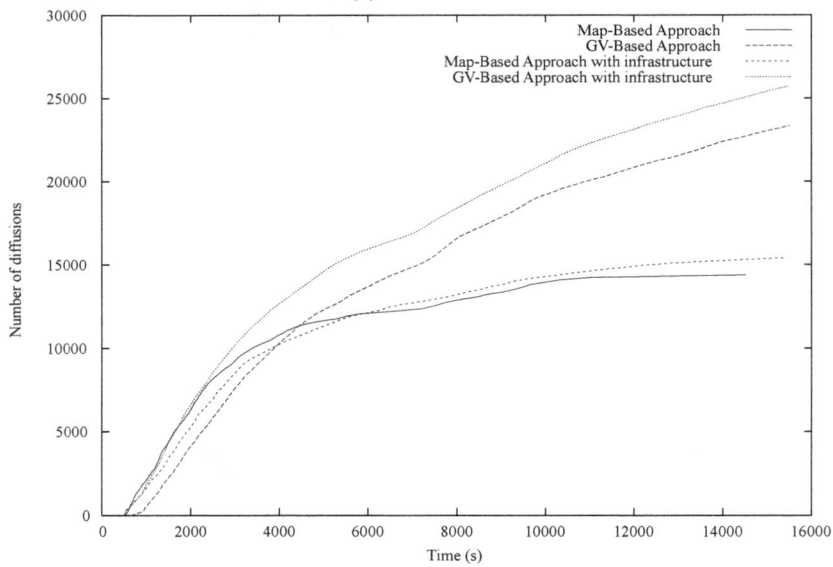

(b) Rural area

Fig. 12. Comparison of the evolution of the number of messages diffused over time between maps and vectors.

of messages diffused while routing query results towards the mailbox. Figure 12 shows a comparison between the results obtained for maps (*Map-based approach*) and vectors (*GV-based approach*). We considered and compared the best configurations. The exploitation of digital maps leads to a reduction of the number of messages diffused whatever the environment and independently of whether an infrastructure is deployed or not. This is due to the fact that loops are avoided and thus messages reach the destination with a lower number of hops.

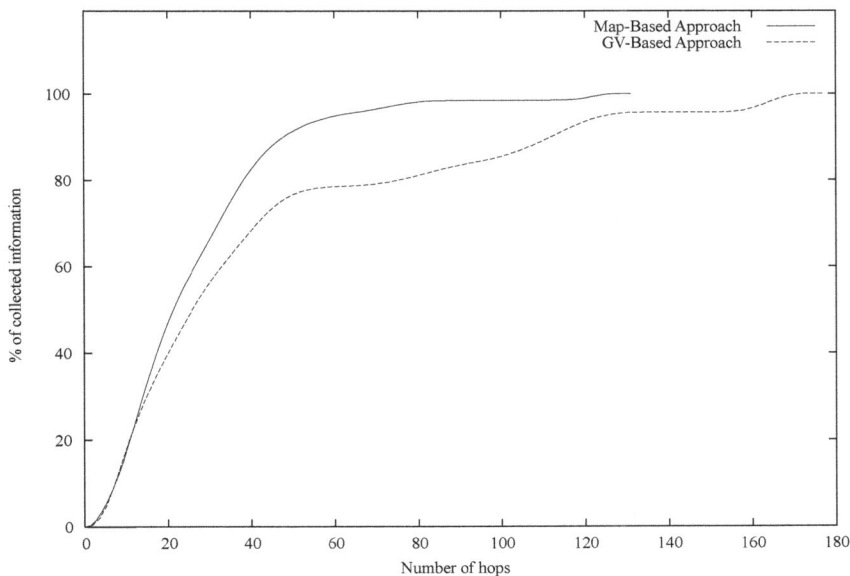

(a) Urban area

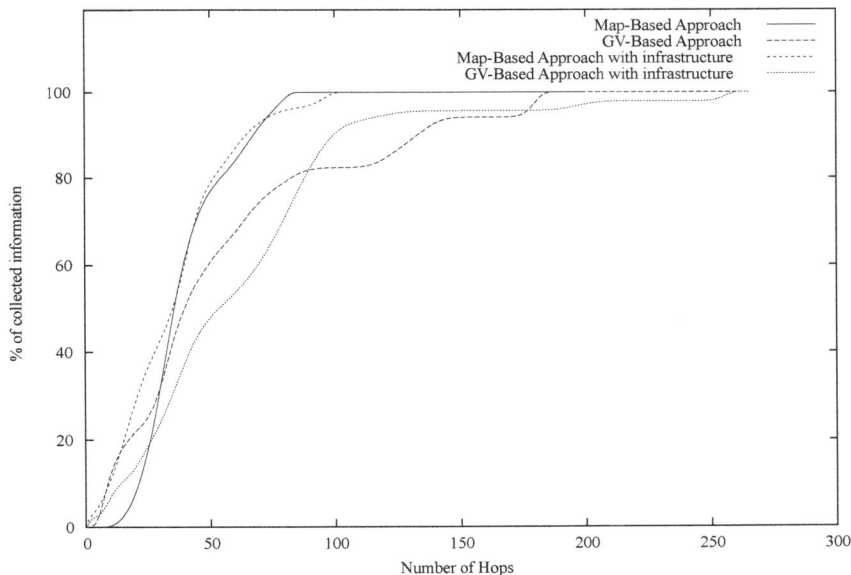

(b) Rural area

Fig. 13. Comparison of the percentage of recovered information according to the number of hops between maps and vectors.

Figure 13 shows the number of hops needed to reach the mailbox considering the use of digital maps and geographic vectors. We kept the same parameters as for our previous study on the number of messages. We can observe that the number of hops is strongly reduced by using the information provided by digital maps, which also confirms our assumptions regarding the reduction of the number of messages diffused observed in Fig. 12, whatever the environment.

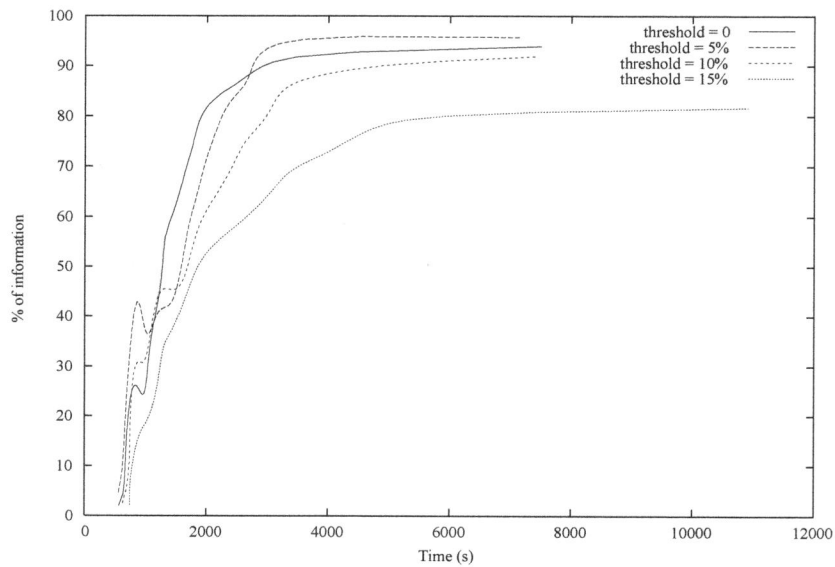

(a) Urban area

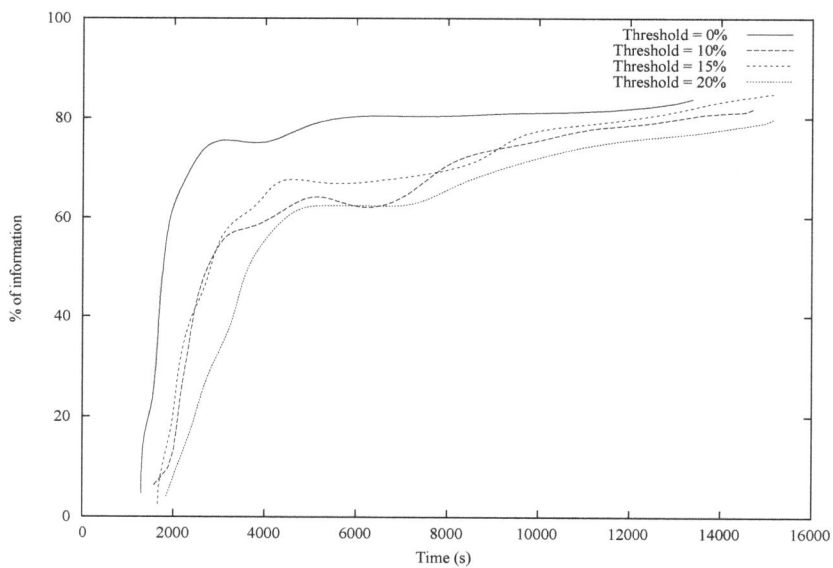

(b) Rural area

Fig. 14. Percentage of information collected over time by duplicating the carriers.

5.3.3. Evaluation of another possible optimization: Multiple carriers

Obviously, one solution to improve the percentage of results collected in the mailbox is to increase the number of message carriers at each hop. Figure 14 shows that, with the approach based on maps, up to 90% in the urban area and 80% in the rural area of the results available can be collected if the two closest vehicles driving in the right direction are used as carriers (instead of only one carrier as in our basic approach). However, this also leads to a severe increase in the network load, as depicted in Fig. 15, and

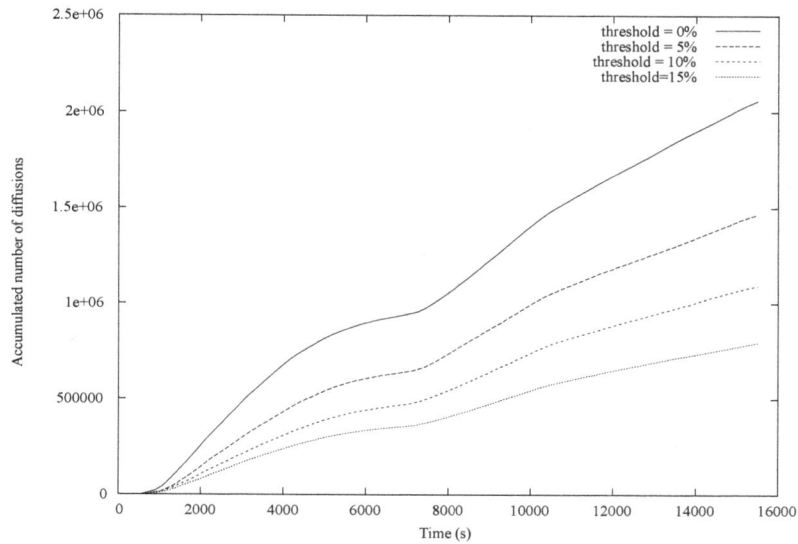

(a) Urban area

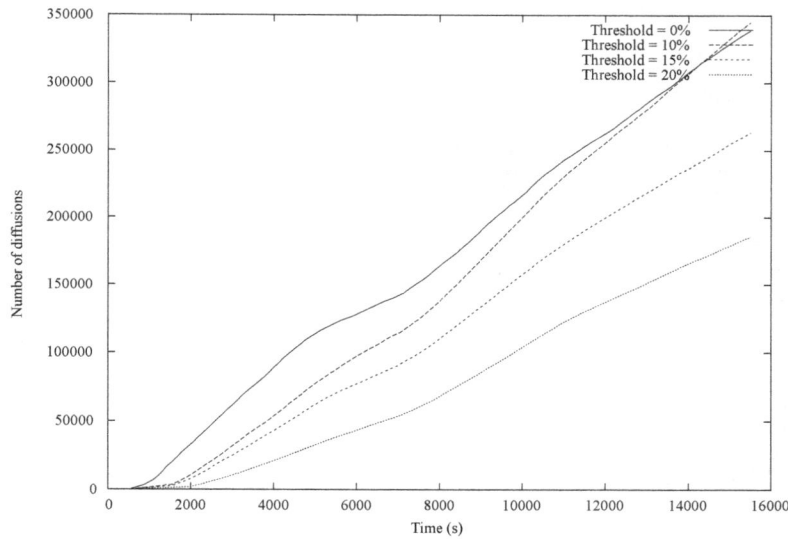

(b) Rural area

Fig. 15. Evolution of the number of messages diffused over time by duplicating the carriers.

to many duplicates received at the mailbox, especially in the urban area (where the density of vehicles is high). Therefore, this optimization should be used in case the density of vehicles is low and retrieving a very high percentage of results is needed.

6. Conclusions and future work

In this paper, we presented GeoVanet, an approach that supports query processing in vehicular networks by providing an efficient and effective protocol for routing the results of the queries. GeoVanet therefore allows users to obtain the interesting data items within a bounded time. Through numerous simulations, we have proved the effectiveness of our solution in several environments and with different assumptions. The performance of Geovanet is high, especially when its implementation exploits the information available on digital road maps.

In the future, we plan to study in detail the impact of the number of mailboxes and their location. We would also like to study the potential interest of partitioning and distributing a single mailbox in different locations. Another interesting research direction will be to consider mixed environments where some vehicles may be equipped with digital maps and others are not.

Acknowledgments

This work was partially supported by the French ANR agency in the scope of the OPTIMACS project and by the CICYT project TIN2010-21387-C02-02.

References

[1] S. Androutsellis-Theotokis and D. Spinellis, A survey of peer-to-peer content distribution technologies, *ACM Computing Surveys* **36**(4) (2004), 335–371.

[2] F. Aurenhammer, Voronoi diagrams – a survey of a fundamental geometric data structure, *ACM Computing Surveys* **23** (1991), 345–405.

[3] P. Belanovic, D. Valerio, A. Paier, T. Zemen, F. Ricciato and C.F. Mecklenbrauker, On wireless links for vehicle-to-infrastructure communications, *IEEE Transactions on Vehicular Technology* **59**(1) (2011), 269–282.

[4] A. Bonifati, P.K. Chrysanthis, A.M. Ouksel and K.-U. Sattler, Distributed databases and peer-to-peer databases: past and present, *SIGMOD Record* **37**(1) (2008), 5–11.

[5] M. Burmester, E. Magkos and V. Chrissikopoulos, Strengthening privacy protection in VANETs. In *2008 IEEE International Conference on Wireless & Mobile Computing, Networking & Communication (WiMob'08)*, pages 508–513. IEEE Computer Society, 2008.

[6] J. Cao, X. Feng, J. Lu and S.K. Das, Mailbox-based scheme for designing mobile agent communication protocols, *Computer* **35**(9) (2002), 54–60.

[7] A.A.V. Castro, G.D.M. Serugendo and D. Konstantas, Hovering information – self-organising information that finds its own storage, In *2008 IEEE International Conference on Sensor Networks, Ubiquitous, and Trustworthy Computing (SUTC'08)*, pages 193–200. IEEE Computer Society, 2008.

[8] P. Cencioni and R.D. Pietro, A mechanism to enforce privacy in vehicle-to-infrastructure communication, *Computer Communications* **31** (2008), 2790–2802.

[9] N. Cenerario, T. Delot and S. Ilarri, A content-based dissemination protocol for VANETs: Exploiting the encounter probability, *IEEE Transactions on Intelligent Transportation Systems*, 2011. 12 pages, to appear, DOI: 10.1109/TITS.2011.2158821.

[10] P. Costa, D. Frey, M. Migliavacca and L. Mottola, Towards lightweight information dissemination in inter-vehicular networks, In *Third International Workshop on Vehicular Ad Hoc Networks (VANET'06)*, pages 20–29. ACM, 2006.

[11] T. Delot, N. Cenerario and S. Ilarri, Vehicular event sharing with a mobile peer-to-peer architecture, *Transportation Research Part C: Emerging Technologies* **18**(4) (2010), 584–598.

[12] T. Delot, S. Ilarri, N. Cenerario and T. Hien, Event sharing in vehicular networks using geographic vectors and maps, *Mobile Information Systems* **7**(1) (2011), 21–44.

[13] T. Delot, S. Ilarri, M. Thilliez, G. Vargas-Solar and S. Lecomte, Multi-scale query processing in vehicular networks, *Journal of Ambient Intelligence and Humanized Computing*, 2011. 14 pages, to appear, published online (SpringerLink): 05 June 2011, DOI: 10.1007/s12652-011-0058-y.

[14] T. Delot, N. Mitton, S. Ilarri and T. Hien, Decentralized pull-based information gathering in vehicular networks using GeoVanet, In *12th International Conference on Mobile Data Management (MDM 2011)*, pages 174–183. IEEE Computer Society, 2011.

[15] N. Dutta, A peer to peer based information sharing scheme in vehicular ad hoc networks, In *11th International Conference on Mobile Data Management (MDM 2010)*, pages 309–310. IEEE Computer Society, 2010.

[16] M. Gerla and L. Kleinrock, Vehicular networks and the future of the mobile internet, *Computer Networks* **55**(2) (2011), 457–469.

[17] G. Ghinita, P. Kalnis, A. Khoshgozaran, C. Shahabi and K.-L. Tan, Private queries in location based services: anonymizers are not necessary, In *ACM SIGMOD International Conference on Management of Data (SIGMOD'08)*, pages 121–132. ACM, 2008.

[18] A.M. Hanashi, I. Awan and M. Woodward, Performance evaluation with different mobility models for dynamic probabilistic flooding in MANETs, *Mobile Information Systems* **5**(1) (2009), 65–80.

[19] H. Huang, J.H. Hartman and T.N. Hurst, Efficient and robust query processing for mobile wireless sensor networks, *International Journal of Sensor Networks* **2**(1/2) (2007), 99–107.

[20] S. Ilarri, C. Bobed and E. Mena, An approach to process continuous location-dependent queries on moving objects with support for location granules, *Journal of Systems and Software* **84**(8) (2011), 1327–1350.

[21] S. Ilarri, E. Mena and A. Illarramendi, Location-dependent query processing: Where we are and where we are heading, *ACM Computing Surveys* **42**(3) (2010), 1–73.

[22] J. Jeong, S. Guo, Y. Gu, T. He and D.H. Du, TSF: Trajectory-based statistical forwarding for infrastructure-to-vehicle data delivery in vehicular networks. In *30th IEEE International Conference on Distributed Computing Systems (ICDCS 2010)*, pages 557–566. IEEE Computer Society, 2010.

[23] E. Kang, M.J. Kim, E.-J. Lee and U.-M. Kim, DHT-based mobile service discovery protocol for mobile ad hoc networks, In *Fourth International Conference on Intelligent Computing (ICIC'08): Advanced Intelligent Computing Theories and Applications – With Aspects of Theoretical and Methodological Issues*, volume 5226 of *Lecture Notes in Computer Science*, pages 610–619. Springer, 2008.

[24] E. Kulla, M. Hiyama, M. Ikeda, L. Barolli, V. Kolici and R. Miho, MANET performance for source and destination moving scenarios considering OLSR and AODV protocols, *Mobile Information Systems* **6**(4) (2010), 325–339.

[25] U. Lee, J.-S. Park, E. Amir and M. Gerla, FleaNet: A virtual market place on vehicular networks, *IEEE Transactions on Vehicular Technology* **59**(1) (2010), 344–355.

[26] I. Lequerica, P.M. Ruiz and V. Cabrera, Improvement of vehicular communications by using 3G capabilities to disseminate control information, *IEEE Network: The Magazine of Global Internetworking* **24**(1) (2010), 32–38.

[27] F. Li and Y. Wang, Routing in vehicular ad hoc networks: A survey, *IEEE Vehicular Technology Magazine* **2**(2) (2007), 12–22.

[28] T.-H. Lin, H.-C. Chao and I. Woungang, An enhanced MPR-based solution for flooding of broadcast messages in OLSR wireless ad hoc networks, *Mobile Information Systems* **6**(3) (2010), 249–257.

[29] C. Lochert, B. Scheuermann, M. Caliskan and M. Mauve, The feasibility of information dissemination in vehicular ad-hoc networks, In *Fourth Annual Conference on Wireless On demand Network Systems and Services (WONS'07)*, pages 92–99. IEEE Computer Society, 2007.

[30] C. Maihöfer, A survey of geocast routing protocols, *IEEE Communications Surveys & Tutorials* **6**(2) (2004), 32–42.

[31] G.K. Mitropoulos, I.S. Karanasiou, A. Hinsberger, F. Aguado-Agelet, H. Wieker, H.-J. Hilt, S. Mammar and G. Noecker, Wireless local danger warning: Cooperative foresighted driving using intervehicle communication, *IEEE Transactions on Intelligent Transportation Systems* **11**(3) (2010), 539–553.

[32] Y. Morgan, Notes on DSRC & WAVE standards suite: Its architecture, design, and characteristics, *IEEE Communications Surveys & Tutorials* **12**(4) (2010), 504–518.

[33] M. Motani, V. Srinivasan and P.S. Nuggehalli, PeopleNet: engineering a wireless virtual social network. In *11th Annual International Conference on Mobile Computing and Networking (MobiCom'05)*, pages 243–257. ACM, 2005.

[34] V. Namboodiri, M. Agarwal and L. Gao, A study on the feasibility of mobile gateways for vehicular ad-hoc networks, In *First ACM International Workshop on Vehicular Ad Hoc Networks (VANET'04)*, pages 66–75. ACM, 2004.

[35] S.-Y. Ni, Y.-C. Tseng, Y.-S. Chen and J.-P. Sheu, The broadcast storm problem in a mobile ad hoc network. In *Fifth Annual ACM/IEEE International Conference on Mobile Computing and Networking (MobiCom'99)*, pages 151–162. ACM, 1999.

[36] S. Olariu and M.C. Weigle, editors, *Vehicular Networks: From Theory to Practice*, Chapman & Hall/CRC, 2009.

[37] M. T. Özsu and P. Valduriez, *Principles of Distributed Database Systems*, chapter 1 "Introduction", pages 1–40. Springer, New York, NY, USA, 2011. Third edition. See Section 1.3 "Data Delivery Alternatives" (pages 5–7).

[38] M.T. Özsu and P. Valduriez, *Principles of Distributed Database Systems, Third Edition*, chapter "Peer-to-Peer Data Management", pages 611–655. Springer, 2011.

[39] M.G.C. Resende and P.M. Pardalos, *Handbook of Optimization in Telecommunications*, Springer, New York, NY, USA, 2006.

[40] R. Saqour, M. Shanuldin and M. Ismail, Prediction schemes to enhance the routing process in geographical GPSR ad hoc protocol, *Mobile Information Systems* **3**(3–4) (2007), 203–220.

[41] E. Spaho, L. Barolli, G. Mino, F. Xhafa, V. Kolici and R. Miho, Implementation of CAVENET and its usage for performance evaluation of AODV, OLSR and DYMO protocols in vehicular networks, *Mobile Information Systems* **6**(3) (2010), 213–227.

[42]   L. Sweeney, k-anonymity: a model for protecting privacy, *International Journal of Uncertainty, Fuzziness and Knowledge-Based Systems* **10** (2002), 557–570.

[43]   R. Trillo, S. Ilarri and E. Mena, Comparison and performance evaluation of mobile agent platforms, In *Third International Conference on Autonomic and Autonomous Systems (ICAS'07)*, pages 41–46. IEEE Computer Society, 2007.

[44]   O. Urra, S. Ilarri, T. Delot and E. Mena, Mobile agents in vehicular networks: Taking a first ride, In *Eight International Conference on Practical Applications of Agents and Multi-Agent Systems (PAAMS 2010)*, volume 70 of *Advances in Intelligent and Soft Computing*, pages 118–124. Springer, 2010.

[45]   A.B. Waluyo, B. Srinivasan and D. Taniar, Research in mobile database query optimization and processing, *Mobile Information Systems* **1**(4) (2005), 225–252.

[46]   A.B. Waluyo, D. Taniar, B. Srinivasan and W. Rahayu, An enhanced global index for location-based mobile broadcast services. In *International Conference on Advanced Information Networking and Applications (AINA 2010)*, pages 1173–1180. IEEE Computer Society, 2010.

[47]   B. Xu, F. Vafaee and O. Wolfson, In-network query processing in mobile P2P databases. In *17th ACM SIGSPATIAL International Conference on Advances in Geographic Information Systems (GIS'09)*, pages 207–216. ACM, 2009.

[48]   Q. Xu, H.T. Shen, Z. Chen, B. Cui, X. Zhou and Y. Dai, Hybrid retrieval mechanisms in vehicle-based P2P networks. In *Ninth International Conference on Computational Science (ICCS'09)*, volume 5544 of *Lecture Notes in Computer Science*, pages 303–314. Springer, 2009.

[49]   K. Xuan, D. Taniar, M. Safar and B. Srinivasan, Time constrained range search queries over moving objects in road networks, In *Eight International Conference on Advances in Mobile Computing and Multimedia (MoMM 2010)*, pages 329–336. ACM, 2010.

[50]   K. Xuan, G. Zhao, D. Taniar, W. Rahayu, M. Safar and B. Srinivasan, Voronoi-based range and continuous range query processing in mobile databases, *Journal of Computer and System Sciences* **77**(4) (2011), 637–651.

[51]   Y. Zhang, J. Zhao and G. Cao, Roadcast: A popularity aware content sharing scheme in VANETs. In *29th IEEE International Conference on Distributed Computing Systems (ICDCS'09)*, pages 223–230. IEEE Computer Society, 2009.

[52]   R. Zimmermann, W.-S. Ku and H. Wang, Spatial data query support in peer-to-peer systems. In *28th Annual International Computer Software and Applications Conference (COMPSAC'04) – Workshops and Fast Abstracts*, volume 2, pages 82–85. IEEE Computer Society, 2004.

Thierry Delot (http://www.univ-valenciennes.fr/ROI/SID/tdelot/) is an Associate Professor at the University of Valenciennes since 2002. He is a member of the LAMIH laboratory (FRE CNRS 3304). He got a PhD in Computer Science at the University of Versailles in 2001. His research interests mainly concern mobile data management and query processing. Since 2007, Thierry is particularly interested in vehicular ad hoc networks.

Nathalie Mitton received the MSc and PhD. degrees in Computer Science from the INSA de Lyon in 2003 and 2006 respectively. She is currently an INRIA full researcher. Her research interests are mainly focused on theoretical aspects of self-organization, self-stabilization, energy efficient routing and neighbour discovery algorithms for wireless sensor networks as well as RFID middlewares. She is involved in several program and organization committees such as MASS 2011, AdHocNow 2011, MASS 2011, WiSARN and WiSARN-Fall 2011&2010, WWASN 2010&2009, SANET 2008&2007.

Sergio Ilarri received his B.S. and his PhD in Computer Science from the University of Zaragoza in 2001 and 2006, respectively. Now, he is an Associate Professor in the Department of Computer Science and Systems Engineering. For a year, he was a visiting researcher in the Mobile Computing Laboratory at the Department of Computer Science at the University of Illinois in Chicago, and he has also cooperated (through several research stays) with the University of Valenciennes and with IRIT in Toulouse. His research interests include data management issues for mobile computing, vehicular networks, mobile agents, and the Semantic Web.

Thomas Hien is the holder of a Master's degree in Computer Science from the University of Valenciennes, France, obtained in 2009. Since then, he is a member of the LAMIH laboratory (FRE CNRS 3304) devoted to the VESPA project as a research engineer. His work mainly concerns the evaluation of inter-vehicle communication systems in realistic environments.

11

Seamless MANET autoconfiguration through enhanced 802.11 beaconing

M. José Villanueva, Carlos T. Calafate*, Álvaro Torres, José Cano, Juan-Carlos Cano and Pietro Manzoni

Department of Computer Engineering, Universitat Politècnica de València, Valencia, Spain

Abstract. The deployment of mobile ad-hoc networks involves several configuration steps, which complicate research efforts and hinder user interest. This problem prompts for new approaches offering full autoconfiguration of terminals at the different network layers involved. In this paper we propose a novel solution for the autoconfiguration of IEEE 802.11 based MANETs that relies on SSID parameter embedding. Our solution allows users to join an existing MANET without resorting to any additional technology, and even in the presence of encrypted communications. Experimental testbed results using a real implementation of the proposed solution show that it introduces significant improvements compared to other existing solutions, allowing nearby stations to be configured in about two seconds, and also enabling multi-hop dissemination of configuration data to take place quickly and efficiently.

Keywords: IEEE 802.11, MANET autoconfiguration, SSID, bootstrap problem

1. Introduction

Mobile Ad Hoc Networks (MANETs) [9] are a networking paradigm where terminals communicate wirelessly and in a multi-hop fashion, not requiring any infrastructure of support. Their characteristics in terms of flexibility and cost have made them the candidate technology for different applications such as rescue and military scenarios [14], information dissemination in vehicular networks [22], multimedia databases [1], or even video communication between peers [18]. However, even after a decade of research efforts, their ease-of-use is still quite low, making it a technology only accessible for experts.

One of the main reasons hindering a large scale deployment of these networks is the initial configuration phase [10]. For a MANET to be fully operational all stations must be configured using compatible layer-2 and layer-3 parameters. In particular, if the IEEE 802.11 technology [7] is adopted for the physical (PHY) and medium access control (MAC) layers, there are some basic parameters that all terminals must share. The common PHY parameters are mainly the modulation type, the frequency, and the synchronization timestamp; notice that these parameters are typically set automatically by the wireless interface without user intervention. Concerning MAC parameters, stations must share: (i) the service set identifier (SSID), (ii) the power-saving mode, (iii) the encryption mode, and (iv) the encryption key. Notice that the power-saving mode and the encryption mode are usually detected automatically, while the other two parameters must be set manually by the user. At layer-3 several parameters must also be

---

*Corresponding author: Carlos T. Calafate, Department of Computer Engineering, Universitat Politècnica de València, Camino de Vera S/N, 46022 Valencia, Spain. E-mail: calafate@disca.upv.es.

set: (i) the IP version used – IPv4 or IPv6-, (ii) the station's IP address, (iii) the network mask, (iv) the routing protocol used, and (v) the gateway to the Internet. Concerning the latter two parameters, notice that routing protocols are essential in MANETs to make multi-hop communication possible [2,4]; additionally, these protocols usually offer gateway information either automatically or upon user request.

The aforementioned list of parameters evidence the complexity in configuring MANET stations. Also, since MANETs lack any sort of centralized server to handle configuration, all terminals involved must share this task in a distributed manner. Overall, we consider that there are mainly two barriers preventing distributed node configuration to be effective: on one hand, a wireless link must be established to share all the configuration parameters required to configure the wireless link itself; on the other hand, the fact that wireless communications are easy to intercept typically requires encryption to be adopted, which further complicates the configuration process if the encryption key itself is one of the configuration parameters required. Notice that in both cases we have a variant of the bootstrapping problem, which is typically complex to solve. Up to now this problem has remained mostly untackled by the research community, and no real alternative to manual setup has been found.

In this paper we propose a solution that is able to solve the MANET autoconfiguration problem described above in a very efficient and straightforward manner, setting up all the different parameters associated with the network layers involved in the process (i.e., PHY, MAC and network layers). Our solution assumes that the IEEE 802.11 technology is used, and thus relies on the only unencrypted piece of information that a user can modify at layer-2, the SSID, to accomplish the goals set. Since the SSID is embedded into beacon frames, periodically broadcasted by all MANET participants, high efficiency is achieved with no cost in terms of additional network traffic.

The paper is organized as follows: in the next section we refer to some related works in this research field. Section 3 briefly introduces BlueWi [10], one of the few autoconfiguration approaches available in the literature addressing both layer-2 and layer-3 requirements. An overview of the proposed solution is then presented in Section 4. Section 5 offers details about an actual implementation of our proposal on a GNU/Linux platform. In Section 6 we offer some performance results obtained in a real-life testbed. A comparison between our solution and BlueWi is then performed in Section 7. Finally, in Section 8, we present our conclusions along with some guidelines for future work.

2. Related works

In the literature we can find several proposals that focus on the IP address assignment problem in MANETs. Mohsin and Prakash [17] propose a proactive scheme for dynamic allocation of IP addresses in MANETs. Their solution uses the concept of binary split, and takes into consideration issues like network partitioning and merging, as well as abrupt departure of nodes from the system. Weniger [16] presents PACMAN, a novel approach for efficient and distributed address autoconfiguration of mobile ad hoc networks. Special features of PACMAN are the support for frequent network partitioning and merging, and very low protocol overhead. This is accomplished by using cross-layer information derived from ongoing routing protocol traffic, e.g., address conflicts are detected in a passive manner based on anomalies in routing protocol traffic. Sheu et al. [11] propose a scheme to assign IP addresses to newly-joined nodes. In their proposal some nodes are selected as coordinators, which are organized in a tree topology by exchanging *hello* messages. New nodes are able to obtain an IP address by listening to the exchanged *hello* messages and contacting the closest coordinator.

More proposals on this topic are addressed in the survey by Weniger and Zitterbart [15], which illustrates the different approaches for solving the IP address autoconfiguration problem in MANETs, highlighting the major challenges involved.

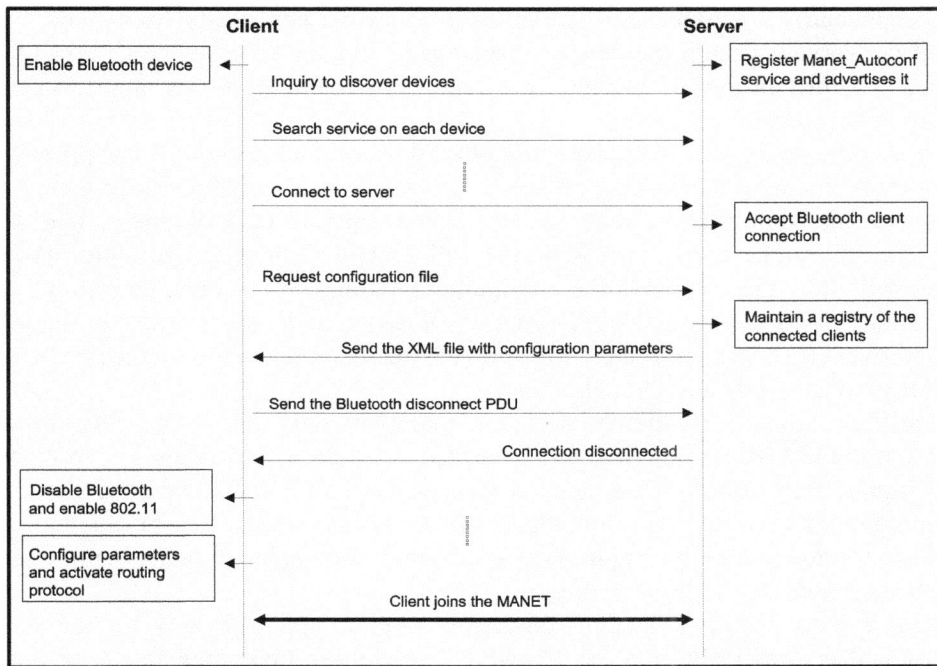

Fig. 1. BlueWi client/server interaction: message interchange diagram.

The main drawback of all the aforementioned proposals is that, for a fully functional MANET to be created, IP address assignment is not the only problem to solve. Thus, a solution offering full configuration of the different network protocols involved, both layer-2 and layer-3, is required.

One of the few works in the literature offering full MANET configuration is the solution proposed by Reyes et al. [10], which relies on Bluetooth to deliver the different configuration parameters required to setup an IEEE 802.11 based MANET. We describe this solution in more detail in the next section, since it will be used for comparison against our own.

In this work we propose a solution offering full MANET configuration that is decentralized, does not require any additional technology besides IEEE 802.11 itself, and does not introduce any extra traffic overhead into the network. Our solution is novel since it addresses both layer-2 and layer-3 configuration (which very few do), while avoiding the limitations of other related works in this field.

3. The BlueWi approach

BlueWi [10] is a solution that relies on Bluetooth [8] wireless interfaces to automate the MANET autoconfiguration process. This solution assumes that all nodes attempting to join the MANET are endowed with both a Wi-Fi and a Bluetooth interface to perform all the required tasks.

Initially, one of the nodes must act as a BlueWi server. This server will register the autoconfiguration service to make it available to all nodes. The rest of Bluetooth devices will function as clients, searching for that service so as to retrieve the MANET configuration parameters, and automatically applying that configuration afterwards. Figure 1 describes this process in more detail.

Every station that wants to join the MANET must first connect to the configuration server via Bluetooth, possibly competing with other stations also waiting to be configured. To do that, stations must perform an *inquiry* action to discover nearby Bluetooth devices. Afterwards they must check the different devices found sequentially until the server offering the desired service (i.e., the *MANET_Autoconf* service) is found. Stations can then establish an L2CAP or RFCOMM connection with the server and download the desired configuration parameters.

The configuration server must make sure it is visible by other devices, and listen to the appropriate L2CAP or RFCOMM port for incoming connections. When a client successfully establishes a connection with the server and requests the configuration data, the server must generate an XML file with all the required information and send it to the client. This XML file will contain all the necessary information for that station to successfully join the MANET. The configuration parameters include the station's IP address and mask, the routing protocol used (e.g. DSR, AODV, OLSR) and all the information required to configure the Wi-Fi interface (SSID, channel, etc.). By allowing the server to determine the IP address of each client we are able to avoid duplicated IP addresses.

Notice that, in the BlueWi solution, the Bluetooth interface is merely used to retrieve the configuration parameters required to join the MANET, while the Wi-Fi interface will allow the station to participate actively in the MANET immediately after the parameters have been received. Thus, after a client station receives its configuration data, it must switch automatically to the Wi-Fi mode. This means that the Bluetooth interface is disconnected and the Wi-Fi interface is activated, allowing to reduce to a minimum the interference between Bluetooth and Wi-Fi technologies.

When the Wi-Fi card is enabled the client station can then proceed to apply the new configuration settings. By doing so it will automatically join the MANET, being able to communicate with other mobile stations that have also configured themselves previously.

Overall, we consider that, despite this solution is able to address both layer-2 and layer-3 configuration requirements, it suffers from some limitations such as requiring all nodes to be endowed with a Bluetooth wireless card, and being centralized, thus suffering from scalability limitations.

4. Overview of the proposed solution

In the field of Wireless Local Area Networks (WLANs), the IEEE 802.11 standard has gained much popularity over the past few years. In fact, its presence is now nearly ubiquitous, although most of the networks are access protected.

The deployment of mobile ad-hoc networks (MANETs) also relies mostly on the IEEE 802.11 standard for the physical and MAC layers. However, differently from WLANs, the lack of access points or any sort of centralized management entity complicates the configuration process for terminals attempting to join the network. In particular, the users must be able to achieve a successful configuration in terms of both layer-2 and layer-3 parameters to enable communication. The characteristics of MANETs – i.e. variable topology, short-lived, decentralized – further complicate the configuration process since the network participants and the different layer-2 parameters may change frequently. Additionally, the support for multi-hop communications requires the same MANET routing protocol to be running on all network nodes.

Due to all the aforementioned issues, the startup of a MANET involves a complex and time-consuming configuration process that may even hinder scalability. Hence, we seek a solution that makes the configuration of MANET stations as simple as possible, so that even those users that are not experts in wireless networking may join and participate in the MANET in a quick, transparent and satisfactory manner.

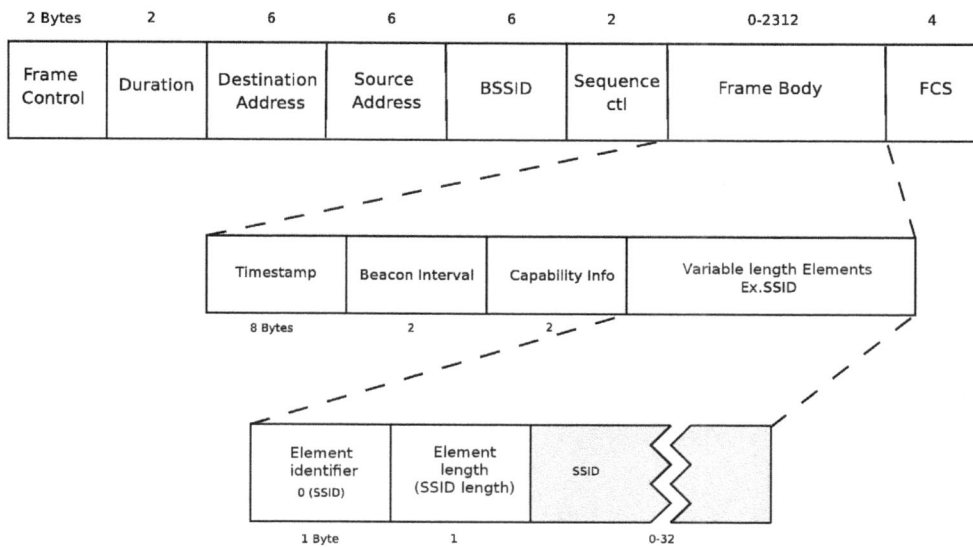

Fig. 2. Details of an IEEE 802.11 beacon frame.

The envisioned decentralized configuration solution takes into consideration that 802.11 is the technology of choice for most of the MANETs created and that, even when communications are encrypted, beacon frames are not. Thus, our proposal relies on beacon frames as potential carriers of the vital information that allows a station to gain awareness of critical configuration parameters.

By analysing the structure of an IEEE 802.11 beacon frame (see Fig. 2) we may observe that all the frame fields are automatically set by the 802.11 MAC layer without user intervention, except for the SSID field. This field is set by the user and carries the network's name, having a maximum size of 32 bytes according to the 802.11 standard.

In our proposed autoconfiguration system the SSID will be used not only to include the network's name, but also to inform stations about configuration details which will allow them to be transparently configured. Such duality is not expected to cause any drawback since most SSIDs in use are characterized by a low byte count. To justify this statement we have taken a large database including about 8 million samples corresponding to the top 1000 SSIDs used worldwide [24], and then plotted the cumulative distribution function for this 1000 SSID sizes. The result of this analysis is presented in Fig. 3. We can see that 92% of the SSIDs in use have a length between 4 and 9 characters, being very large sizes (> 16) quite scarce and lacking any additional benefits. Thus, we consider that limiting the SSID to a smaller size would not represent any significant limitation, especially when targeting ad-hoc networks where the SSID must be defined every time a new network is created.

Taking the previous analysis into consideration, we propose a strategy to partition the SSID into four different blocks as shown in Fig. 4. The name of the network, that is, the legacy SSID, will be in the first block. The second block includes basic network properties: whether it is an IPv4 or IPv6 network, which encryption mode is used, and also which MANET routing protocol should be running. The third block identifies the subnetwork prefix (also including the network mask when IPv4 is used). Finally, the fourth block holds a random value to be used as seed when attempting to derive the session key used for 802.11 MAC encryption.

The proposed solution requires that users attempting to connect to an autoconfigurable MANET parse the SSID to extract all the information required to be connected to the MANET. All data is

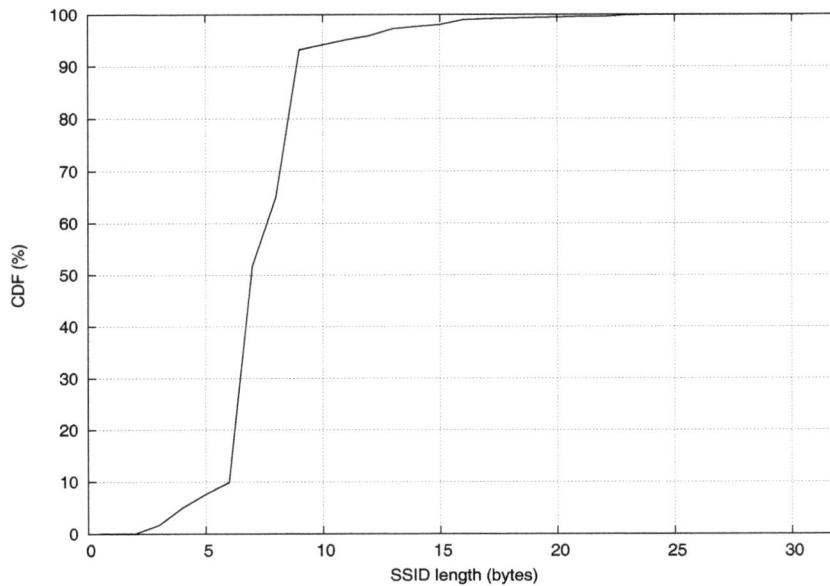

Fig. 3. Cumulative distribution function for the SSID set analysed.

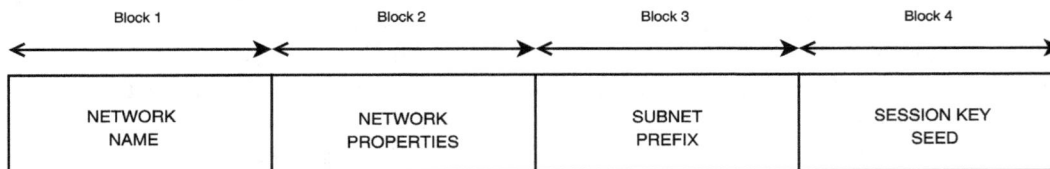

Fig. 4. Generic SSID partitioning strategy.

retrieved automatically through client software, allowing to fully and transparently configure the network connection in terms of both layer 2 and layer 3 parameters.

In terms of advantages offered, our configuration strategy is: (i) scalable, (ii) decentralized, (iii) robust, (iv) efficient, and (v) fully automatic. The arguments that support this statement are the following:

  i) Since beacons are periodically generated at a controlled rate by all MANET stations according to a randomization algorithm, the configuration information is made available to the whole MANET independently of its size, which makes the solution scalable.

 ii) Any station attempting to join the MANET is able to obtain configuration data just by listening to the beacons from any nearby MANET member, avoiding the need for a central coordinator.

iii) As long as a single MANET member remains active, the proposed configuration strategy remains immune to the loss of participating stations, which makes the process robust to failures.

 iv) The proposed strategy does not generate additional network traffic, and allows achieving full node configuration in a short period of time (see Section 6), thus offering high efficiency.

  v) Since the entire setup process is automatic and transparent, it does not require any technical skills from the user, and even inexpert users are able to take full advantage of MANETs in a seamless manner.

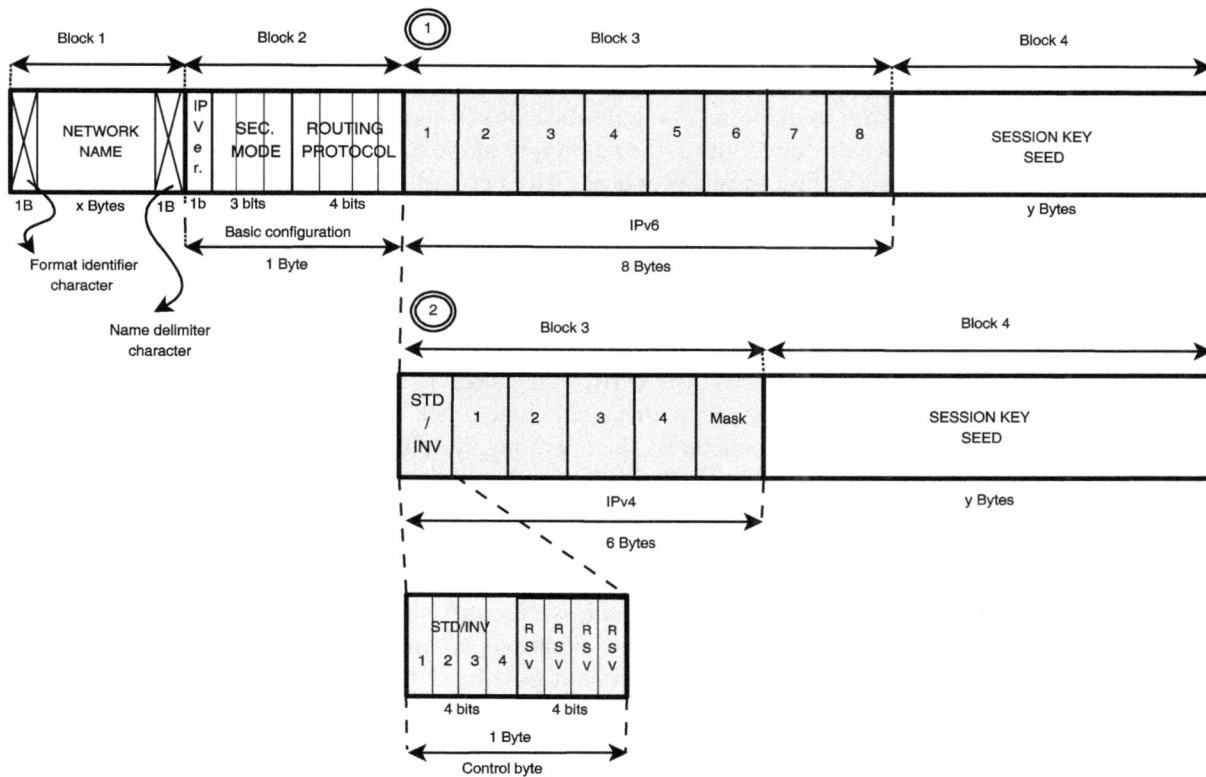

Fig. 5. Detailed SSID partitioning strategy.

5. Implementation details

In this section we describe how the different parameters required for configuration are embedded into the SSID string and later parsed. We will also offer details of an actual implementation of our approach in a Linux-based testbed. The support for 802.11 wireless cards in the Linux operating system has been available since the late nineties through the Wireless Extensions API [12] developed by Jean Tourrilhes, along with a set of wireless tools [13] accessible through the command line that were developed by the same author.

For our endeavour, we also developed command line applications that allow a user to join an existing MANET with autoconfiguration support, as well as starting such a MANET. The latter option requires the user to define the value of the different parameters required to fully configure a network interface card, which includes both layer-2 and layer-3 setup.

Concerning layer-2 parameters, deploying an IEEE 802.11 based MANET basically requires defining the operation mode (ad-hoc), the SSID, the channel used, and some security details. The latter include the security protocol used (WEP, WPA, or WPA2) and the shared key used for authentication and/or encryption.

At layer-3 we must define which version of the IP protocol is used and, for that IP version, the subnetwork used through a network ID and a network mask. To support multi-hop communication, the routing protocol used (e.g., AODV[2], OLSR [23], DYMO [6]) must also be defined.

5.1. Proposed SSID partitioning strategy

Figure 5 illustrates the proposed SSID partitioning strategy, which has been implemented and validated in a real-life testbed. Notice that, in order to distinguish regular beacons from our formatted beacons, a special (non-printable) character has been inserted just at the beginning of the SSID block, thus allowing to quickly identify SSIDs formatted according to our proposal. The network name, that is, the SSID according to its original definition, appears next, followed by another non-printable character that ends block 1. Block 2 is composed by a single byte where the first bit indicates which IP version is in use (IPv4 or IPv6), followed by 3 bits that indicate which 802.11 security mode is active (0 = open access, 1 = WEP-64, 2 = WEP-128, 3 = WPA-PSK, 4 = WPA2-PSK, 5–7 = reserved for future extensions); concerning the last 4 bits in block 2, they are used to identify the MANET routing protocol used (0 = forbidden value, 1 = OLSR, 2 = AODV, 3 = DYMO, 4 = DSR, 5–15 = reserved for future extensions). Notice that value 0 is forbidden to avoid a situation where this single byte block is set to the NULL value, which would be considered as an *end of string* character by the operating system, thus causing and error.

Depending on which IP version was defined in block 2, block 3 will contain an IPv6 network address field (8 bytes) or an IPv4 network address field. In case IPv6 is used, these 8 bytes represent the first half of the address within the Unique Local Unicast [19] range of addresses (FC00::/7); the latter 8 bytes (Interface ID) are derived from the MAC address of the wireless network interface according to the strategy defined in RFC 4291 [20]. If IPv4 is used instead, we identify the network using 4 bytes plus an extra byte to set the network mask. To avoid those situations where one or more bytes are zero (NULL character), we use the first byte (STD/INV) to invert possible NULL values in any of the four bytes that define the IP address, thus converting any 0x00 value into 0xFF. Stations attempting to configure themselves must reverse the inverted bytes to recover the original values. Notice that this strategy was not required for IPv6 since we rely on the Unique Local Unicast range of addresses, which allows picking any value for the 7 bytes following the first, which means we can easily discard any 0x00 values appearing and pick other values instead.

Concerning the last block, it includes the session key seed, which is used to derive the actual session key that will be used to perform MAC layer encryption.

5.2. Deriving the session key

When a new MANET is generated, the value for the session key seed is picked randomly. This seed allows deriving the session key by supposing that all users are aware of a fixed pre-shared key (PSK). When relying on standard 802.11 this shared key is used directly for MAC layer encryption; however, with our solution, this shared key is replaced by a variable session key. This strategy complicates the discovery of the MAC layer encryption key by a potential attacker by making it different every time a new ad hoc network is created.

The size of the seed itself is variable, and depends on the number of bytes used to identify the network (x). In our solution we will restrict the size of this network identifier to a maximum of 10 characters, which is not considered a restriction since 10 characters are enough to uniquely identify an ad-hoc network in any plausible scenario. Once the network identifier is defined, the size of the session key seed is picked so as to fill up the SSID size, thus reaching the maximum length for the SSID field. Although in theory there are 32 bytes available for the SSID, the fact that the operating systems handle it as a string ending with the NULL character reduces it to 31 bytes. This way the size of the session key seed will be either 20-x or 22-x bytes (10 bytes in the worst case), depending on whether IPv6 or IPv4 is used, respectively.

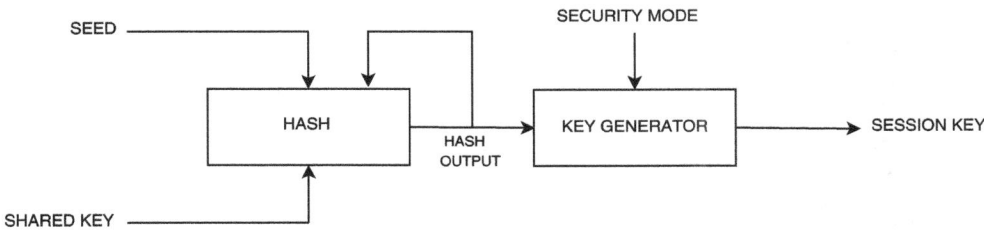

Fig. 6. Session key generation process.

One of the limitations of having the seed embedded into the SSID has to do with the handling of NULL values, as mentioned above. This means that the number of possible combinations will be slightly reduced by this restriction. Thus, the original space of $256^{(20-x)}$ combinations ($256^{(22-x)}$ for IPv4) is reduced to $255^{(20-x)}$ ($255^{(22-x)}$ for IPv4). In the worst case conditions (if the network name uses all 10 characters) there are still about 10^{24} possible seeds ($\sim 7 \times 10^{28}$ for IPv4); this means that the chances that the same seed repeats for a same group of users becomes negligible.

Figure 6 offers more details about the process of session key generation. Initially, the key shared by all MANET users is combined with the seed made available in the SSID (block 4) by using a hash mechanism such as MD5 [21], SHA-1 [3], or RIPMED-160 [5]. Since the number of bits in the hash may be shorter than the one required by the selected security mode, the hash output is fed back to generate a new hash until the key generator module gathers enough bits. Depending on the security mode selected, the key generator module may have to chop part of the input in order to obtain the correct number of bits for encryption. For example, if MD5 is used for hashing and WEP-64 is used for encrypting, a single hashing round suffices since the 128 bit output is enough to obtain the 40 bits required for a valid session key. On the contrary, if WPA-PSK is used, we need two MD5 hashing rounds to generate a session key of 256 bits, as required.

5.3. Methodology of use

In our scheme, we suppose that there is a group of users that regularly creates and joins an ad-hoc network with a specific goal. An example can be a firefighting unit, where on every mission a same group of firemen creates a MANET for communicating among themselves. All users use a same key for accessing the network, referred to as *shared key* in the previous section.

When these users intend to join the same ad-hoc network using our proposed solution, one of them (e.g., the head of the fire squad) creates the ad-hoc network, defining all the parameters required; among them we have the network name, the security mode, the routing protocol used, the IP addressing information and the *seed* used for session key generation. When the terminals used by the other users listen to the beacons generated, they will immediately parse the SSID field to retrieve the configuration details and successfully attach themselves to the MANET. This means that first there will be a layer-2 connection establishment in order to become a member of the Independent Basic Service Set (IBSS) created, followed by IP parameter definitions and the launching of the appropriate MANET routing daemon.

From that point on, any subsequent MANET generated would be quite similar, except that the seed used will always differ, and thus also the session key. This requires a potential attacker to find the session key used, and to launch the attack within the lifetime of a specific ad-hoc network (i.e., for a specific seed). Compared to the default solution, where the session key would be used over and over again, this strategy significantly reduces the effectiveness and interest at performing malicious activities.

Table 1
Average time overhead associated with autoconfiguration tasks

Autoconguration tasks	No security	WEP (64 & 128 bits)	WPA (256 bits)
Obtain configuration	4/20 μs	4 / 20 μs	4 / 20 μs
Generate key	–	23 μs	46 μs
Establish SSID	4,400 μs	4,400 μs	4,400 μs
Apply security key	–	200 μs	17,500 μs
Set IP address	4,000 μs	4,000 μs	4,000 μs
Start routing protocol	5,600 μs	5,600 μs	5,600 μs
Total time	$\sim$14 ms	$\sim$14.2 ms	$\sim$ 31.5 ms

6. Validation and performance analysis

In the previous section we described the implementation details of the proposed autoconfiguration system, which includes two components: one that allows creating the ad-hoc network, executed only by the first station, and another that allows joining an existing autoconfigurable MANET. In this section we present some performance results obtained when validating our solution in our ad-hoc network testbed.

Performance measurements were made using five middle-range laptops with similar hardware, running at 1.6 GHz with a single CPU and with 1 GB of RAM. The wireless cards used were Intel embedded devices supporting the IEEE 802.11g standard, and all terminals are within transmission range of the terminal that initially creates the ad-hoc network, unless stated otherwise.

6.1. Assessing the overhead introduced per task

Table 1 shows the average time overhead results obtained for the different security strategies. We do not include WPA2 encryption since it is not yet supported by the Linux OS in ad-hoc mode.

With respect to the first task, the MANET creation component requires parsing the user's input and generating an SSID string with autoconfiguration information (according to the strategy shown in Fig. 5), while the autoconfiguration component must merely parse the beacon received to extract configuration information. Thus, while the latter task is achieved in just 4 μs, the former (SSID generation) requires 20 μs.

The remaining tasks are similar for both components developed. In particular, the second task is related to key generation, which is achieved according to the strategy shown in Fig. 6; obviously, this step is skipped if security is disabled. The third and fourth tasks consist in setting the layer-2 parameters, such as the SSID and the encryption key. In case WPA is used, a configuration file must be created before launching the *wpa_supplicant* tool, which is responsible for WPA/WPA2 configuration tasks in Linux. This causes the time associated to that task to account for more than half of the total configuration time.

The last two tasks – IP definition and launching the MANET routing daemon – are related to network layer configuration, being common in all cases.

Overall, the autoconfiguration times can be considered quite low, although we have to take into account that the measurements presented in this section refer to the tasks taking place at the application layer. Since the dissemination of configuration information requires beacons to be received, and autoconfiguration tasks to be completed, prior to start generating new beacons, the total autoconfiguration time is usually higher. In the next section we will focus on these issues.

6.2. Autoconfiguration times in a multi-hop environment

When attempting to autoconfigure different stations in a wireless multi-hop environment, two different issues must be taken into account: (i) there is a delay from the time the first station creates the MANET

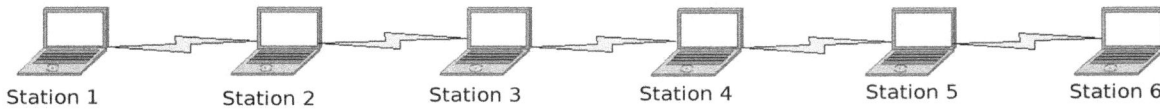

Fig. 7. Chain topology used to measure the propagation time for autoconfiguration beacons.

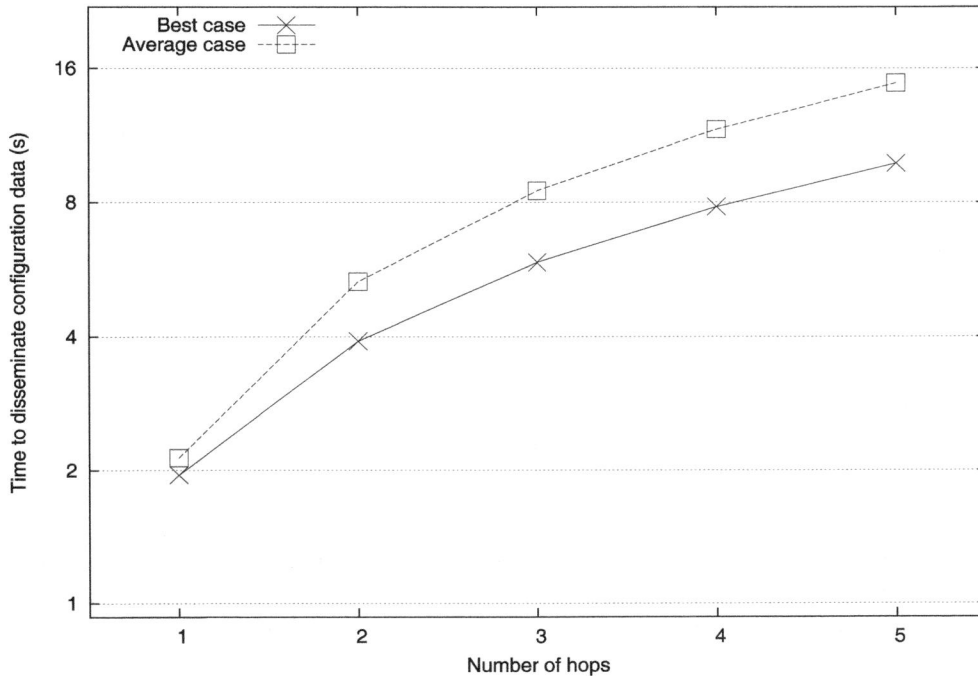

Fig. 8. Autoconfiguration times for nodes at multiple hop distances from the initiating node.

to the time nearby stations are able to receive the first beacon with autoconfiguration data embedded into the SSID; and (ii) when a new station wants to join the MANET. it must scan the different channels for beacons containing autoconfiguration data. Since in the ad-hoc mode the beacon generation process is distributed and follows a random algorithm, the actual time required to detect the beacons may vary.

To study the multi-hop propagation behaviour of autoconfiguration data, we devised a scenario (see Fig. 7) where nodes are arranged according to a chain topology. Distances between nodes are high enough to assure that radio communications are only possible with one-hop neighbours.

In our setting, Station 1 is responsible for starting the MANET. So, at the beginning of our experiment, Station 1 uses the autoconfiguration application to create a new ad-hoc network, while Stations 2 to 6 attempt to connect to the existing network by starting the autoconfiguration application in the *join* mode.

Figure 8 shows how the autoconfiguration information propagates at multiple hops. We can see that Station 2 gets configured in about 2 seconds since it is very close to the station that initiates the ad-hoc network. In particular, most of this time is associated with detecting the beacon, being configuration parameters applied in just a few milliseconds, as shown earlier (see Table 1).

As we increase the number of hops, it would be desirable to experience a linear increase of this auto-configuration time, being such linear increase represented as *best case* in Fig. 8. However, experimental

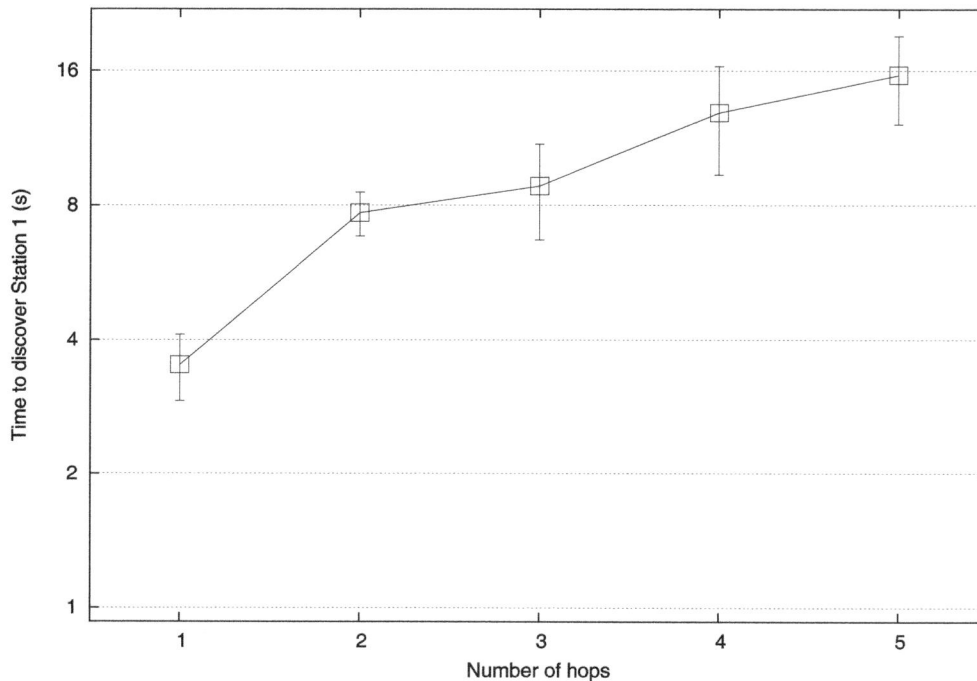

Fig. 9. OLSR topology updating time when joining the network at different hop counts from Station 1.

results show that the average propagation times are associated with a more than proportional increase, which is explained by the random beacon generation process. Remember that the IEEE 802.11 standard establishes that, in the ad-hoc network operation mode, beacons are generated by all stations involved in a distributed fashion by following a randomization algorithm. Thus, in our example, Station 2 would only generate a beacon about half the times, while Station 1 would generate beacons in the other half of the cases. As more stations get involved, the chances that a particular station generates a beacon become smaller, which slows down beacon dissemination. In our scenario, we find that the station at 5 hops from the first one (Station 6) must wait on average 14.8 seconds to detect the first autoconfiguration beacon. As a final remark, we should emphasize that such beacon propagation times are the typical times for multi-hop ad-hoc network environments, being that our solution does not impose a significant additional delay to the process.

In terms of scalability, we consider that our solution is scalable by design since the configuration information data propagates at the beacon propagation rate, which becomes highly effective even in large-sized and highly disperse MANET environments.

Concerning new nodes intending to join the MANET, they can initiate the configuration process as soon as the ad-hoc network is detected (after any periodic beacon is received), usually waiting for only a few seconds on average. In this context, we also measure the delays introduced by routing protocols, that is, the routing topology dissemination time. Notice that, once a station becomes configured and connected, there will be an additional delay introduced by the routing protocol to update the network topology. In our testbed, the routing protocol adopted was OLSR, using the standard parameter values defined in [23]. Thus, we performed a second group of tests where we measured the time elapsed from the instant when the autoconfiguration application completes its tasks, until a valid route to the first node

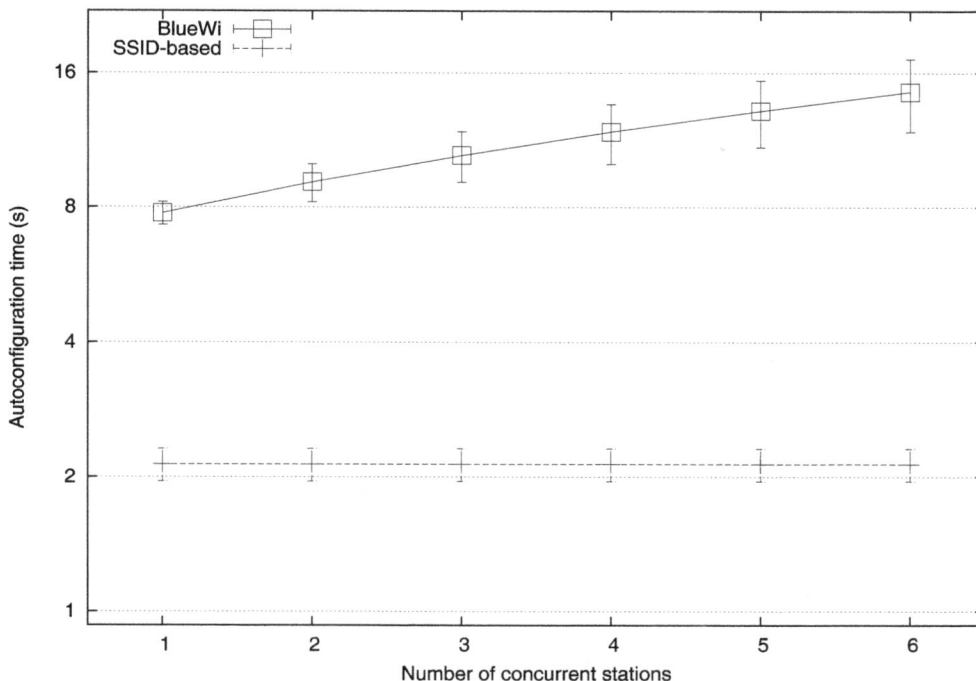

Fig. 10. Autoconfiguration time when varying the number of terminals being configured at one hop.

in the chain topology becomes available. Notice that, after the beacon is detected, each station will apply the autoconfiguration parameters, which also includes launching the OLSR protocol daemon.

In the tests that follow, all the previous nodes in the chain are configured and connected from a routing perspective when the new station arrives.

Figure 9 presents our experimental results assuming that Stations 2 to 6 will gradually join the network, creating the topology shown in Fig. 7. The values represented in the Fig. 9 show that routing information dissemination with OLSR imposes a significant time overhead, especially at more than one hop. This is expected since, in the scope of OLSR, communication with one-hop neighbours only requires neighbour detection procedures, while higher hop counts require topology updating procedures to be triggered. Thus, when Station 2 attempts to join the network and contact Station 1, OLSR takes between 3 and 4 seconds to provide a route to this station, while for Station 6 it will take OLSR between 12 and 19 seconds to provide a valid route.

By combining the results of Figs 8 and 9, we find that the time required for a MANET to be fully connected and operational can be reduced to less than one minute if the proposed autoconfiguration strategy is adopted, even with stations located several hops away from the station starting the MANET, and even when using a routing protocol with a relatively slow responsiveness (e.g. OLSR).

7. Comparison between BlueWi and the proposed solution

In this section we perform a comparison between our solution and the BlueWi solution introduced in Section 3.

Table 2
Comparison between BlueWi and our SSID-based autoconfiguration technique

Characteristic	BlueWi	SSID-based
Configuration strategy	Centralized	Distributed
Autoconfiguration server	Required	Not required
Multi-hop configuration dissemination	Not supported	Supported
Wireless technologies	Wi-Fi, Bluetooth	Wi-Fi
Number of simultaneous users serviced	7	No limit
IPv4 support	Yes (DHCP-like)	Yes (MAC address based)
IPv6 support	No	Yes (MAC address based)
WEP/WPA/WPA2 support	Yes	Yes
Rotating encryption keys	Yes (manually)	Yes (random seed)
Routing protocol	Any	Any
User control and logging	Yes	No
User access control	Bluetooth pin	Pre-shared Key
Best-case configuration time	7.31 s	1.95 s

As referred before, BlueWi requires clients to establish a Bluetooth channel with a BlueWi server to retrieve all the configuration parameters required. The Wi-Fi interface is then configured according to that parameter set. Our proposal significantly differs from BlueWi since it does not assume any sort of server. In fact, any station can start the ad-hoc network, and as long as a single station keeps that network alive, other stations can autoconfigure themselves and join the network.

Figure 10 shows the total autoconfiguration time when attempting to simultaneously configure different numbers of terminals. All terminals are assumed to be at one hop from either the Bluetooth server (BlueWi) or the station that starts the MANET (SSID-based proposal) for comparison. In terms of radio range, the BlueWi solution limits the maximum distance between the Bluetooth server and the stations being configured to 10 meters (default Bluetooth range) or 100 meters, depending on the Bluetooth device class. For our SSID-based solution, stations at one hop from the station starting the MANET are able to detect its beacons for distances up to 250 meters although, as shown in the previous section, multi-hop configuration is possible and does not suppose any impediment.

From Fig. 10 we can see that the autoconfiguration time for our SSID-based solution is independent of the number of stations involved. This is expected since beacons are broadcasted, being received by all wireless devices within range. Concerning BlueWi, we find that autoconfiguration tasks require several seconds more. This additional time is mostly associated with Bluetooth device discovery procedures (*Inquiry*), which takes about 5.12 seconds to complete, and that are a prerequisite before attempting to contact the Bluetooth server. Also, the number of concurrent stations retrieving configuration parameters will reduce the channel capacity dedicated to each station, thereby increasing the total time involved.

To complete our comparison between BlueWi and our proposal, Table 2 summarizes the main differences between both solutions. Overall, we find that the proposed autoconfiguration system based on SSID parameter embedding offers significant improvements over a pre-existent solution (BlueWi), representing a significant step forward in the state-of-the-art within the field of MANET autoconfiguration.

8. Conclusions and future work

Despite all on-going efforts, the issue of MANET usability is still an important research topic since the complexity when attempting to configure MANET terminals remains high. Besides complexity itself, other issues such as the need to rely on encrypted communications further complicate the configuration problem.

In this paper we propose a novel solution for terminal autoconfiguration that is able to fully configure both layer-2 and layer-3 parameters that are critical to join an 802.11-based MANET. Our solution relies on the SSID field that is present on the periodic beacons generated by IEEE 802.11 compliant stations to announce basic configuration data. By listening to beacons and parsing the SSID field, new stations are able to determine all the information required to successfully join the MANET.

To validate our proposal we developed two software components, one that allows creating a new autoconfigurable MANET, and another one that allows joining an existing autoconfigurable MANET. Experimental results show that both software components are able to perform all the configuration tasks required in a very short period of time. In particular, the total time required is below 15 ms if security is basic (WEP) or disabled, and it is below 32 ms if WPA is used instead.

By deploying a small scenario using a chain topology we showed that multi-hop configuration dissemination can be performed in an efficient manner, introducing on average a delay of about 3.2 seconds per hop. Also, experimental results have showed that, after the configuration process is completed, additional time may be required to allow the chosen routing protocol to update the topology. This is particularly true for proactive routing protocols such as OLSR, which require several seconds to detect new stations and update the network topology.

To complete our analysis, we compared our proposal against BlueWi, a similar solution available in the literature, showing that our strategy offers significant benefits and improvements with respect to the latter.

Overall, we consider that the proposed solution can fill-in the gap between regular users and ad-hoc network technologies, allowing to accelerate the adoption of distributed communication paradigms to a wider range of application scenarios.

As future work we plan to develop a similar set of tools to other operating systems besides GNU/Linux, thus embracing a greater number of potential users.

Acknowledgments

This work was partially supported by the *Ministerio de Educación y Ciencia*, Spain, under Grant TIN2011-27543-C03-01.

References

[1] B. Yang and A.R. Hurson, Similarity-based clustering strategy for mobile ad hoc multimedia databases, *Mobile Information Systems* **1** (December 2005), 253–273.

[2] C.E. Perkins, E.M. Belding-Royer and S.R. Das, Ad hoc on-demand distance vector (AODV) routing, Request for Comments 3561, MANET Working Group, http://www.ietf.org/rfc/rfc3561.txt, July 2003. Work in progress.

[3] D. Eastlake and P. Jones, US Secure Hash Algorithm 1 (SHA1), IETF RFC 3174, September 2001.

[4] E. Kulla, M. Hiyama, M. Ikeda, L. Barolli, V. Kolici and R. Miho, MANET performance for source and destination moving scenarios considering OLSR and AODV protocols, *Mobile Information Systems* **6** (2010), 325–339.

[5] H. Dobbertin, A. Bosselaers and B. Preneel, RIPEMD-160, a strengthened version of RIPEMD, In *Fast Software Encryption, LNCS 1039*, 1996, pp. 71–82, .

[6] I. Chakeres and C. Perkins, Dynamic MANET On-demand (DYMO) Routing, Internet Draft, MANET Working Group, draft-ietf-manet-dymo-12.txt, February 2008, Work in progress.

[7] IEEE 802.11 WG, International Standard for Information Technology – Telecom. and Information exchange between systems – Local and Metropolitan Area Networks – Specific Requirements – Part 11: Wireless Medium Access Control (MAC) and Physical Layer (PHY) Specifications, ISO/IEC 8802-11:1999(E) IEEE Std. 802.11, 1999.

[8] IEEE 802.15.1(tm) IEEE Standard for Information technology – Telecommunications and information exchange between systems – Local and metropolitan area networks–Specific requirements Part 15.1: Wireless Medium Access Control (MAC) and Physical Layer (PHY) Specifications for Wireless Personal Area Networks (WPANs(tm)), 2002.

[9] IETF, MANET Working Group Charter, http://www.ietf.org/html.charters/manet-charter.html.

[10] J.C. Reyes, E. Burgoa, C.T. Calafate, J.-C. Cano and P. Manzoni, A MANET autoconfiguration system based on Bluetooth technology. In *3rd International Symposium on Wireless Communication Systems (ISWCS)*, Valencia, Spain, September 2006.

[11] J.-P. Sheu, S.-C. Tu and L.-H. Chan, A distributed IP address assignment scheme in ad hoc networks, *Int J Ad Hoc Ubiquitous Comput* 3(1) (2007), 10–20.

[12] J. Tourrilhes, Wireless extensions for linux, Hewlett Packard Laboratories, Palo Alto, 1996, Available at the author's home page at http://www.hpl.hp.com/.

[13] J. Tourrilhes, Wireless tools for linux, Hewlett Packard Laboratories, Palo Alto, 2008, Available at: http://www.hpl.hp.com/personal/Jean_Tourrilhes/.

[14] J.-Z. Sun, Mobile ad hoc networking: an essential technology for pervasive computing, In *International Conferences on Info-tech and Info-net*, volume 3, 2001, pp. 316–321.

[15] K. Weniger and M. Zitterbart, Mobile ad hoc networks – current approaches and future directions, *Network, IEEE* 18(4) (2004), 6–11.

[16] K. Weniger, Pacman: passive autoconfiguration for mobile ad hoc networks. *Selected Areas in Communications, IEEE Journal on*, 23(3): 507–519, March 2005.

[17] M. Mohsin and R. Prakash, IP Address Assignment in a Mobile Ad Hoc Network. In *Proceedings of Military Communications Conference (MILCOM 2002)*, volume 2, pages 856–861, Anaheim, California, USA, 2002.

[18] N. Qadri, M. Altaf, M. Fleury and M. Ghanbari, Robust video communication over an urban VANET, *Mobile Information Systems* 6 (2010), 259–280.

[19] R. Hinden and B. Haberman, Unique Local IPv6 Unicast Addresses, IETF RFC 4193, October 2005.

[20] R. Hinden and S. Deering, IP Version 6 Addressing Architecture, IETF RFC 4291, February 2006.

[21] R. Rivest, The MD5 Message-Digest Algorithm, IETF RFC 1321, April 1992.

[22] S. Manvi, M. Kakkasageri and J. Pitt, Multiagent based information dissemination in vehicular ad hoc networks, *Mobile Information Systems* 5 (2009), 363–389.

[23] T. Clausen and P. Jacquet, Optimized link state routing protocol (OLSR). Request for Comments 3626, MANET Working Group, http://www.ietf.org/rfc/rfc3626.txt, October 2003. Work in progress.

[24] wigle.net, Wireless Geographic Logging Engine, http://www.wigle.net/gps/gps/main/ssidstats, June 2010.

Maria José Villanueva graduated in Computer Science at the Technical University of Valencia in 2009. She is currently a Ph.D. student financed by the Science and Innovation Public Department (MICINN) under an FPU Spanish research and teaching fellowship. Maria José Villanueva's research is focused on the application of sofware engineering and information systems principles in the bioinformatics domain, as well as on the development of health services for personalized medicine.

Carlos T. Calafate is an associate professor of the Department of Computer Engineering at the Technical University of Valencia (UPV), Spain. He graduated with honors in Electrical and Computer Engineering at the University of Oporto (Portugal) in 2001. He received his Ph.D. degree in Computer Engineering from the Technical University of Valencia in 2006. He is a member of the Computer Networks research group (GRC). His research interests include mobile and pervasive computing, security and QoS on wireless networks, as well as video coding and streaming.

Álvaro Torres graduated in Computer Science at the Technical University of Valencia in 2009. He is currently a Ph.D. student financed by the Science and Innovation Public Department (MICINN) under an FPU Spanish research and teaching fellowship. His research is focused on the video delivery and quality of services issues in vehicular network environments.

José Cano received his M.S. degree and Ph.D. in Computer Science from the Technical University of Valencia (UPV), Spain, in 2004 and 2012, respectively. He is a member of the Computer Networks Group (GRC) at the UPV since 2005. He is also member of the Parallel Architectures Group (GAP) at UPV since 2009. His current research interests include ubiquitous applications, wireless technologies, and networks-on-chips.

Juan Carlos Cano received the MS and PhD degrees in computer science from the Technical University of Valencia, Spain in 1994 and 2002 respectively. He was a visiting researcher at the Computer Engineering Department of the Baskin School of Engineering at University of California, Santa Cruz, USA, in 2002, where he worked in the area of mobile ad hoc networks under the supervision of Professor J.J. Garcãa-Luna-Aceves. He has served as Technical Program Committee member of many IEEE Conferences and serves as a regular reviewer for leading IEEE Conferences and Journals such as ICC, Globecomm,

WCNC, ISCC, and IEEE Transaction on Mobile Computing. Currently he is full professor of computer engineering in the Department of Computer Engineering at the Technical University of Valencia, Spain. His current research interests include wireless networks protocols design, context aware applications and ubiquitous computing. From 1995 to 1997 he worked at IBM's manufacturing division in Valencia, Spain.

Pietro Manzoni is a full professor of computer science at the "Universidad Politecnica de Valencia" (SPAIN). He received the M.S. degree in Computer Science from the "Universita' degli Studi" of Milan (ITALY) in 1989 and the Ph.D. in Computer Science from the Polytechnic University of Milan (ITALY) in 1995. His research activity is related to wireless networks protocol design, modeling and implementation.

Permissions

List of Contributors

Francesco Palmieri
Dipartimento di Ingegneria dell'Informazione, Seconda Università degli Studi di Napoli, Aversa (CE), Italy

Aniello Castiglione
Dipartimento di Informatica, Università degli Studi di Salerno, Fisciano (SA), Italy

Admir Barolli
Department of Computers and Information Science, Seikei University, Tokyo, Japan

Evjola Spaho
Graduate School of Engineering, Fukuoka Institute of Technology (FIT), Fukuoka, Japan

Leonard Barolli
Department of Information and Communication Engineering, Fukuoka Institute of Technology (FIT), Fukuoka, Japan

Fatos Xhafa
Department of Languages and Informatics Systems, Technical University of Catalonia, Jordi Girona 1-3, Barcelona, Spain

Makoto Takizawa
Department of Computers and Information Science, Seikei University, Tokyo, Japan

Jiro Iwashige
Department of Information and Communication Engineering, Fukuoka Institute of Technology (FIT), 3-30-1Wajiro-Higashi, Higashi-Ku, Fukuoka 811-0295, Japan

Leonard Barolli
Department of Information and Communication Engineering, Fukuoka Institute of Technology (FIT), 3-30-1Wajiro-Higashi, Higashi-Ku, Fukuoka 811-0295, Japan

Motohiko Iwaida
Graduate School of Engineering, Fukuoka Institute of Technology (FIT), 3-30-1Wajiro-Higashi, Higashi-Ku, Fukuoka 811-0295, Japan

Saki Kameyama
Graduate School of Engineering, Fukuoka Institute of Technology (FIT), 3-30-1Wajiro-Higashi, Higashi-Ku, Fukuoka 811-0295, Japan

Rana Asif Rehman
Department of Electronics and Computer Engineering, Graduate School, Hongik University, Sejong City 337-701, Republic of Korea

Jong Kim
Seon Architects and Engineering Group, Cheongju, Chungbuk 360-020, Republic of Korea

Byung-Seo Kim
Department of Computer and Information Communication Engineering, Hongik University, Sejong City 337-701, Republic of Korea

David Chunhu Li
Department of Computer Science and Information Engineering, National Central University, No. 300 Jhongda Road, Jhongli City, Taoyuan County 32001, Taiwan

Li-Der Chou
Department of Computer Science and Information Engineering, National Central University, No. 300 Jhongda Road, Jhongli City, Taoyuan County 32001, Taiwan

Li-Ming Tseng
Department of Computer Science and Information Engineering, National Central University, No. 300 Jhongda Road, Jhongli City, Taoyuan County 32001, Taiwan

Yi-Ming Chen
Department of Information Management, National Central University, No. 300 Jhongda Road, Jhongli City, Taoyuan County 32001, Taiwan

Kai-Wei Kuo
Department of Computer Science and Information Engineering, National Central University, No. 300 Jhongda Road, Jhongli City, Taoyuan County 32001, Taiwan

Yuka Komai
Department of Multimedia Engineering, Graduate School of Information Science and Technology, Osaka University, Osaka, Japan

Yuya Sasaki
Department of Multimedia Engineering, Graduate School of Information Science and Technology, Osaka University, Osaka, Japan

Takahiro Hara
Department of Multimedia Engineering, Graduate School of Information Science and Technology, Osaka University, Osaka, Japan

Shojiro Nishio
Department of Multimedia Engineering, Graduate School of Information Science and Technology, Osaka University, Osaka, Japan

Federico Mari
DI, University of Roma "La Sapienza", Roma, Italy

Igor Melatti
DI, University of Roma "La Sapienza", Roma, Italy

Enrico Tronci
DI, University of Roma "La Sapienza", Roma, Italy

Alberto Finzi
DSF, University of Napoli "Federico II", Complesso Universitario di Monte Sant'Angelo, Napoli, Italy

Thierry Delot
University Lille North of France, Valenciennes, France

Sergio Ilarri
University of Zaragoza, Zaragoza, Spain

Nicolas Cenerario
University Lille North of France, Valenciennes, France

Thomas Hien
University Lille North of France, Valenciennes, France

Shangguang Wang
State Key Laboratory of Networking and Switching Technology, Beijing University of Posts and Telecommunications, Beijing, China

Zibin Zheng
Shenzhen Research Institute, The Chinese University of Hong Kong, Hong Kong, China

Zhengping Wu
Department of Computer Science and Engineering, University of Bridgeport, Bridgeport, CT, USA

Qibo Sun
State Key Laboratory of Networking and Switching Technology, Beijing University of Posts and Telecommunications, Beijing, China

Hua Zou
State Key Laboratory of Networking and Switching Technology, Beijing University of Posts and Telecommunications, Beijing, China

Fangchun Yang
State Key Laboratory of Networking and Switching Technology, Beijing University of Posts and Telecommunications, Beijing, China

Tarek R. Sheltami
Computer Engineering Department, King Fahd University of Petroleum and Minerals, Dhahran 31216, Saudi Arabia

Thierry Delot
University Lille North of France, LAMIH FRE CNRS, Valenciennes, France

Nathalie Mitton
INRIA Lille-Nord Europe, Villeneuve, France

Sergio Ilarric
IIS Department, University of Zaragoza, Zaragoza, Spain

Thomas Hien
University Lille North of France, LAMIH FRE CNRS, Valenciennes, France

M. José Villanueva
Department of Computer Engineering, Universitat Politècnica de València, Valencia, Spain

Carlos T. Calafate
Department of Computer Engineering, Universitat Politècnica de València, Valencia, Spain

Álvaro Torres
Department of Computer Engineering, Universitat Politècnica de València, Valencia, Spain

José Cano
Department of Computer Engineering, Universitat Politècnica de València, Valencia, Spain

Juan-Carlos Cano
Department of Computer Engineering, Universitat Politècnica de València, Valencia, Spain

Pietro Manzoni
Department of Computer Engineering, Universitat Politècnica de València, Valencia, Spain